Topics in Mining, Metallurgy and Materials Engineering

"Topics in Mining, Metallurgy and Materials Engineering" welcomes manuscripts in these three main focus areas: Extractive Metallurgy/Mineral Technology; Manufacturing Processes, and Materials Science and Technology. Manuscripts should present scientific solutions for technological problems. The three focus areas have a vertically lined multidisciplinarity, starting from mineral assets, their extraction and processing, their transformation into materials useful for the society, and their interaction with the environment.

**** Indexed by Scopus (2020) ****

Raúl Pérez-Bustamante

Fundamental, Practical and Disruptive Applications of Powder Metals

Springer

Raúl Pérez-Bustamante
SECIHTI-Centro de Investigación en Materiales Avanzados A.C. (CIMAV)
Av. Miguel de Cervantes Saavedra, Complejo Industrial Chihuahua
Chihuahua, Chihuahua, Mexico

ISSN 2364-3293 ISSN 2364-3307 (electronic)
Topics in Mining, Metallurgy and Materials Engineering
ISBN 978-3-032-27129-7 ISBN 978-3-032-27130-3 (eBook)
https://doi.org/10.1007/978-3-032-27130-3

This Springer imprint is published by the registered company Springer Nature Switzerland AG
The registered company address is: Gewerbestrasse 11, 6330 Cham, Switzerland

This book is dedicated to all those who, through research, teaching, and professional practice, have chosen powder metallurgy as a means to understand and transform matter into solutions that serve society. It is addressed to those who recognize that the development of new materials and the integration of emerging technologies not only expand the possibilities of processing and manufacturing, but also redefine the ways in which we learn, teach, and share technical knowledge. In a context where innovation advances rapidly, new technologies become indispensable tools for education, analysis, and creation, enabling complex concepts to be conveyed to new generations and fostering more dynamic, accessible, and collaborative forms of learning. May these pages serve as a meeting point between experience and innovation, as a recognition of the sustained effort behind each advance, and as an invitation to continue exploring, with rigor and curiosity, the many opportunities offered by the materials and processes of the present and the future.

Foreword

You can tell this was written by someone who has spent a long time working with materials, not just reading about them. The explanations take their time and do not try to clean up processes that are known to be sensitive, sometimes stubborn, and rarely forgiving. Powder metallurgy comes across as it is lived in practice, as a field where understanding grows by watching closely, running experiments, and going back to the basics when things do not work out as expected. That way of thinking carries through the discussion, keeping what happens at the particle level tied to what later shows up in real parts. The way powders are handled, pressed, and heated plays a key role in how the final piece will work. These steps are not just part of the process; they directly influence functional capabilities of the material. How they are carried out depends on the type of material, the setup, and the conditions in which the final product will be used.

More established routes and newer processing approaches are discussed together, without trying to place one above the other. Pressing and sintering, hot isostatic pressing, injection molding, and powder-based additive processes are all presented as different ways of dealing with the same basic problem, turning loose powder into material you can count on. Porosity, interfaces, and microstructure are handled as everyday concerns rather than theoretical ideas, the kind of things that guide choices during processing and later show up in service behavior. There is also room to point out limits and sensitivities that come with real materials and real processes, things that are well known to people working in the field even if they are not always written down. What gives the text its strength is a plain way of talking about technical work, shaped by someone comfortable moving between the lab and real-world applications, and willing to share not just results, but the thinking behind them.

José Manuel Mendoza Duarte

Centro de Investigación en Materiales Avanzados A.C. (CIMAV), Av. Miguel de Cervantes Saavedra, Complejo Industrial Chihuahua, Chihuahua, Chihuahua, Mexico

Declaration of Competing Interests

The author declares that no known competing financial interests or personal relationships exist that could have appeared to influence the work reported in this book.

Preface

Powder metallurgy has become one of the most relevant processing routes in modern materials engineering, particularly in applications requiring precise control over microstructure, composition, and component geometry. Unlike conventional melt-based processes, powder consolidation establishes a direct link between particle-scale phenomena and the mechanical, thermal, and functional behavior of the final material, which has supported its widespread use across metallic, ceramic, and composite systems and its integration with advanced manufacturing technologies. In this book, powder metallurgy is presented as a coherent framework for microstructural design rather than as a collection of isolated forming techniques. From powder production and conditioning to sintering mechanisms, it is shown that particle size, morphology, surface chemistry, and thermomechanical history govern densification behavior, microstructural evolution, and material stability, making an understanding of diffusion, plastic deformation, and mass transport essential for interpreting the performance of powder-based materials.

The discussion extends to different shaping and consolidation strategies, including press-and-sinter processing, hot isostatic pressing, metal injection molding, and powder-based additive manufacturing, each of which imposes distinct mechanical and thermal conditions that influence pore distribution, interparticle bonding, and phase stability. Despite their technological maturity, these processes remain sensitive to density gradients, interfacial reactions, and defects associated with the particulate nature of the feedstock, particularly in reactive or multicomponent systems. This sensitivity becomes evident in the analysis of industrial applications, where powder metallurgy demonstrates its ability to compete with wrought or cast materials when porosity, phase distribution, and residual stress states are properly controlled. Tribological components, magnetic materials, energy-related systems, and biomedical applications illustrate how porosity may function either as a structural limitation or as a functional asset, depending on its morphology and connectivity, reinforcing the view of microstructure as a design variable rather than a by-product of processing.

Recent developments that extend the traditional scope of powder metallurgy are also addressed, including solid-state processes, hybrid manufacturing routes, and sustainability-oriented strategies. Powder reuse, metallic scrap recycling, and energy-efficient sintering are presented as natural extensions of a technology that inherently promotes material efficiency and waste reduction. Throughout the chapters, advanced characterization techniques and modeling tools are employed in response to practical needs rather than as part of a rigid methodological framework. Direct observation of powders, interfaces, and microstructures, combined with diffusion-based and microstructural evolution models, enables interpretation beyond empirical description, while the combined use of experimental and computational approaches helps connect length scales and clarify the actual effects of processing conditions on material behavior. In this context, powder metallurgy emerges as a field in which processing, microstructure, and performance are continuously interrelated, informing decisions in material design, manufacturing, and application.

Chihuahua, Chihuahua, Mexico Raúl Pérez-Bustamante

Acknowledgments

This book is the result of a long process shaped by time, reflection, and dialogue, and it would not have been possible without the support of people and institutions that, in different ways, made its development possible. Scientific research grows not only from ideas and results, but also from settings that allow questions to take shape, work to continue over time, and attention to be sustained. I would like to express my deepest gratitude to the Program of Investigadoras e Investigadores por México of SECIHTI, whose support has been essential throughout this work. This program provides the conditions that make sustained research possible, encouraging commitment, continuity, and genuine integration within host institutions. Thanks to this support, it has been possible to pursue long-term research lines, work closely with students, build collaborations, and devote the necessary time to the thinking and writing required for a book of this kind. The confidence extended through this program has been a steady source of encouragement over the course of this project.

I would also like to thank Meltio and Renishaw for kindly providing images that form part of the visual material included here. Their willingness to share this material helped strengthen the connection between research, technology, and industrial practice, and is sincerely appreciated. I am equally grateful to the colleagues with whom I have shared discussions, ideas, and collaborations over the years. Many of those conversations, whether brief or extended, have left both visible and subtle traces across these pages. I also wish to acknowledge the students who, through their questions, curiosity, and engagement, have helped keep the discussion active and grounded, reminding me of the shared and formative nature of academic work.

Finally, I extend my heartfelt thanks to my family and to those close to me who accompanied this process with patience, understanding, and steady support. The time devoted to research and writing is often quiet and extended, and that everyday presence made it possible to carry this work through to completion.

Competing Interests The author has no competing interests to declare that are relevant to the content of this manuscript.

Contents

Abbreviations and Acronyms

AAM	Autonomous additive manufacturing
ADAM	Atomic diffusion additive manufacturing
AI	Artificial intelligence
AM	Additive manufacturing
AMCs	Aluminum matrix composites
ANN	Artificial Neural Network
APS	Atmospheric plasma spray
AS	Arc spraying
ASMCs	Amorphous soft magnetic composites
BCC	Body-centered cubic
BMD	Bound Metal Deposition™
BTBM	Bonding treatment and ball milling
CAC	Calcium aluminate cement
CCCs	Conformal cooling channels
CE	Circular economy
CHD	Case hardening depth
CIP	Cold isostatic pressing
CP	Ceramic powder
CPM	Croom Precision Medical
CS	Cold spray
CT	Computer tomography
CTE	Coefficient of thermal expansion
CZ	Clad zone
DED	Directed energy deposition
DFT	Density functional theory
DMLS	Direct metal laser sintering
DSC	Differential scanning calorimetry
DZ	Dilution zone
EBM	Electron beam melting

EB-PBF	Electron beam powder-bed fusion
EBSD	Electron backscatter diffraction
EDS	Energy dispersive spectroscopy
EP	Epoxy resin
FCC	Face-centered cubic
FDA	Food and Drug Administration
FDM	Fused deposition modeling
FEA	Finite element analysis
FEM	Finite element method
FFF	Fused filament fabrication
FRAM	Friction rolling additive manufacturing
FWHM	Full width at half maximum
GFA	Glass forming ability
GNP	Graphene nanoplatelets
HA	Hydroxyapatite
HAZ	Heat-affected zone
hBN	Hexagonal boron nitride
HEAs	High-entropy alloys
HEBM	High-energy ball milling
HIP	Hot isostatic pressing
HPM	High-performance powder metallurgy
HVC	High-velocity compaction
HVOF	High-velocity oxy-fuel
IACS	International Annealed Copper Standard
IAM	Intelligent Additive Manufacturing
kMC	Kinetic Monte Carlo
LAM	Laser additive manufacturing
LBAM	Laser-based additive manufacturing
LCA	Life cycle assessment
LCC	Life cycle costing
LDED	Laser direct energy deposition
LHEAs	Lightweight high-entropy alloys
LLDPE	Linear low-density polyethylene
LMD	Laser metal deposition
L-PBF	Laser powder bed fusion
LPIM	Low-pressure injection molding
MA	Mechanical alloying
MAS	Microwave-assisted sintering
MD	Molecular dynamics
MD/MC	Molecular dynamics/Monte Carlo
MEX	Material extrusion
MF3	Metal fused filament fabrication
MIM	Metal injection molding

MIs	Mold inserts
MJS	Multiphase jet solidification
ML	Machine Learning
MLR	Multiple linear regression
MM	Mechanical milling
MMCs	Metal matrix composites
MOFs	Metal organic frameworks
NCP	Non-aqueous chemical precipitation
ODS	Oxide dispersion strengthening
OPC UA	Open Platform Communications Unified Architecture
OPH	Oxide precipitation hardened
PBF	Powder bed fusion
PEG	Polyethylene glycol
PIM	Powder injection molding
PIMDMLS	DMLS-grade powder in the PIM process
PIMPIM	PIM quality powder
PM	Powder metallurgy
PMMA	Polymethyl methacrylate
POM	Polyoxymethylene
PPC	Polypropylene carbonate
PPX	Poly para xylene
PTFE	Polytetrafluoroethylene
RA	Reactive alumina
RHEAs	Refractory high-entropy alloys
SEM	Scanning electron microscopy
SH	Space holder
SLM	Selective laser melting
SMAs	Shape memory alloys
SMCs	Soft magnetic composites
SPS	Spark plasma sintering
SRO	Short-range order
SSP	Severe shot peening
SVMR	Support vector machine regression
TEM	Transmission electron microscopy
TGA	Thermogravimetric analysis
TMCs	Titanium matrix composites
VEC	Valence electron concentration
VPP	Vat photopolymerization
WAAM	Wire and arc additive manufacturing
WBM	Wet ball milling
WLMD	Wire laser metal deposition
XPS	X-ray photoelectron spectroscopy

XRD	X-ray diffraction
XRI	X-ray imaging
YAG	Yttrium aluminum garnet
Zr-MOFs	Zirconium-based metal-organic frameworks

Chapter 1
Foundations of Powder Metallurgy

Abstract The deliberate coordination of powder production, consolidation, and thermally activated bonding provides a robust means to control microstructure, properties, and final component geometry. Characteristics imparted during powder preparation, such as particle morphology, size distribution, and surface chemistry, whether arising from mechanical milling, chemical synthesis, or atomization, strongly influence flow behavior, packing efficiency, and sintering response, thereby shaping densification kinetics and the development of defects. Subsequent consolidation methods, including uniaxial and isostatic pressing, powder injection molding, and layerwise additive fusion, introduce distinct stress states that determine green density uniformity and govern shrinkage during thermal processing. During sintering, mass transport proceeds through surface, grain boundary, and lattice diffusion mechanisms, while field-assisted approaches enhance diffusion rates and help limit grain growth and oxide retention. Additional steps such as infiltration, heat treatment, and surface finishing further adjust interfacial quality, residual stress states, and overall functional performance. Applicable to metals, ceramics, and composite systems, powder metallurgy enables tailored porosity, phase constitution, and reinforcement architectures, allowing combinations of strength, fatigue resistance, thermal stability, and functionality that are difficult to achieve through melt-based routes. When combined with near-net-shape processing, powder reuse, and compatibility with additive manufacturing, this approach links microstructural design flexibility with material efficiency and the broader goals of sustainable manufacturing.

Keywords Powder metallurgy · Microstructure control · Sintering mechanisms · Sustainable manufacturing

1.1 Introduction

Powder metallurgy (PM) is a manufacturing technology that integrates materials science, thermomechanical processing, and design engineering to produce components with controlled microstructures and high precision; its foundation lies in the

R. Pérez-Bustamante, *Fundamental, Practical and Disruptive Applications of Powder Metals*, Topics in Mining, Metallurgy and Materials Engineering,
https://doi.org/10.1007/978-3-032-27130-3_1

transformation of fine powders into functional solids through compaction, sintering, and, more recently, additive and hybrid routes [1–4]. From its origins as a method to process refractory and difficult to cast metals, PM has evolved into a versatile technique enabling near-net shape manufacturing with minimal waste and material efficiency and whose advances in powder production, through atomization, mechanical alloying, and chemical synthesis, have made it possible to control particle morphology, particle size, and purity, directly influencing flow behavior, packing density, and final part properties [5–9].

Compaction and shaping technologies, including uniaxial and isostatic pressing, metal injection molding, and high velocity compaction, provide the means to achieve homogeneous density, while sintering mechanisms such as solid-state and spark plasma sintering (SPS) govern diffusion, phase transformation, and grain growth that define the mechanical response of a material [7, 10, 11]. The versatility of PM provides the development of high performance alloys, ceramics, and composites, enabling lightweight structures in automotive and aerospace industries, biocompatible implants with a controlled porosity, and functional materials for energy storage and thermal applications [7, 12–15].

The integration of PM with additive manufacturing (AM) has further expanded its scope, allowing precise control of composition and geometry to fabricate complex, multimaterial, and functionally graded components [16–21]. At the same time, PM aligns naturally with sustainable manufacturing principles through reduced raw material consumption, lower energy requirements, and the capacity to recycle metallic scrap and reuse powders without degradation. Its compatibility with circular economy models positions PM as a key enabler of environmentally responsible production while maintaining high structural and functional performance. As a result, PM stands as a mature yet continuously advancing technology that bridges scientific innovation, industrial scalability, and ecological sustainability in modern materials engineering [22].

1.2 Principles of Powder Metallurgy Processes

1.2.1 Fundamentals and Historical Evolution

The foundations of PM trace back to prehistoric metalworking, a time before the advent of high temperature smelting. As Pickens [4] denotes, early civilizations, such as the Egyptians, forged iron-rich particles below their melting point through hammering, a primitive form of pressing and sintering that marked the earliest practice of PM. The subsequent evolution of furnace technologies in the Industrial Revolution diminished the relevance of PM until the nineteenth century, when metals with high melting points like platinum and tungsten required non-melting fabrication routes. PM thus reemerged as an indispensable process, later becoming central to the development of cemented carbides and tungsten filaments.

In the post-war industrial expansion of the 1950s, PM found renewed prominence, especially within the automotive and aerospace industries. The industrial consolidation of technology was driven by its ability to efficiently form complex parts with minimal waste and superior microstructural control [23]. This scalability made PM particularly suited for the fabrication of structural and tribological components in transportation systems. The rise of PM coincided with increasing environmental concerns, leading to its integration into energy-efficient systems, lightweight materials, and electric vehicle technologies.

The fundamental principle of PM involves three major stages: powder production, cold consolidation, and sintering. In this process, small particles are consolidated into dense solids. Early advancements in atomization enabled rapid cooling rates in the range of 10^3–10^6 $K{\cdot}s^{-1}$, fostering microstructural refinement through reduced grain and dendrite sizes. These fine microstructures contributed to improved mechanical properties by hindering dislocation motion and promoting the formation of nonequilibrium metastable phases. Pickens' observations show that such rapid solidification processes extended alloying limits and allowed dispersion of otherwise insoluble elements, leading to enhanced corrosion resistance and mechanical performance [4].

Technological refinement of metal powder materials continued with mechanical alloying and oxide dispersion strengthening (ODS). The pioneering work on sintered aluminum powder demonstrated that ball milling in oxygen-rich atmospheres could produce oxide dispersions within aluminum matrices, dramatically increasing high temperature strength and creep resistance. This microstructural engineering became the conceptual precursor for the development of advanced composite systems. Parallel to the evolution of PM processing, material design also advanced. Aluminum matrix composites (AMCs), as reported by Ujah and Kallon [24], represent a natural extension of PM principles. The introduction of ceramic reinforcements such as alumina (Al_2O_3), silicon carbide (SiC), and aluminum nitride (AlN) within aluminum alloy matrices provided promising results in wear resistance, thermal stability, and strength-to-weight ratios. These composites produced through PM routes and technologies such as SPS exhibit fine microstructural uniformity and superior interfacial bonding between reinforcement and matrix, overcoming many of the limitations of liquid route processing. Their observations highlighted the relationship between particle size, porosity, and mechanical properties, where finer powders reduce porosity and enhance stiffness and hardness.

A more recent milestone in PM evolution is the emergence of high entropy alloys (HEAs), characterized by multi-element compositions that stabilize single-phase solid solutions through configurational entropy. HEAs fabricated by PM techniques exhibit refined grain structures, an improved diffusion resistance, and enhanced high temperature performance. Mechanical alloying provides a homogeneous mixing of elements at the atomic scale, with subsequent sintering, whether by hot pressing or advanced methods like high-frequency induction sintering, achieving near-full densification. Such processing advances represent the latest stage in the historical continuum from ancient sintering to modern nanostructured materials.

1.2.2 Powder Production Techniques (Mechanical, Chemical, Atomization)

The production of metallic powders represents the first and most fundamental stage in PM, determining the physical and chemical characteristics that influence compaction behavior, sintering kinetics, and the final properties of consolidated materials (Fig. 1.1). Various powder production techniques exist: mechanical, chemical, and atomization, each governed by distinct thermodynamic and kinetic mechanisms that control particle morphology, purity, and microstructural homogeneity [26].

Mechanical methods constitute a versatile and energy-efficient approach for powder production, particularly suitable for recycling metallic waste. Rojas-Díaz et al. [27] investigated the transformation of aluminum saw chips into powders through mechanical grinding in a ball mill, aiming to evaluate recyclability and the influence of grinding time on particle morphology and powder characteristics. Using a rotational speed of 55 rpm and grinding times between 18 and 96 h, the researchers observed a progressive morphological transition, where flattened flakes formed at shorter times due to plastic deformation and dislocation accumulation, whereas extended milling times induced cold welding and fragmentation, leading to more rounded particles. Particle size, morphology, and specific surface area were correlated with grinding time, with prolonged milling producing crystallite refinement and an increase in dislocation density. However, excessive grinding led to particle agglomeration, affecting flowability and packing behavior.

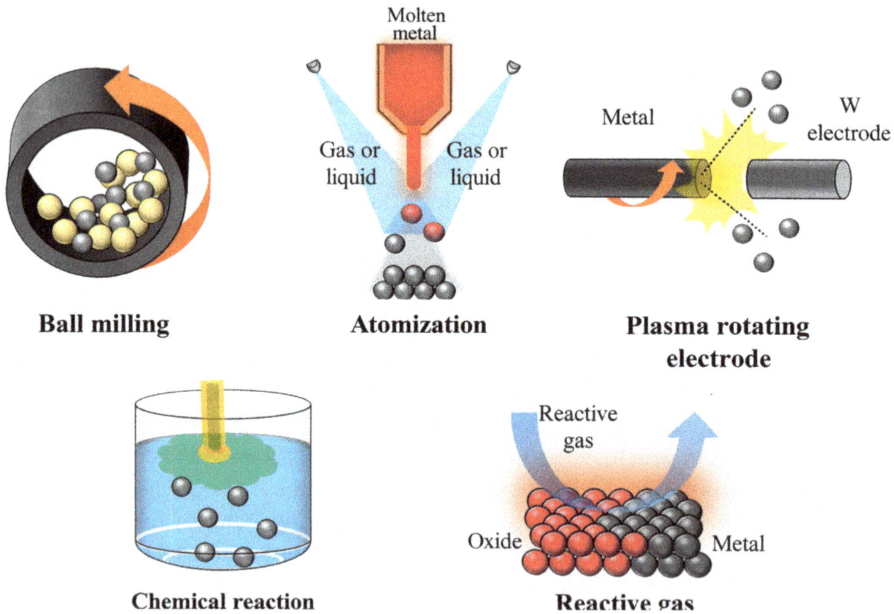

Fig. 1.1 Scheme of the most common powder production methods [25]

Rojas-Díaz et al. [27] also highlighted the challenges associated with the high oxygen affinity of aluminum, which leads to the formation of Al_2O_3 surface layers during milling. This oxide film increases powder stability but hinders interparticle bonding during sintering. Sintering at 620 °C for 1 h under argon increased densification by reducing surface oxides and promoting diffusion bonding. The successful recycling of saw chips demonstrated the feasibility of mechanical routes for sustainable powder production while maintaining acceptable purity and microstructural control. However, beyond mechanical techniques, chemical processes, such as reduction of metal oxides, electrolytic deposition, and precipitation, offer fine particle sizes and high purity but at higher cost and lower throughput. It is necessary to observe that among powder production routes, atomization techniques are the most widely used for large-scale, high quality metallic powder synthesis. Kassym and Perveen [28] classified atomization processes into gas, water, centrifugal, and plasma atomization, emphasizing their role in AM and PM applications. In these methods, molten metal is disrupted into fine droplets by high-velocity fluid jets, followed by rapid cooling and solidification. The morphology, size distribution, and microstructural features depend on parameters such as melt superheat, gas pressure, and atomization medium. Gas atomization produces powders with high sphericity and smooth surfaces under inert atmospheres, yielding narrow size distributions ideal for both pressing and additive routes. It was reported that atomization of metal and alloy powders process parameters, such as melt temperature, atomizing gas type, and nozzle geometry, affect droplet breakup dynamics and cooling rates [29]. High gas pressure and smaller nozzle diameters were correlated with finer powder sizes, while the use of inert gases like argon or nitrogen reduced oxidation and improved powder flowability. It was noted that water atomization, though economical and capable of high production rates, tends to produce irregular, oxidized particles, making it more suitable for ferrous materials or applications where surface oxidation is less critical. In contrast, plasma atomization achieves nearly perfect sphericity and low oxygen content, generating powders particularly suitable for biomedical and aerospace alloys, even though at higher energy cost.

Comparative analysis by Kassym and Perveen and by Comparison of Ultrasonic and Other Atomization Methods showed that ultrasonic atomization produced smaller and more uniform particles through cavitation-induced droplet fragmentation although its application remained restricted to low-viscosity melts and small-scale processing [28, 30]. This contrasts with centrifugal atomization, where high speed rotation disperses molten droplets radially, providing high production rates and excellent particle purity but with a wider particle size distribution. The ultrasonic technique was shown to be particularly effective for producing reactive powders and fine alloys, as the localized high-frequency oscillations minimize oxidation and allow precise control of particle morphology.

1.2.3 Compaction and Shaping Methods (Uniaxial Pressing, Isostatic Pressing, Injection Molding, Additive Routes)

The transformation of metallic powders into dense, shaped components constitutes the essence of PM where compaction and shaping are decisive stages in defining the final density, microstructure, and performance of the material. These methods have evolved from simple uniaxial pressing to more sophisticated routes, including isostatic compaction, metal injection molding (MIM), and powder-bed AM, each governed by distinct mechanisms of pressure transmission, particle rearrangement, and densification kinetics.

1.2.4 Uniaxial and High Velocity Compaction (HVC)

Conventional uniaxial pressing remains the foundational method in PM, utilizing rigid dies to apply pressure along a single axis. However, modern adaptations such as high velocity compaction (HVC) have significantly advanced the process (Fig. 1.2). HVC introduces kinetic energy via impact or vibration, leading to instantaneous pressure transmission that enhances particle rearrangement and localized plastic deformation [32, 33]. The dynamic loading minimizes internal friction and results in higher green densities compared to static compaction. Their experimental investigations demonstrated that powder particle deformation is dominated by adiabatic shear and localized strain accumulation, leading to reduced porosity and improved uniformity. The use of HVC is particularly effective for high strength

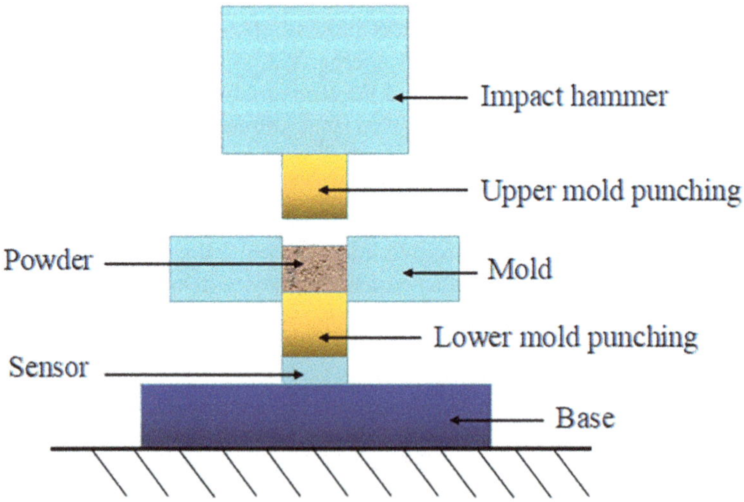

Fig. 1.2 Experiment device schematic diagram of HVC [31]

ferrous and nonferrous powders, achieving relative densities up to 98% with shorter cycle times and reduced tooling wear.

1.2.5 Isostatic Pressing and Hot Isostatic Pressing (HIP)

While uniaxial pressing is limited by frictional gradients along the die walls, isostatic pressing applies pressure uniformly in all directions through a fluid medium. This principle is exploited in both cold isostatic pressing (CIP) and hot isostatic pressing (HIP, Fig. 1.3). Wang and coworkers [35] described the use of HIP in fabricating titanium alloys and their composites, demonstrating near-wrought densities exceeding 99% and the elimination of microstructural defects common in pressureless sintering. It is reported that powders were encapsulated and subjected to simultaneous heat and pressure, promoting diffusion bonding and grain boundary healing. The HIPed Ti-6Al-4V alloys exhibited fine α/β lamellar structures and mechanical properties equivalent to forged counterparts. Moreover, the authors highlighted that the combination of CIP preforming followed by HIP sintering produces homogeneous densification and minimizes segregation, particularly for powders with irregular morphologies.

Near-net-shape HIP of Inconel 625 reinforced with SiC and TiB_2 powders produced fully dense Ni-based metal matrix composites with a continuous network of ceramic reinforcements along prior particle boundaries for space propulsion applications [36]. Microhardness and wear resistance increased markedly with

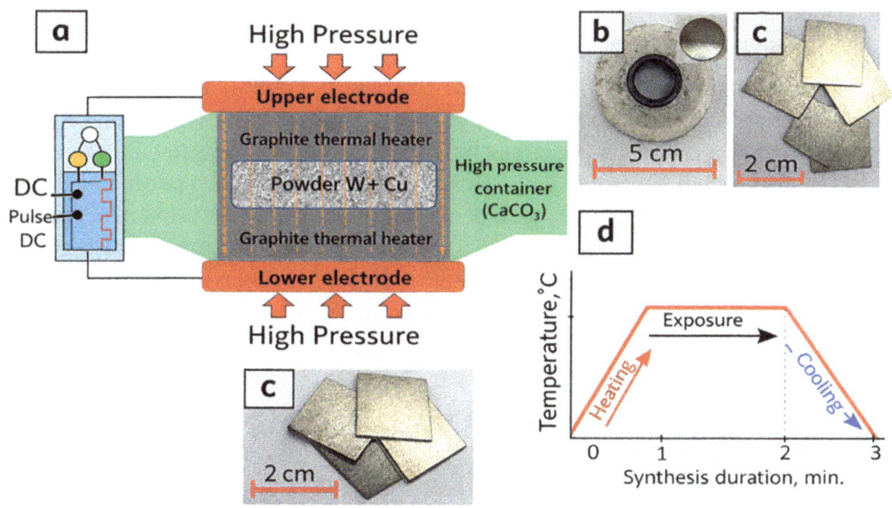

Fig. 1.3 Schematic view of an isostatic hot pressing (**a**), sample preparation container with graphite heating element and W-Cu samples (insert) (**b**), radiation shields based on W-Cu composite with different thicknesses (**c**), and a schematic view of the sintering process (**d**) [34]

reinforcement volume fraction, confirming that HIP not only consolidates but also homogenizes complex composite systems. The feasibility of producing functional, near-net mechanical seals through PM HIP highlighted its industrial applicability in high stress aerospace environments.

1.2.6 Metal Injection Molding

The versatility of powder shaping is further extended by metal injection molding, which integrates polymer processing principles into PM. Moon et al. [10] observed that the progress and challenges of iron-based powders in MIM (Fig. 1.4), emphasizing its capacity for mass production of complex geometries with high dimensional accuracy and minimal post-machining. The process involves the preparation of feedstock composed of fine metal powders and thermoplastic binders, which are injection molded, debound, and sintered. MIM allows for near-net-shape production of intricate components used in automotive and biomedical applications. Experimental advancements in binder systems and powder size optimization have enabled densities above 96%, rivaling conventional PM compaction.

Furthermore, prealloyed powders, typically produced by gas atomization, offer superior flowability and mechanical performance compared to mixed or master-alloy powders, though at higher cost. The combination of carbonyl iron powders and hybrid alloy systems enhances sinterability and surface finish, expanding the scope of MIM into magnetic and structural applications, which indicates that the evolution of shaping technologies has culminated in AM, where powder compaction and fusion occur simultaneously in a layerwise fashion. Moreover, powder bed fusion (PBF) is the most advanced AM route, encompassing laser (L-PBF) and electron beam (EB-PBF) processes [38]. In L-PBF, a laser selectively melts powder layers according to a digital model, enabling near-full density (≥99%) and intricate geometries impossible through conventional pressing. EB-PBF operates under vacuum and offers reduced residual stresses and oxidation, particularly beneficial for reactive metals such as Ti and Ni-based alloys. However, L-PBF provides superior resolution and surface finish due to its finer melt pools. Improper parameter combinations can lead to porosity and residual stress accumulation. The authors compared PBF with directed energy deposition (DED), noting that while DED achieves faster deposition rates, PBF delivers higher precision and mechanical consistency.

1.2.7 Mechanisms, Stages, and Influencing Factors of Sintering

The sintering stage in powder metallurgy governed densification kinetics, microstructural evolution, and the resulting mechanical performance of consolidated

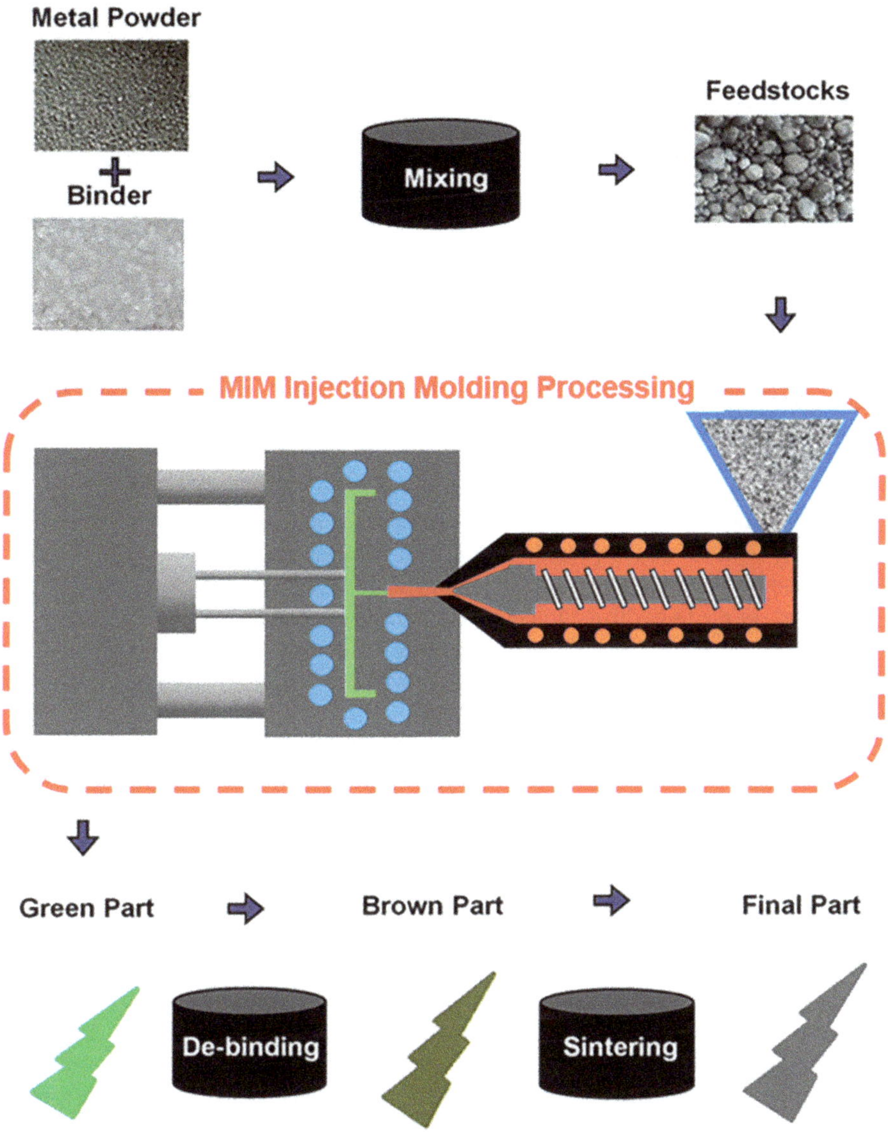

Fig. 1.4 MIM processing flow diagram [37]

materials (Fig. 1.5). Three sintering stages were identified, beginning with an initial regime characterized by interparticle neck formation [11]. The intermediate stage was marked by densification and pore coalescence. The final stage is where residual porosity is minimized and a stable, dense microstructure is achieved. In the initial stage, atoms migrate from regions of high surface energy toward particle contact zones, driven by gradients in chemical potential and the reduction of total interfacial

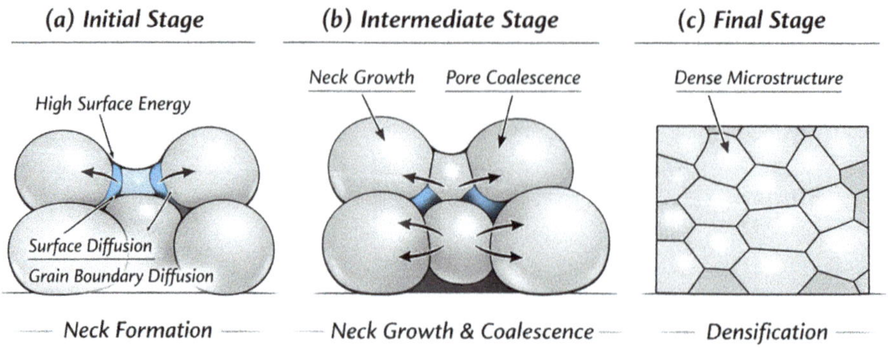

Fig. 1.5 Schematic 2D representation of the sintering process in PM showing (**a**) the initial stage with neck formation, (**b**) the intermediate stage with neck growth and densification, and (**c**) the final stage characterized by a dense, stable microstructure

energy. This atomic transport results in the formation and growth of sintering necks, whose rate depends on temperature, time, and particle characteristics. The research group identified surface and grain boundary diffusion as dominant mechanisms at early stages, while bulk diffusion became predominant during the intermediate stage as necks coalesce and densification accelerates. These mechanisms define pore morphology and grain boundary formation, which are directly correlated with the mechanical integrity of the final material.

In composite systems, sintering behavior is further affected by the interaction between the metallic matrix and reinforcing particles. Kumar et al. [39] investigated this behavior in an AA7075-B_4C composite fabricated by conventional PM routes, reporting that densification occurs through a combination of grain boundary diffusion and thermally activated plastic flow. Increasing sintering temperature increased diffusion rates and matrix reinforcement bonding up to an optimum level, beyond which grain coarsening reduced hardness and strength, with excessive thermal exposure promoting coarsening and porosity entrapment. A detailed analysis of SPS described the contribution of an electric field-assisted mechanism relative to conventional thermal sintering [14, 15, 32]. In SPS, a pulsed direct current generates localized plasma discharges and Joule heating at particle contact points, producing highly localized temperature gradients [14]. These effects enhance atomic diffusion, induce partial surface melting, and significantly accelerate neck formation. The combination of simultaneous heating and applied uniaxial pressure promotes plastic flow and bonding between particles, reducing the required sintering temperature by 200–300 °C compared to traditional methods. It was also demonstrated that the SPS process enables heating rates up to 1000 °C/min, effectively suppressing grain growth while achieving near-theoretical densities [14, 15]. Therefore, densification is highly dependent on current amplitude, pulse duration, and dwell time: moderate sintering durations improved microstructural homogeneity, while prolonged exposures resulted in grain coarsening and reduced ductility.

At the microscopic scale, the mass transport phenomena governing sintering involve multiple diffusion paths. Six major mechanisms contribute to atomic movement during sintering, of which three, grain boundary diffusion, lattice diffusion from the boundary, and dislocation-mediated plastic flow, directly enhance densification [11]. The predominance of each mechanism depends on process conditions: in conventional sintering, lattice diffusion dominates, whereas in SPS or other field-assisted methods, surface diffusion and electromigration play more significant roles due to electric activation and local plasma formation. During pulsed discharge, transient plasmas develop between particles, promoting evaporation, localized melting, and rapid homogenization [32]. These events accelerate bonding and reduce the sintering time required to achieve densities above 99%. Experiments on pure titanium and Cu-Ni alloys revealed that precise control of pulse parameters and applied pressure prevents abnormal grain growth while maintaining high densification efficiency.

1.2.8 Infiltration, Heat Treatment, Machining, and Surface Finishing as Secondary Operations

Secondary operations in PM play an important role in enhancing mechanical performance, dimensional precision, and surface integrity of sintered components (Fig. 1.6). These post-sintering treatments, including infiltration, heat treatment, machining, and surface finishing, are designed to refine microstructure, optimize density, and tailor mechanical behavior according to service requirements. Secondary processes, though historically less explored than primary PM operations, are

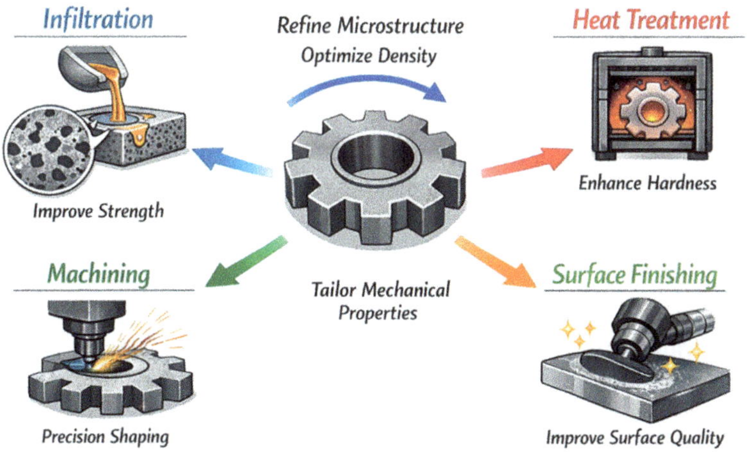

Fig. 1.6 Secondary post-sintering operations in PM, such as infiltration, heat treatment, machining, and surface finishing, used to refine microstructure, improve density, and tailor the mechanical performance of PM components

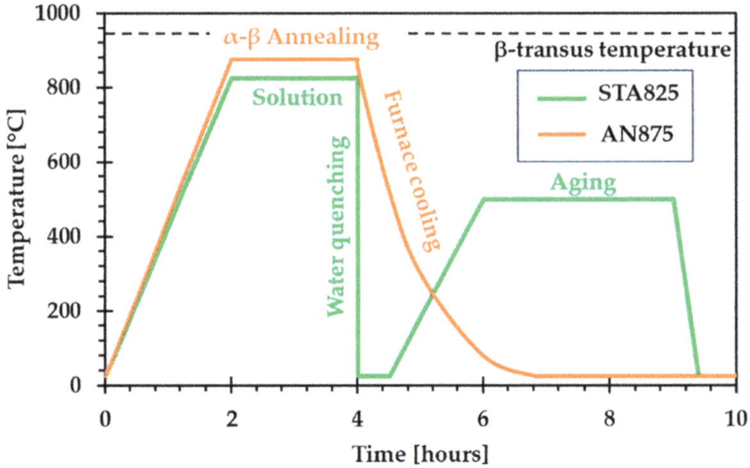

Fig. 1.7 Representation of STA825 (green) and AN875 (orange) thermal cycles [42]

decisive in determining the final mechanical response of metallic and composite systems due to their influence on microstructural homogenization and stress distribution [40]. Infiltration, particularly melt infiltration, is commonly employed to improve density and load-bearing capacity in porous PM components. Melt infiltration with metallic alloys such as bronze or copper increased bonding strength and wear resistance in porous iron substrates, producing a quasi-dense metallic network with higher hardness and lower friction coefficients [6, 41]. The infiltration process promotes the filling of interconnected porosity and establishes metallurgical continuity between the infiltrant and the host matrix, effectively reducing microvoids that act as stress concentrators. Moreover, the infiltration temperature and infiltration rate were identified as key parameters; excessive temperature caused interfacial segregation, while insufficient temperature led to incomplete pore filling. The optimized infiltration regime achieved by the authors resulted in a 40% increase in yield strength and a 25% improvement in wear resistance compared to the non-infiltrated control samples.

Heat treatment following sintering modified microstructure through precipitation hardening, grain refinement, and phase transformation (an example of heat-treated STA825 and AN875 is presented in Fig. 1.7). In PM titanium alloys, tailored thermal schedules were used to refine the α/β morphology and to improve the balance between strength and ductility [43]. Post-sintering solution treatment followed by aging significantly refines β grain size and enhances dislocation density, leading to a 15% increase in tensile strength and improved fatigue life. Combining rapid heat treatment with subsequent aging on near-β titanium alloys improved the strength-to-weight ratio by promoting uniform α phase precipitation within a refined β matrix [44].

Machining and surface finishing operations are often necessary to achieve the required dimensional tolerances and surface smoothness (Fig. 1.8), especially for

Fig. 1.8 An example of a Ti-MIM component, in this case a tripod base [45]

functional components exposed to cyclic or tribological loading. In metal matrix composites (MMCs) processed via PM, machining-induced residual stresses and surface deformation can influence performance, particularly in aluminum-based systems where reinforcement particles harden the surface but also increase tool wear [46]. It was noted that advanced machining approaches, such as high speed cutting and cryogenic-assisted machining, reduce microcrack formation at the matrix reinforcement interface. Furthermore, secondary finishing treatments, including polishing and coating, were reported to enhance fatigue resistance and tribological stability by minimizing surface asperities that act as crack initiation sites.

Optimization of these secondary processes has been approached through statistical and computational methods to enhance overall performance. Taguchi and response surface methodologies have been successfully applied to identify optimal conditions for secondary treatments, significantly reducing the experimental complexity and cost while improving repeatability and process control [40], and have emphasized the potential of multi-response optimization to simultaneously enhance multiple mechanical outcomes such as strength, hardness, and wear resistance.

1.3 Key Materials and Their Properties (Metals, Ceramics, and Composites)

1.3.1 Common Alloys, Particle Morphology, and Mechanical Behavior of Metallic Powders

Metallic powders controlled both processing behavior and performance in powder metallurgy and additive manufacturing (Fig. 1.9). Aluminum, titanium, and nickel-based alloys were most widely used due to their favorable strength-to-weight ratios, corrosion resistance, and suitability for high temperature operation. Their properties and behaviors are profoundly influenced by particle morphology, production method, and microstructural stability during consolidation. The high reflectivity of aluminum to laser radiation and its poor powder flowability result in processing instabilities, porosity, and crack formation in the final parts. Gas atomization, the most common method for producing metallic powders, yields near-spherical particles with high packing density and consistent flowability, essential for both laser powder bed fusion and traditional PM compaction processes [47]. However, deviations from ideal morphology, such as irregular or elongated particles, lead to poor layer uniformity and inhomogeneous densification.

The relationship between powder morphology and mechanical behavior is especially pronounced in composite systems, where the morphology of the composite

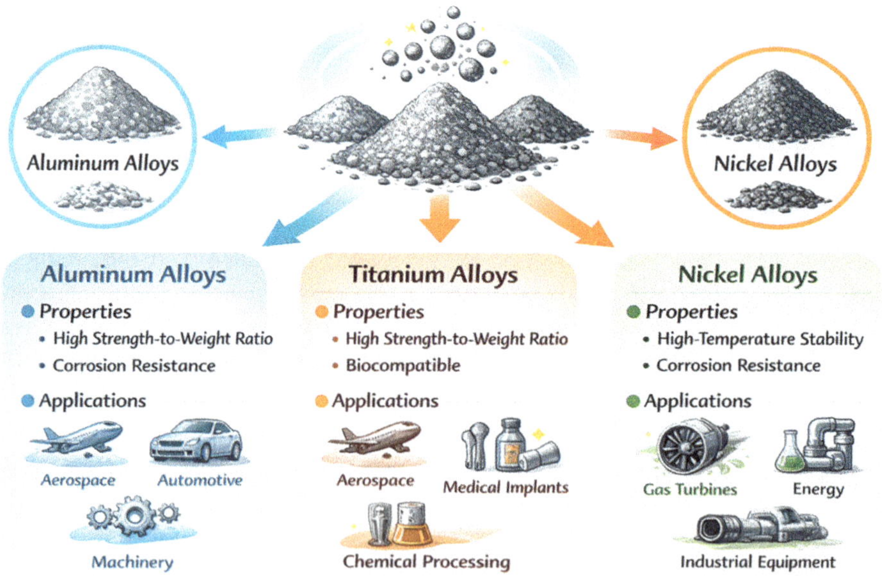

Fig. 1.9 Metallic powders used in PM and AM, highlighting aluminum, titanium, and nickel alloys and their key properties for lightweight, corrosion-resistant, and high temperature applications

powder could be optimized for laser powder bed fusion despite its non-spherical shape [19]. Their experiments demonstrated that mechanical alloying, although traditionally associated with poor flowability, can produce powders with homogeneously distributed ceramic reinforcements when properly processed and sieved. The AlSi10Mg-TiB_2 composite powders exhibited flow and packing behavior comparable to that of gas-atomized aluminum, indicating that careful control of particle size distribution and surface texture can mitigate flow issues.

Further investigations into the reuse and recyclability of metal powders have revealed that repeated thermal exposure and oxidation during AM cycles can alter surface chemistry and reduce flowability. Aluminum powders show a high tendency for moisture absorption, leading to hydrogen-induced porosity upon melting [18, 35, 47]. This phenomenon compromises mechanical integrity and highlights the necessity of controlled environments with low oxygen partial pressure. Gas-atomized spherical powders retained high flowability, whereas progressive oxidation led to thickening of surface films that modified energy absorption during laser interaction and altered melt pool dynamics. Reused powders therefore required screening and surface conditioning to maintain consistent processing behavior.

Mechanically, metallic powders used in AM and PM display distinctive performance trends depending on alloy type. Aluminum alloys, particularly AlSi10Mg and AlSi12, show high ductility and moderate strength due to fine microstructures formed during rapid solidification [19]. However, their mechanical response is sensitive to pore formation and microsegregation. In contrast, titanium alloys produced via powder routes exhibit excellent strength and fatigue resistance, attributed to their inherent α-β phase balance and refined grain structures [44]. Nickel-based alloys, with higher density and superior oxidation resistance, remain indispensable for high temperature applications. However, they demand careful control of powder chemistry to prevent contamination [18]. Moreover, a key insight lies in the mechanical response of composite powders under consolidation [19]. The AlSi10Mg-TiB_2 systems demonstrated that increasing the volume fraction of ceramic particles enhances stiffness and wear resistance but requires a balance to avoid agglomeration and crack initiation [16]. Powders with homogeneously embedded reinforcements between 0.5 and 10 μm exhibited superior isotropy and load transfer efficiency compared to systems with nanoscale particles prone to clustering.

1.3.2 Ceramic Powders: Brittleness, High Temperature Resistance, and Structural Role

Ceramic powders contributed to the development of structural and functional materials through high temperature stability, hardness, and reinforcement capability (See secondary electron SEM micrographs presented in Fig. 1.10). Covalent and ionic bonding produced high elastic modulus and hardness, while limiting plastic deformation and fracture toughness. These characteristics make ceramics suitable for high

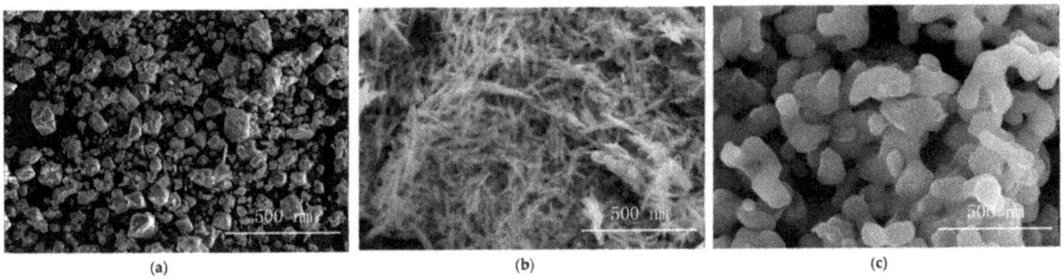

Fig. 1.10 SEM images of (**a**) Al_2O_3-NPs, (**b**) Al_2O_3-NWs, and (**c**) Al_2O_3-NPLs [48]

temperature and wear-resistant applications, but their processing requires careful control of powder morphology, distribution, and sintering behavior to mitigate defects. Silicon nitride (Si_3N_4) ceramics demonstrate outstanding mechanical and thermal properties, reporting applications in aerospace and energy systems [49]. However, their inherent brittleness and difficulty in machining complex geometries necessitate the use of AM or powder modification strategies to improve their processability. In this regard, a non-aqueous chemical precipitation (NCP) process was applied to coat Si_3N_4 powders with yttrium aluminum garnet (YAG), enabling uniform distribution of sintering additives and reducing viscosity by 42% compared to traditional wet ball milling (WBM) [15]. This coating enhanced the rheological and curing behavior of Si_3N_4 slurries used in vat photopolymerization, leading to ceramics with improved densification and bending strength of 833 MPa, approximately 18% higher than WBM-prepared samples. The uniform YAG coating facilitated elongated grain formation and homogeneous sintering, which mitigated stress concentration and enhanced the Weibull modulus, demonstrating that powder surface engineering can overcome the limitations imposed by brittleness.

Beyond nitrides, ceramic powders also contribute significantly to the development of composite systems where their refractory nature imparts thermal stability and dimensional integrity. The inclusion of ceramic powders in cementitious matrices has been explored to replace traditional calcium aluminate cement (CAC), combining sustainability with high temperature resistance. Koňáková et al. [49] reported that substituting up to 60% of CAC with ceramic powder (CP) and reactive alumina (RA) produced composites capable of withstanding temperatures up to 1400 °C without degradation of strength. The ternary system A15-C15, containing 15% RA and 15% CP, exhibited compressive strength of 107.6 MPa and maintained hydration activity at 280 J/g, reflecting a synergistic enhancement between the pozzolanic reactivity of ceramic powders and the high alumina content of reactive fillers. Microstructural analysis revealed reduced porosity and formation of stable calcium aluminosilicate hydrates, which promoted long term structural integrity. The high temperature resistance of ceramic powders derives primarily from their stable crystalline phases, such as Si_3N_4, Al_2O_3, and mullite, which maintain strength and stiffness at elevated temperatures. Their brittleness, while a drawback under mechanical stress, is offset by their ability to maintain dimensional stability and resist creep, oxidation, and softening beyond 1000 °C.

1.3.3 Composites: Combining Phases for Tailored Properties

The design and performance of composite materials rely on the synergistic combination of two or more distinct phases, typically a continuous metallic or polymeric matrix and a discontinuous reinforcement phase that can be ceramic, carbonaceous, or metallic. Within PM, AMCs are the most representative examples of this concept, combining lightness, strength, and thermal stability. Composites are distinguished by the nature of their matrix and reinforcement, and aluminum alloys are often chosen as the base phase because of their low density and good castability, while reinforcements such as SiC, B_4C, Al_2O_3, or TiC provide the stiffness and hardness required for high performance applications [39]. These reinforcements, when finely dispersed within the metallic matrix, modify its load-bearing capacity and resistance to deformation, producing tailored mechanical and tribological properties. A schematic representation of the microstructures of composites obtained by the consolidation of ceramic and composite powders is presented in Fig. 1.11.

Among the various fabrication routes, solid-state processing such as PM offers a notable advantage in achieving uniform dispersion and clean interfaces between phases. Sadhu et al. [51] report that the PM route allows a high reinforcement fraction, up to 75%, while maintaining good particle bonding and homogeneity. It was observed that SiC particles enhance load-bearing capability and induce dislocation density through the mismatch in thermal expansion with the aluminum matrix, strengthening the composite by mechanisms such as load transfer and dislocation hardening. For instance, the addition of SiC resulted in a maximum improvement in hardness and tensile strength of 198% and 295%, respectively, compared to pure aluminum, while also reducing wear rate and friction coefficient by over 60%. Similarly, graphene additions improved both strength and ductility owing to its exceptional stiffness, high aspect ratio, and lubricating properties, which facilitate stress transfer and crack deflection. The PM route also mitigates the undesirable interfacial reactions that can occur in liquid routes, which often compromise bonding and increase porosity.

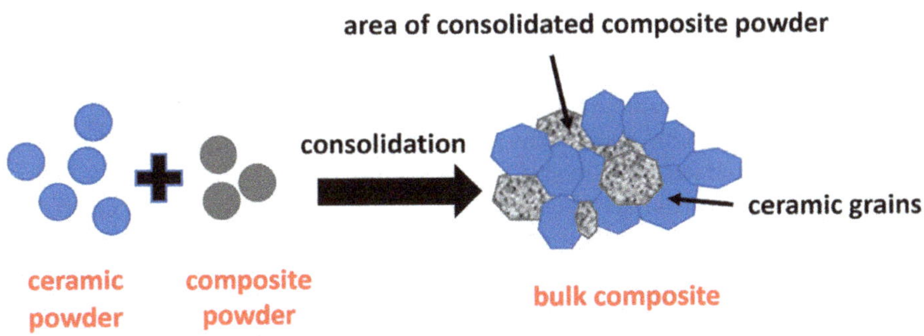

Fig. 1.11 Schematic representation of the microstructures of composites obtained by the consolidation of ceramic and composite powders [50]

Additional research shows that the mechanical response of composites is closely tied to reinforcement morphology and distribution. A uniform dispersion achieved through solid-state routes ensures a more homogeneous microstructure than that obtained via casting, which is prone to segregation or clustering of the ceramic phase. The particle size ratio between aluminum and SiC (PSR $\approx$ 1) optimizes density and hardness by promoting better packing and minimizing voids. Moreover, the addition of bimodal reinforcement distributions, where large and fine SiC particles coexist, enhances both the packing density and thermal conductivity of the composite. In graphene-reinforced systems, similar control of morphology determines the efficiency of load transfer; however, excessive graphene leads to agglomeration, acting as a defect source that reduces ductility [39, 51].

From a mechanical perspective, Al-SiC and Al-graphene composites exhibit distinctive strengthening mechanisms. SiC primarily contributes through Orowan looping and dislocation pinning, while graphene promotes strain hardening and crack bridging across the matrix. The combined reinforcement of SiC and graphene in hybrid composites can balance stiffness and toughness by merging ceramic and carbonaceous strengthening mechanisms. The observed improvements in tensile and compressive properties derive from effective stress sharing between the matrix and reinforcements, as well as refinement of the aluminum grain structure during sintering. Additionally, the electrical and thermal conductivity of these composites are highly dependent on the integrity of the interface. Dense, well bonded microstructures facilitate electron and phonon transport, while voids and weak interfaces act as scattering centers that lower conductivity.

1.4 Relevance of Powder Metallurgy in Modern Manufacturing

1.4.1 Advantages over Conventional Processing Methods

PM represents a transformative approach to materials processing, offering distinct advantages over conventional manufacturing methods, especially in terms of material efficiency, near-net shape capability, and the fabrication of geometrically complex components. The process fundamentally minimizes material wastage by utilizing powders as the primary feedstock, ensuring that nearly all material contributes to the final part, unlike subtractive techniques such as machining. PM enables near-net shape production, reducing post-processing requirements and promoting cost-effective mass manufacturing with tight dimensional control, particularly in metal injection molding and sintering applications [1]. This material utilization efficiency becomes crucial when processing expensive or rare metals such as titanium, nickel-based alloys, and high performance magnesium systems, where raw material conservation directly impacts economic feasibility and sustainability. Furthermore, the PM route allows the development of intricate and complex geometries that are challenging or even impossible to achieve via casting or forging. The introduction of AM techniques within the PM framework, particularly selective laser melting (SLM) and laser direct energy deposition (LDED), has revolutionized the design possibilities of metallic components [8]. These techniques facilitate the creation of components with internal channels, graded porosities, and lattice architectures without the need for extensive tooling or machining, drastically reducing lead times and material waste. The direct correlation between process parameters, such as laser power, scanning speed, and powder morphology, and the resultant microstructure underscores the adaptability of PM in tailoring mechanical and functional properties according to application specific demands.

In the context of functional alloys like NiTi, the precision and flexibility of PM processes contribute to improved microstructural uniformity and enhanced shape memory behavior. HIP and SPS enable the production of dense, homogeneous NiTi components with controlled porosity and minimal impurity phases, outperforming conventionally cast materials in terms of fatigue resistance and surface integrity. Moreover, AM-based PM routes allow the fabrication of NiTi components with highly complex geometries suitable for biomedical implants and aerospace actuators, further extending the scope of PM beyond traditional limitations. The ability to integrate multiple materials in a single build, as achieved in direct energy deposition, provides an additional layer of design freedom for functionally graded materials and multiphase composites [8].

PM enabled control of microstructure in addition to geometric complexity and material efficiency. Adjustment of compaction pressure, sintering temperature, and particle size distribution produced microstructures with improved mechanical response and isotropic behavior, including in materials with limited machinability [1]. Lower sintering temperatures and controlled atmospheres in powder metallurgy limited oxidation and contamination, improving fatigue and corrosion resistance in reactive systems such as magnesium alloys, while Ti-50Ta processed by this route exhibited a diffusion-controlled microstructure with refined grains and increased hardness without cast-related macrosegregation. Near-net-shape processing reduced machining, energy consumption, and material waste, with unused powders recoverable and recyclable in powder bed systems, supporting material conservation and cost efficiency in circular manufacturing cycles [8].

1.4.2 Limitations and Challenges

PM faces three primary categories of limitations, powder cost, part size constraints, and defect sensitivity, that condition its scalability despite its advantages in material efficiency and near-net shape forming, as can be observed in Fig. 1.12 in a component produced by MIM. Powder cost is dominated by the production of spherical, low-oxygen particles by gas or plasma atomization, which imposes a high energy and capital burden; handling requirements for reactive alloys (e.g., Ti, Al, Mg) further raise costs due to inert storage, sieving, and safety protocols [3]. In parallel,

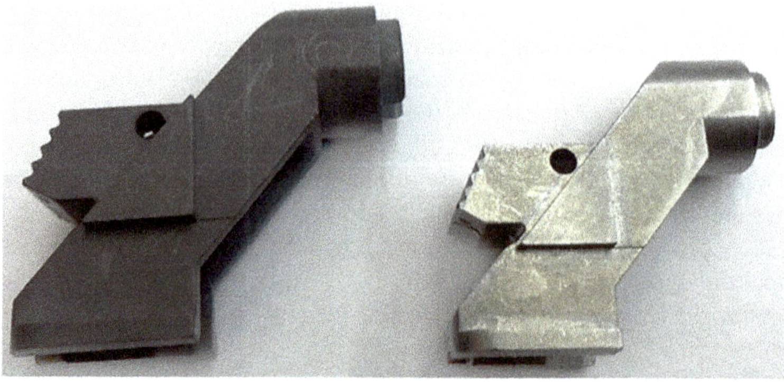

Fig. 1.12 Green part (**left**) and "guidepost" sintered part made of MIM-4140 steel (**right**) [52]

when PM is applied to light alloy design, the powder route itself adds a premium relative to ingot metallurgy, even though it enables chemistries and microstructures inaccessible at equilibrium; this economic balance is explicitly noted for Al-based systems where gas-atomized feedstocks and specialized blending/sintering sequences are required.

Part size limits arise from both tooling/pressing physics and thermal management during layerwise fusion. In conventional PM, uniaxial compaction of high aspect ratio shapes generates density gradients along the pressing axis due to frictional losses at die walls, which propagate into nonuniform shrinkage and property anisotropy after sintering; isostatic routes mitigate but do not eliminate the cost/complexity for large parts. In powder bed additive routes, build volume and heat accumulation constrain the maximum component envelope; residual stresses from steep thermal gradients promote warping and cracking in tall builds, particularly in high strength Ni and Ti alloys, so geometry and scan strategy must be engineered within machine limits.

Defects constitute a third challenge that couples directly to fatigue and fracture performance (Fig. 1.13). In laser/electron powder bed fusion, lack-of-fusion porosity, keyhole pores, and unmelted particles originate from mismatches between power density, scan speed, hatch spacing, and layer thickness that perturb melt pool stability; spatter and oxide films exacerbate lack of fusion by disrupting wetting and interlayer bonding [20]. In press and sinter PM of light alloys, surface oxides (e.g., Al_2O_3) on powder particles inhibit metallurgical bonding at necks, leaving interparticle films that act as crack initiators under cyclic loading; diffusion distances and thermodynamics of deoxidation during solid state or supersolidus liquid phase sintering govern whether these films are ruptured or retained.

Process parameter windows that avoid these pitfalls are consequently narrow. For AM, strategies such as tailored scan vectors, preheating, and controlled atmosphere limit oxygen pickup and stabilize melt pools, but they add time and equipment complexity; even then, post-processing like HIP may be necessary to close defects at the expense of cost and throughput. For press and sinter PM, careful control of

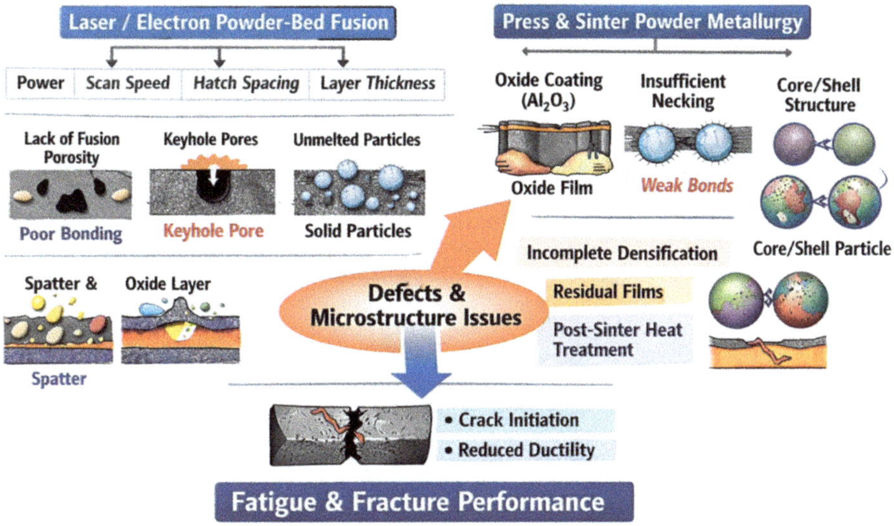

Fig. 1.13 Schematic overview of defect formation in powder-bed fusion and press-and-sinter PM, highlighting how process-induced porosity, oxide films, and incomplete densification couple microstructure to reduced fatigue resistance, fracture performance, and ductility

particle size distribution, dispersants, pressing route, and sintering schedule is required to minimize gradients and oxide persistence, with the recognition that these controls themselves contribute to processing overhead [14, 53].

1.4.3 Applications Across Industries

PM has found broad applicability across major industrial sectors due to its capacity to deliver components with tailored microstructures, high dimensional accuracy, and efficient material utilization [2]. In the automotive industry (Fig. 1.14), PM is a mature technology that provides lightweight, high strength components with minimal machining requirements. PM enables the large-scale production of gears, valve guides, connecting rods, and bushings with excellent wear resistance and cost efficiency through the optimization of powder composition, compaction pressure, and sintering temperature [54]. Fe-Cu-C systems exhibit improved mechanical performance when lubricants and alloying agents are finely distributed in the powder mixture, allowing superior densification and homogeneous microstructures. Furthermore, in automotive applications of PM, the environmental and economic advantages of PM were highlighted, emphasizing reduced CO_2 emissions and waste generation relative to casting and forging due to near-net shape forming and the reuse of powder feedstock during production [41, 54].

In the aerospace sector (Fig. 1.15), PM technologies enable the fabrication of high performance alloys such as titanium and nickel-based systems that combine low density with exceptional strength and creep resistance [17, 43]. Prealloyed

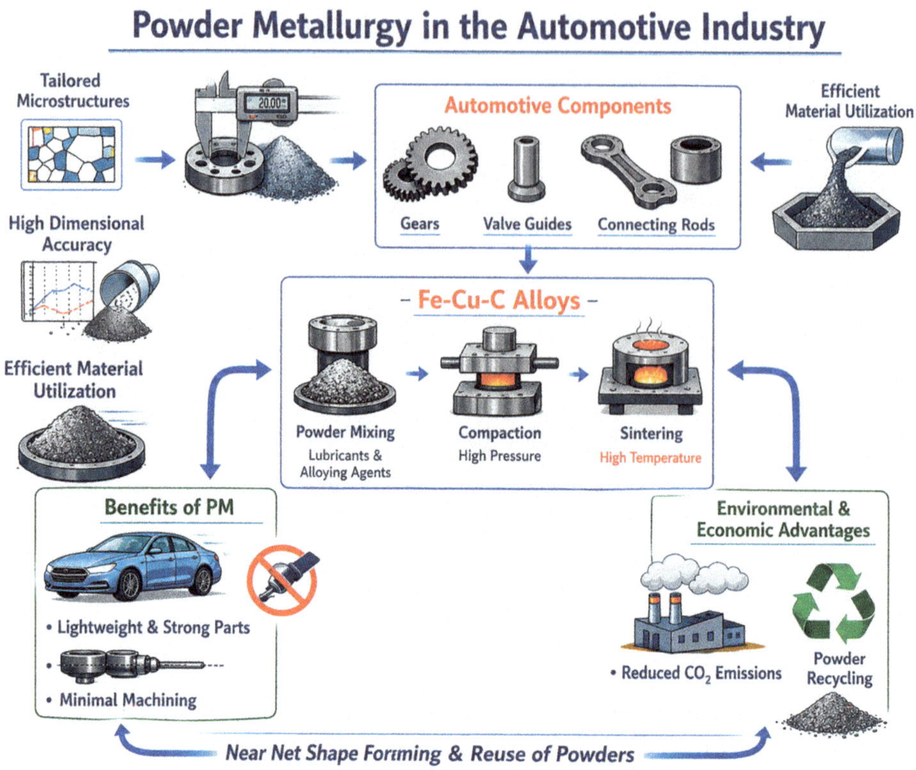

Fig. 1.14 Schematic overview of PM in the automotive industry, highlighting processing steps, optimized Fe–Cu–C systems, key components, and environmental and economic advantages

Ti-6Al-4V powders consolidated by hot isostatic pressing and subsequent thermomechanical treatment yielded components with densities above 99.5%, tensile strengths exceeding 900 MPa, and elongation greater than 10% for aeronautical use. Control of particle size distribution and oxygen content in the feedstock reduced defect populations and increased fatigue reliability under cyclic loading conditions. Rapid solid-state diffusion during sintering promoted α/β phase equilibration, producing a uniform lamellar structure comparable to that of wrought material [44]. In comparison, conventional casting routes often suffer from macrosegregation and coarse grain growth, highlighting the advantage of PM in microstructural refinement and reproducibility for demanding aerospace environments [56].

The adaptability of PM extends to the biomedical field, where the ability to precisely engineer alloy composition and porosity allows the design of biocompatible and load-bearing implants (Fig. 1.16). Research carried out on Ti-Nb-Mg alloys fabricated by PM showed that partial Mg volatilization during sintering induced controlled porosity, leading to elastic moduli close to that of cortical bone and thus reducing stress shielding [32]. The interconnected pore network also enhances osteointegration potential without sacrificing compressive strength, which reached values near 650 MPa. Complementarily, investigations on modified 316L stainless

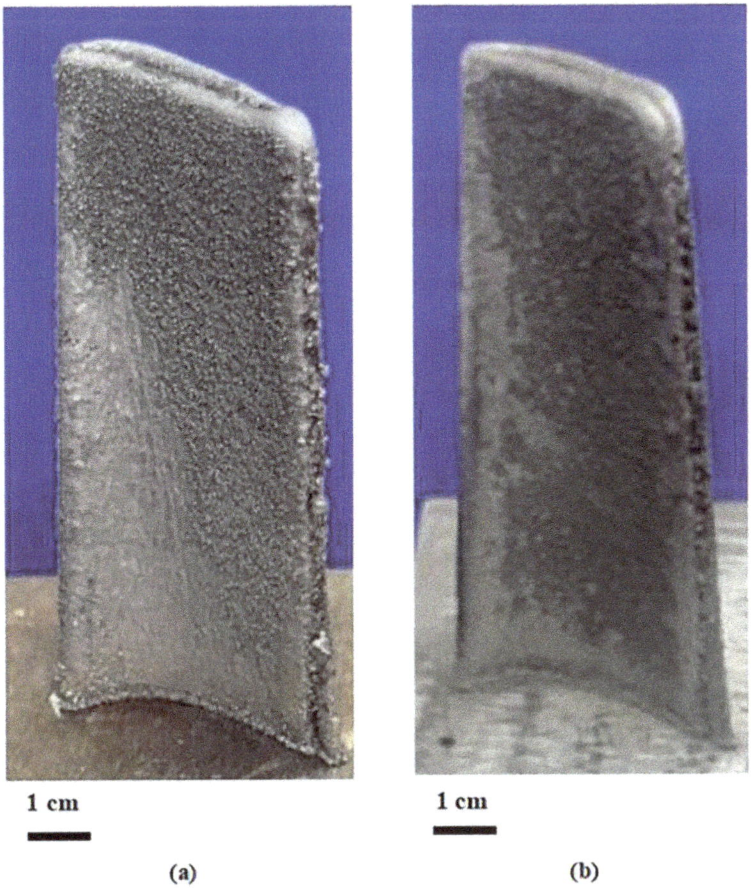

Fig. 1.15 Macroscopic morphology of DED-printed aerospace blades: (**a**) using new powder, (**b**) using mixed powder [55]

steel powders revealed that the addition of Mo and Mn during mechanical alloying refined austenitic grain size and suppressed σ-phase formation, improving both corrosion resistance and ductility. The sintered 316L compacts exhibited hardness values exceeding 200 HV and ultimate tensile strengths above 500 MPa, confirming the capability of PM to adjust microstructural stability and mechanical performance through powder design and sintering control [5].

The energy sector increasingly benefits from the precision of PM in fabricating nanostructured materials with enhanced electrochemical functionality. The development of oxide and carbon-based nanocomposites synthesized via high energy ball milling demonstrates the relevance of PM beyond structural applications. Repeated cold welding and fracture during mechanical milling generated metastable solid solutions with refined grain boundaries that enhanced ionic transport in nanocomposite electrode systems. Mechanically alloyed NiO-C and TiO_2-graphene

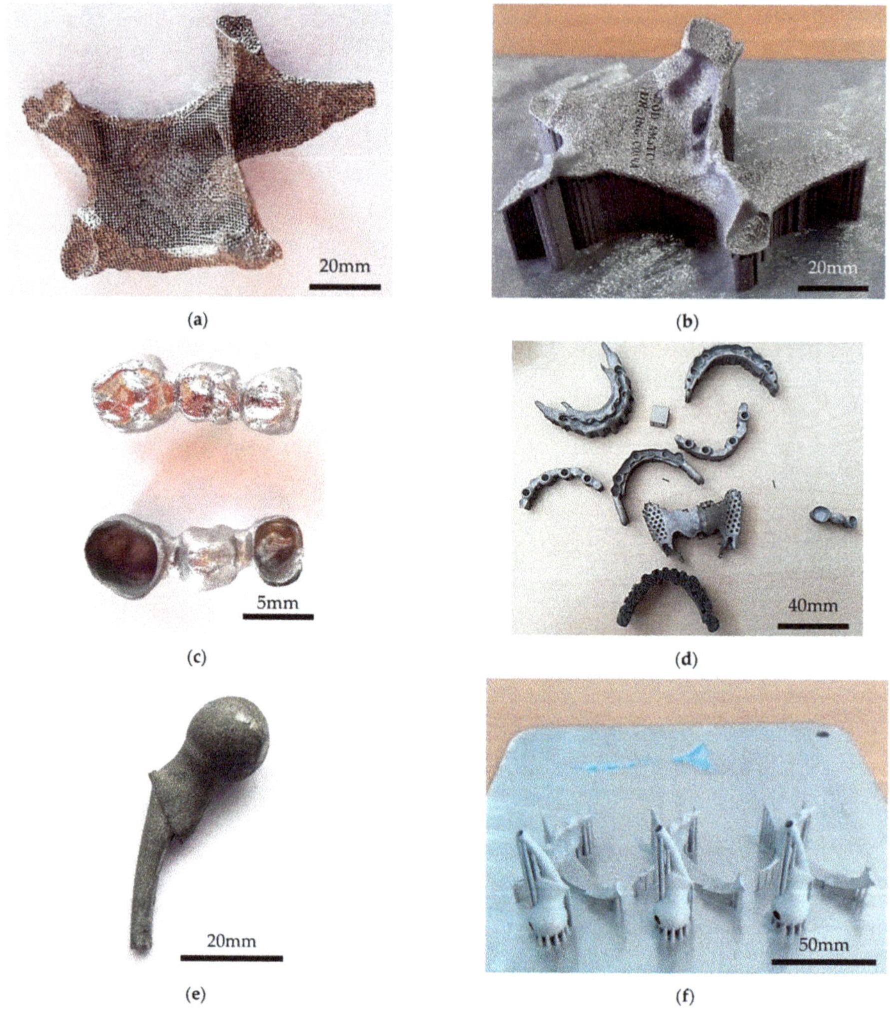

Fig. 1.16 Real parts fabricated by SLM: (**a**, **b**) Maxillofacial implants, pure Ti, (**c**, **d**) dental applications, CoCrWMo, (**e**, **f**) orthopedic implants, Ti6Al7Nb [57]

composites exhibited higher capacitance retention and improved cycling stability than conventionally mixed materials, associated with homogeneous phase dispersion and reduced particle size [7, 18]. This ability to control nanoscale architecture through purely mechanical routes positions PM as a cost-effective technique for electrode and hydrogen-storage materials requiring precise stoichiometry and large surface area.

1.5 Conclusions

The material surveyed in this chapter establishes powder metallurgy as a unified processing framework in which powder characteristics, consolidation methods, and sintering phenomena are intrinsically coupled to the performance of the final material. The evolution of powder production techniques, from mechanical milling and chemical synthesis to gas and plasma atomization, has enabled control over particle morphology, size distribution, purity, and phase stability, which directly determine flow behavior, packing efficiency, and sintering response. These attributes form the foundation upon which all subsequent processing stages operate. In this regard, shaping and consolidation routes translate these powder-scale properties into macroscopic form. Uniaxial and isostatic pressing, metal injection molding, and additive manufacturing impose distinct mechanical and thermal fields that regulate particle rearrangement, plastic deformation, and interparticle contact. When appropriately selected and combined with suitable feedstock design, these routes yield components with high density, dimensional precision, and complex geometries while minimizing material waste. In addition, sintering provides the thermodynamic continuity between discrete powders and bulk solids through diffusion-driven mass transport and microstructural evolution. The balance between surface, grain boundary, and lattice diffusion, as well as field-assisted mechanisms in processes such as SPS, determines the rate of densification, the stability of grain boundaries, and the elimination of porosity. These processes define the mechanical strength, fatigue resistance, and thermal stability of the consolidated material.

References

1. Araoyinbo A et al (2022) Overview of powder metallurgy process and its advantages. AIP Conf Proc 2437:020153
2. Cor E et al (2020) Environmental analysis of the powder metallurgy value chain: a methodology for comparison with conventional manufacturing. In: Euro PM2020—European powder metallurgy conference, Grenoble, France: European Powder Metallurgy Association (EPMA)
3. Kipouros GJ, Caley WF, Bishop DP (2006) On the advantages of using powder metallurgy in new light metal alloy design. Metall Mater Trans A 37(12):3429–3436
4. Pickens JR (1981) Aluminium powder metallurgy technology for high-strength applications. J Mater Sci 16(6):1437–1457
5. Ali S et al (2022) Microstructure and mechanical properties of modified 316L stainless steel alloy for biomedical applications using powder metallurgy. Materials 15(8):2822
6. Awad M, Hassan NM, Kannan S (2021) Mechanical properties of melt infiltration and powder metallurgy fabricated aluminum metal matrix composite. Proc Inst Mech Eng B J Eng Manuf 235(13):2093–2107
7. Manohar G, Pandey KM, Maity SR (2021) Effect of sintering mechanisms on mechanical properties of AA7075/B4C composite fabricated by powder metallurgy techniques. Ceram Int 47(11):15147–15154
8. Parvizi S et al (2021) Effective parameters on the final properties of NiTi-based alloys manufactured by powder metallurgy methods: a review. Prog Mater Sci 117:100739

9. Wang S et al (2025) Diffusion-dependent microstructure and mechanical properties of Ti50Ta alloy via powder metallurgy. Mater Charact 225:115164
10. Moon AP et al (2021) Opportunity and challenges of iron powders for metal injection molding. ISIJ Int 61(7):2015–2033
11. Mazlan MR et al (2023) Necking mechanism under various sintering process parameters—a review. J Mater Res Technol 23:2189–2201
12. Saini RK, Pandel U, Nadakuduru VN (2025) A review on high-entropy alloys synthesized via powder metallurgy route. Phys Met Metallogr 126(5):501–518
13. Venugopal B, Xavior AM (2024) Development of high entropy alloys (HEAs): current trends. Heliyon 10(7):e26464
14. Li XY et al (2021) The development and application of spark plasma sintering technique in advanced metal structure materials: a review. Powder Metall Met Ceram 60(7):410–438
15. Li Y et al (2025) Effects of ceramic powder coating treatment on the curing behavior, mechanical properties, and microstructure of 3D printed silicon nitride ceramics. J Eur Ceram Soc 45(3):116952
16. Karabulut Y, Ünal R (2022) Additive manufacturing of ceramic particle-reinforced aluminum-based metal matrix composites: a review. J Mater Sci 57(41):19212–19242
17. Katz-Demyanetz A et al (2019) Powder-bed additive manufacturing for aerospace application: techniques, metallic and metal/ceramic composite materials and trends. Manuf Rev 6:5
18. Moghimian P et al (2021) Metal powders in additive manufacturing: a review on reusability and recyclability of common titanium, nickel and aluminum alloys. Addit Manuf 43:102017
19. Parsons EM, Shaik SZ (2022) Additive manufacturing of aluminum metal matrix composites: mechanical alloying of composite powders and single track consolidation with laser powder bed fusion. Addit Manuf 50:102450
20. Vafadar A et al (2021) Advances in metal additive manufacturing: a review of common processes, industrial applications, and current challenges. Appl Sci 11(3):1213
21. ZainElabdeen IH et al (2024) Recent advancements in hybrid additive manufacturing of similar and dissimilar metals via laser powder bed fusion. Mater Sci Eng A 909:146833
22. Rane KK, Date PP (2017) Sustainable recycling of ferrous metallic scrap using powder metallurgy process. J Sustain Metall 3(2):251–264
23. Tsutsui T (2010) Recent technology of powder metallurgy and applications. Hitachi chemical technical report, (54), pp 12–20
24. Ujah CO, Kallon DVV (2022) Trends in aluminium matrix composite development. Crystals 12(10):1357
25. Lanzutti A, Marin E (2024) The challenges and advances in recycling/re-using powder for metal 3D printing: a comprehensive review. Metals 14(8):886
26. Soong SZ, Lai WL, Kay Lup AN (2023) Atomization of metal and alloy powders: processes, parameters, and properties. AICHE J 69(11):e18217
27. Rojas-Díaz LM et al (2020) Production and characterization of aluminum powder derived from mechanical saw chips and its processing through powder metallurgy. Powder Technol 360:301–311
28. Kassym K, Perveen A (2020) Atomization processes of metal powders for 3D printing. Mater Today Proc 26:1727–1733
29. Edosa OO, Tekweme FK, Gupta K (2021) A review on the influence of process parameters on powder metallurgy parts. Eng Appl Sci Res 49(3):433–443
30. Bałasz B et al (2023) Comparison of ultrasonic and other atomization methods in metal powder production. J Achiev Mater Manuf Eng 116:11–24
31. Liu M et al (2024) Finite element analysis of densification process in high velocity compaction of iron-based powder. Materials 17(13):3085
32. Li Z et al (2021) Characteristics of Ti-Nb-Mg alloy by powder metallurgy for biomedical applications. Mater Charact 173:110953
33. Bai Y et al (2021) A review on high velocity compaction mechanism of powder metallurgy. Sci Prog 104(2):00368504211016945

34. Tishkevich DI et al (2022) Isostatic hot pressed W–Cu composites with nanosized grain boundaries: microstructure, structure and radiation shielding efficiency against gamma rays. Nanomaterials (Basel) 12(10):1642
35. Wang Z, Tan Y, Li N (2023) Powder metallurgy of titanium alloys: a brief review. J Alloys Compd 965:171030
36. Sergi A et al (2022) Development of Ni-base metal matrix composites by powder metallurgy hot isostatic pressing for space applications. Adv Powder Technol 33(2):103411
37. Lin C-M, Yen P-Y, Tan C-M (2024) Optimization of metal injection molding processing conditions for reducing black lines and meld lines in bone plates. Polymers 16(23):3241
38. Dejene ND, Lemu HG (2023) Current status and challenges of powder bed fusion-based metal additive manufacturing: literature review. Metals 13(2):424
39. Sharma AK, Bhandari R, Pinca-Bretotean C (2021) A systematic overview on fabrication aspects and methods of aluminum metal matrix composites. Mater Today Proc 45:4133–4138
40. Lingam D et al (2024) Optimization methods in powder metallurgy for enhancing the mechanical properties: a systematic literature review. Eng Res Express 6(2):022504
41. Ramakrishnan P (2013) 17—Automotive applications of powder metallurgy. In: Chang I, Zhao Y (eds) Advances in powder metallurgy. Woodhead Publishing, Sawston, pp 493–519
42. Pirro G et al (2025) Effect of heat treatments and related microstructural modifications on high-cycle fatigue behavior of powder bed fusion–laser beam-fabricated Ti-6Al-2Sn-4Zr-6Mo alloy. Metals 15(8):849
43. André V et al (2005) Production of titanium alloys for advanced aerospace systems by powder metallurgy. Mater Res 8:443–446
44. Qiu F et al (2024) Achieving superior performance in powder-metallurgy near-β titanium alloy by combining hot rolling and rapid heat treatment followed by aging. J Mater Sci Technol 171: 24–36
45. German RM (2013) Progress in titanium metal powder injection molding. Materials 6(8): 3641–3662
46. Seikh Z et al (2025) Metal matrix composites processed through powder metallurgy: a brief overview. J Inst Eng India Ser D 106(1):771–778
47. Wang A et al (2021) 3D printing of aluminum alloys using laser powder deposition: a review. Int J Adv Manuf Technol 116(1):1–37
48. Tang Y et al (2020) Effect of Al2O3 with different nanostructures on the insulating properties of epoxy-based composites. Materials 13(19):4235
49. Koňáková D et al (2025) The role of ceramic powder in sustainable thermal-resistant cement-based composites. Constr Build Mater 487:142095
50. Zygmuntowicz J et al (2023) Microstructure and mechanical characterization of novel Al2O3–(NiAl–Al2O3) composites fabricated via pulse plasma sintering. Materials 16(11):4136
51. Sadhu KK, Mandal N, Sahoo RR (2023) SiC/graphene reinforced aluminum metal matrix composites prepared by powder metallurgy: a review. J Manuf Process 91:10–43
52. Kutsbakh A et al (2025) Modeling of feedstock formability to optimize mold design and prevent possible defects during metal injection molding. J Manuf Mater Process 9(6):203
53. Sharma SK et al (2025) Significance of the powder metallurgy approach and its processing parameters on the mechanical behavior of magnesium-based materials. Nanomaterials 15(2):92
54. Tan Z-q et al (2020) New development of powder metallurgy in automotive industry. J Cent South Univ 27(6):1611–1623
55. Liu X et al (2025) Sustainable additive manufacturing of IN718 blades: powder recycling strategies for reduced carbon footprint. Materials 18(6):1344
56. Miko T et al (2023) A novel process to produce Ti parts from powder metallurgy with advanced properties for aeronautical applications. Aerospace 10(4):332
57. Leordean DV et al (2025) Evaluating the reliability of powder bed fusion for biomedical materials: an experimental approach. Appl Sci 15(8):4542

Chapter 2
Core Techniques in Powder Metallurgy

Abstract Powder metallurgy is a diffusion-controlled manufacturing approach operating across multiple length scales, in which densification efficiency and property optimization emerge from coordinated control of powder characteristics, thermomechanical conditions, and mass transport. In conventional press and sinter processing, performance is governed by green density uniformity, friction management, and atmosphere-regulated diffusion to minimize residual porosity and ensure reproducible mechanical properties. Accelerated heating techniques such as induction, microwave, and spark plasma sintering reduce kinetic time scales, enhancing interparticle bonding while limiting detrimental interfacial reactions. Hot isostatic pressing enables near-theoretical density and microstructural refinement through the combined effects of temperature and isostatic pressure, resulting in high ductility and fatigue resistance. Advanced routes including metal injection molding, binder jetting, and powder-based additive manufacturing extend these principles through controlled feedstock rheology and spatially resolved densification, enabling complex geometries with mechanical performance comparable to wrought materials. Overall, densification is governed by diffusion driven by curvature-induced chemical potential gradients, increasingly optimized through the integration of experimental data and computational modeling.

Keywords Diffusion-controlled densification · Advanced sintering techniques · Hot isostatic pressing · Powder-based additive manufacturing

2.1 Introduction

Powder metallurgy (PM) constitutes a technologically established framework for producing structural and functional metallic components through the controlled consolidation of particulate feedstocks, where densification, microstructure, and performance are governed by coupled thermomechanical and diffusional phenomena [1–5]. Traditional press and sinter processing together with hot isostatic pressing constituted the main industrial routes, enabling near-net-shape fabrication of steels, titanium alloys, refractory systems, and particle-reinforced composites through

R. Pérez-Bustamante, *Fundamental, Practical and Disruptive Applications of Powder Metals*, Topics in Mining, Metallurgy and Materials Engineering,
https://doi.org/10.1007/978-3-032-27130-3_2

compaction followed by thermal or thermomechanical consolidation. Moreover, densification during press and sinter is controlled by green density, porosity evolution, and diffusion-driven particle coalescence, while microstructural development is determined by sintering temperature, dwell time, atmosphere, and the interaction between matrix and reinforcement phases in composite systems [6–8].

Experimental evidence across iron, aluminum, titanium, and tungsten systems demonstrates that inappropriate combinations of compaction pressure, lubricant content, and thermal exposure generate density gradients, interfacial reactions, and brittle phases that compromise strength and ductility, whereas optimized conditions yield equiaxed grains, uniform pore distributions, and reproducible mechanical behavior [9–12]. Hot isostatic pressing extends this paradigm by introducing isostatic pressure at elevated temperature, promoting nonlinear plastic flow and diffusion-assisted pore elimination that produces fully dense microstructures with controlled grain boundary character and enhanced homogeneity. Despite these capabilities, both routes exhibit intrinsic constraints in controlling nanoscale features, phase distributions, and thermodynamic nonequilibrium states, motivating the integration of mechanical alloying, hybrid sintering, and advanced powder conditioning methods to expand accessible microstructural regimes [7, 9–14].

Parallel developments in metal injection molding, additive manufacturing, and binder jetting further demonstrate that powder characteristics, binder powder interactions, and post-sintering treatments critically determine density, dimensional stability, and defect populations in complex geometries. Underlying these diverse processing routes is a unifying thermodynamic and kinetic basis in which curvature-driven chemical potential gradients, capillary pressures, diffusion pathways, and field-assisted transport mechanisms govern neck formation, grain growth, and pore elimination from the nanoscale to bulk components [15–19]. This chapter therefore operates on the premise that quantitative control of powder attributes, compaction mechanics, and sintering kinetics provides a deterministic pathway to minimizing defects and stabilizing microstructures across traditional and additive powder metallurgy platforms. Therefore, this chapter examines the coupling between process parameters, diffusion mechanisms, and defect evolution, with the objective of defining conditions that control densification efficiency, microstructural uniformity, and mechanical performance in press and sinter, hot isostatic pressing, and powder-based manufacturing routes.

2.2 Traditional Methods: Press and Sinter and Hot Isostatic Pressing

2.2.1 Fundamentals of Press and Sinter Processing

Powder metallurgy (PM) through the press and sinter route constitutes one of the foundational techniques of traditional metallic powder processing, serving as a

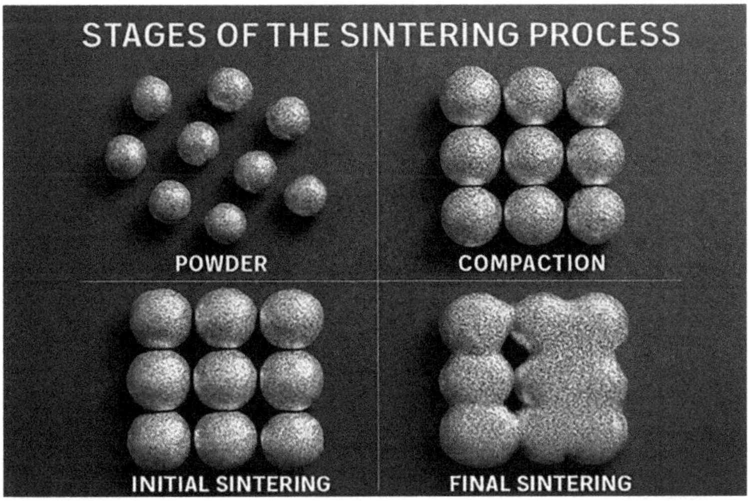

Fig. 2.1 Microstructural evolution during metallic powder sintering: from loose particles to densified solid

benchmark for the development of advanced structural materials (see microstructural evolution during metallic powder sintering: from loose particles to densified solid, Fig. 2.1). This method, defined by the compaction of metallic powders followed by thermal consolidation, has demonstrated remarkable versatility in the fabrication of steels, titanium alloys, and particle-reinforced metal matrix composites. Its effectiveness stems from the ability to precisely control densification, microstructure, and final mechanical properties by adjusting variables such as compaction pressure, sintering temperature, and processing atmosphere.

The densification behavior of a Fe-Cr-C steel was investigated using innovative fast sintering techniques and compared to the traditional press and sinter route, where prealloyed steels such as Astaloy CrM exhibit low compressibility, which limits the achievement of high densities via conventional sintering [20]. The efficient densification depends primarily on reducing porosity and ensuring a homogeneous current flow during field-assisted processes, though the fundamental mechanism remains consistent with thermal sintering, particle coalescence driven by atomic diffusion. The direct correlation between residual porosity and transverse rupture strength highlights the role of green density and thermal cycle control in achieving reproducible mechanical performance.

A comparison between vacuum sintering and induction sintering of titanium alloys produced by elemental press and sinter processing showed that consolidation proceeded by solid-state diffusion under both conditions, with different heating mechanisms. Induction sintering generated internal heating through eddy currents, reaching heating rates of approximately 250 °C min^{-1}, substantially higher than those in conventional furnaces [14]. This method enabled a reduction of sintering time to roughly 5% of that required for vacuum sintering, while maintaining chemical homogeneity and achieving relative densities of 94–95%. Microstructural analysis showed that the residual pores were mostly spherical and evenly distributed, indicating that the compacts reached the final sintering stage. The presence of equiaxed α grains and $\alpha + \beta$ lamellae further confirmed the proper microstructural evolution typical of the press and sinter process [14].

Tabie [10] reported the behavior of press and sinter processing in the fabrication of particle-reinforced titanium matrix composites (TMCs). Consolidation of metallic and reinforcement powders such as SiC, B_4C, or graphene is typically achieved through cold consolidation followed by sintering under vacuum or controlled atmospheres. Reported results emphasized that sintering temperature optimization is essential to avoid undesirable interfacial reactions between titanium and ceramic reinforcements, particularly the excessive formation of TiC or Ti_5Si_3 phases that compromise ductility and promote brittle fracture [10]. Uniform particle dispersion and porosity control were found to be decisive factors in ensuring efficient load transfer and thermal stability. Composites produced via the press and sinter route exhibited significant improvements in hardness, wear resistance, and densification, particularly when reinforced with boron carbide or graphene nanoparticles, with tensile strength increases of up to 52% compared to unreinforced matrices.

2.2.2 *Powder Compaction: Pressure, Lubricants, and Green Strength*

Powder compaction is the first and most critical stage in the PM process, as it determines the uniformity, green density, and mechanical integrity of the compacted part (Fig. 2.2). In this stage, applied pressure, lubrication, and particle morphology play interrelated roles in establishing the strength of the green compact before sintering; and the PM process is governed by mechanical consolidation, where metal powders are pressed below their melting point to create near-net-shape components [21]. The compaction step involves the rearrangement and plastic deformation of particles, which reduces porosity and promotes interparticle bonding. The optimization of process variables such as pressure and lubrication directly affects the powder flow behavior, compaction homogeneity, and the resulting green strength. In mechanical alloying and compaction, ductile and brittle particles experience different deformation behaviors, where welding and fracturing occur simultaneously until a steady-state microstructure is achieved. The degree of densification, particle welding, and eventual fracture balance defines the compact's microstructural integrity and its ability to withstand handling and sintering without cracking.

Compaction pressure controlled density distribution within the green body, with Cu-Sn-Zn-graphite composites pressed at 200 bar to obtain sufficient green strength while retaining the lubricating function of graphite [8]. The results demonstrated that moderate pressures produced sufficient bonding between metallic particles,

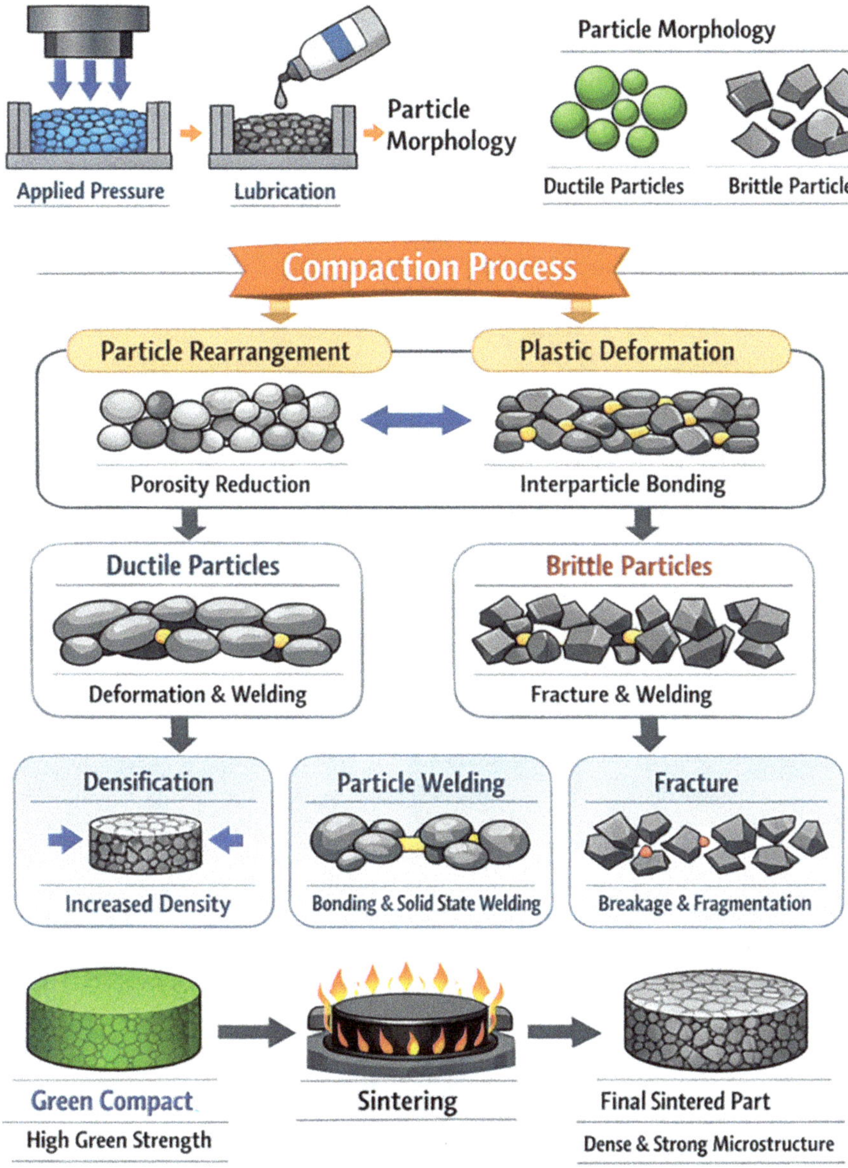

Fig. 2.2 Schematic 2D representation of powder compaction in PM, illustrating the roles of applied pressure, lubrication, and particle morphology on particle rearrangement, plastic deformation

enhancing density and mechanical performance. However, excessive pressure can lead to die wall friction, inhomogeneous density gradients, and lubricant entrapment, which may subsequently cause cracking or delamination during ejection. It was observed that when graphite was incorporated at 1.0 wt%, the compact exhibited optimal green and sintered densities, indicating that adequate interparticle contact was achieved under controlled compaction forces.

2.2.3 Sintering Behavior and Microstructural Development

The sintering behavior in PM governs the evolution of density, grain structure, and mechanical properties through atomic diffusion and phase transformations, which are controlled by sintering temperature and dwell time. These parameters also govern interfacial reactions, phase evolution, and microstructural homogeneity in AA7075-B_4C composites [22]. Conventional sintering conducted at 620 °C with a holding time of 150 min resulted in excessive porosity and pronounced grain coarsening, attributed to prolonged thermal exposure. This condition promoted intensified interfacial diffusion and the formation of brittle reaction products, particularly AlB_2 intermetallic phases, which adversely affected the microstructural integrity of the composite. In contrast, spark plasma sintering (SPS, Fig. 2.3) and microwave-assisted sintering (MAS) achieved superior densification at lower temperatures of approximately 500 °C and shorter times of 5 and 30 min, respectively. These rapid heating techniques enhanced diffusion rates and bonding efficiency

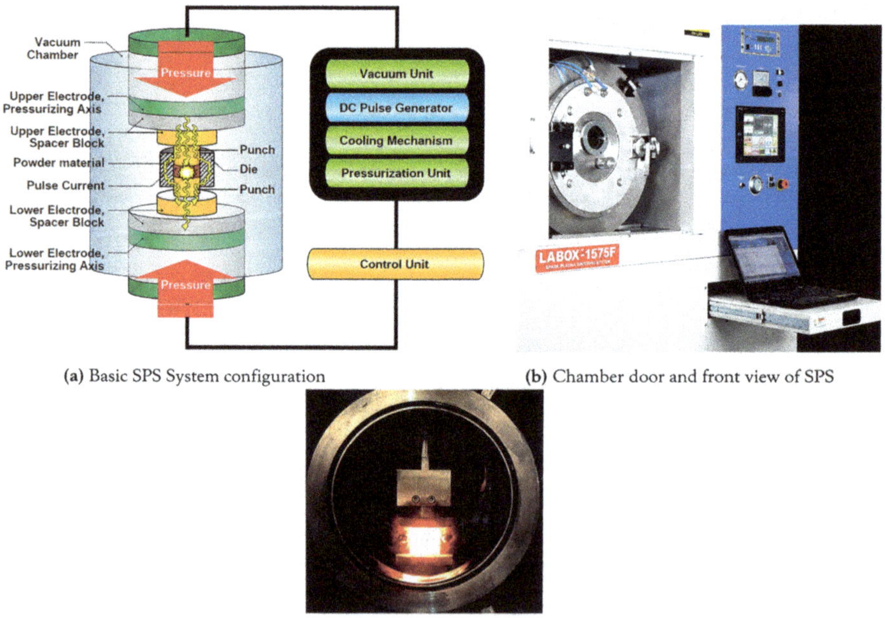

(a) Basic SPS System configuration

(b) Chamber door and front view of SPS

(c) Sintering stage and graphite mold in SPS water cooling vacuum chamber at 1273 K

Fig. 2.3 Basic SPS System configuration: (**a**) system configuration; (**b**) chamber door and front view of SPS apparatus; (**c**) sintering stage and graphite mold in SPS water-cooled vacuum chamber at 1273 K [23]

between aluminum and boron carbide particles without inducing detrimental reactions. Scanning electron microscopy (SEM) and energy dispersive spectroscopy (EDS) analyses confirmed uniform particle distribution and clean interfaces in SPS and MAS samples, while conventional sintering resulted in agglomerations and microvoids that acted as stress concentrators and weakened the interface. It was observed that reduced sintering time and controlled temperature promote the development of dense microstructures with fewer defects and stronger interfacial bonding, which significantly enhanced tensile strength, hardness, and ductility in SPS-processed composites.

The effect of sintering temperature and alloying elements on microstructural refinement was also observed by Dareh Baghi et al. [24], who examined Ti-6Al-4V alloys synthesized by conventional PM [24]. The sintering process conducted at 1100 and 1250 °C under an argon atmosphere led to the formation of dual-phase α and β structures, with α grains and α + β lamellae dominating the microstructure. Increasing the sintering temperature enhanced diffusion and grain coalescence, resulting in higher density but also coarser grains. The porosity decreased with increasing compaction pressure and sintering temperature, directly correlating with improved mechanical strength. Optical and electron microscopy revealed the transformation of elongated α lamellae into more equiaxed grains at higher temperatures, indicating the progress of solid-state diffusion and recrystallization during sintering. X-ray diffraction confirmed the presence of α (hcp) and β (bcc) phases, and hardness testing demonstrated that the α phase retained higher microhardness due to its refined morphology.

The sintering of tungsten heavy alloys further highlights the importance of elemental additions in microstructural evolution. The addition of iron to W-Ni-Cu heavy alloys enhanced densification and grain refinement through the promotion of liquid phase formation at lower temperatures [25, 26]. Sintering was carried out at 1500 °C, and the Fe addition lowered the liquidus temperature, accelerating particle rearrangement and pore elimination. Microstructural analysis revealed a transition from irregular tungsten grains to more spheroidal morphologies, improving wettability between the binder and tungsten phases. Mechanical testing demonstrated that Fe modified alloys exhibited higher tensile strength and hardness due to refined grain boundaries and reduced porosity, confirming the beneficial influence of controlled alloying on sintering kinetics and microstructure stabilization [26].

The addition of nickel and copper facilitated neck growth between tungsten particles at 1200 °C, leading to dense and fine-grained structures without excessive grain growth. X-ray diffraction and transmission electron microscopy (TEM) revealed that nanoscale grain sizes were maintained while achieving high relative density [13]. The refined grains improved mechanical properties by increasing boundary area and impeding dislocation motion, which resulted in higher hardness and yield strength compared to conventionally sintered alloys. The combination of nanocrystallinity and controlled alloying thus enabled low temperature sintering with superior mechanical performance, establishing an efficient route for high density refractory alloys through diffusion-enhanced mechanisms.

2.2.4 Mechanisms and Parameters in Hot Isostatic Pressing

HIP is a thermomechanical consolidation process that combines high temperature and isostatic pressure to achieve near-full density in PM components (Figs. 2.4 and 2.5). The process operates through diffusion-driven bonding and pore elimination, producing microstructures comparable to or superior to those obtained by

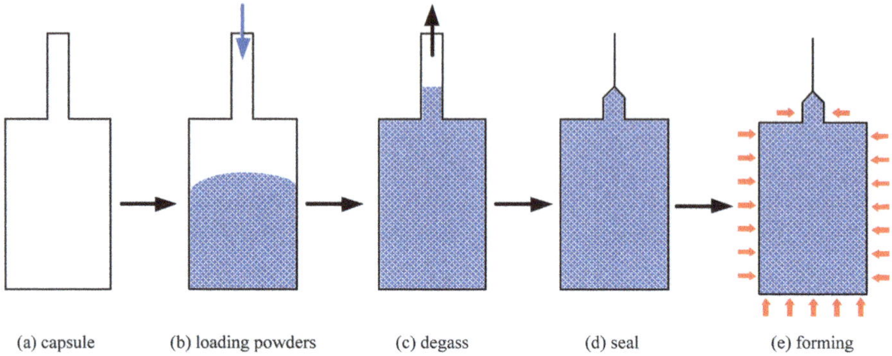

Fig. 2.4 Schematic illustration of the HIP process. (**a**) capsule, (**b**) powder loading, (**c**) degassing, (**d**) sealing, and (**e**) HIP forming [27]

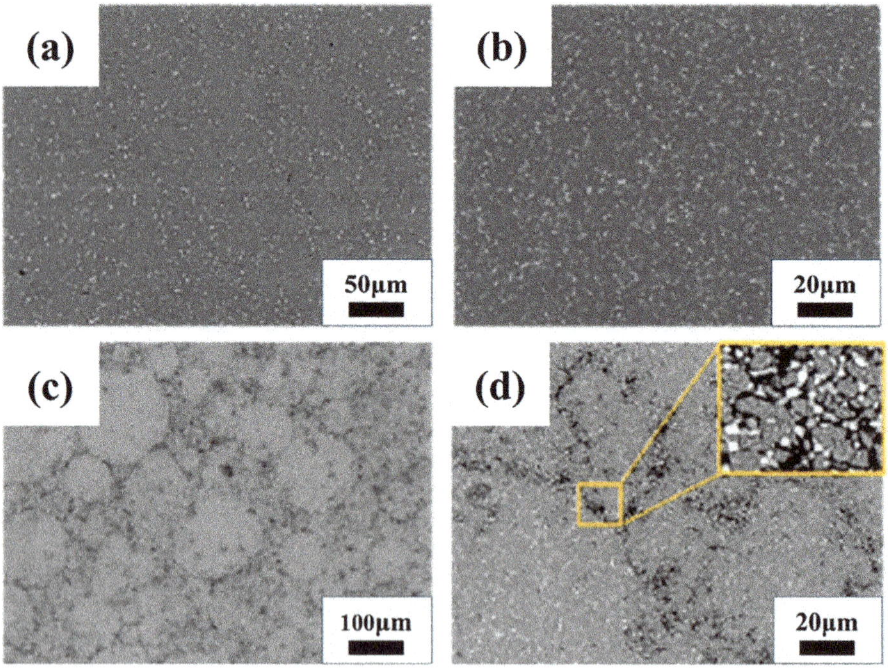

Fig. 2.5 FESEM images of a high speed steel sample: (**a**, **b**) before etching; (**c**, **d**) after etching [28]

conventional casting and forging. Densification during HIP follows nonlinear rheological behavior influenced by coupled temperature, pressure, and strain rate dependencies [5]. Constitutive relations such as the Shima-Oyane formulation were required for accurate prediction of densification kinetics under isostatic loading conditions [29]. Through combined finite element and discrete element simulations, they established a temperature-compensated Shima-Oyane model to describe the yield behavior of nickel-based powders. The model was validated using uniaxial compression experiments at 800–1100 °C and pressures up to 120 MPa, reproducing the measured stress-strain responses with an error below 5%. Results show that, at elevated temperatures, plastic flow and diffusion-assisted deformation dominate, enabling rapid pore closure and uniform densification. The yield stress decreased markedly with temperature, confirming the strong thermal sensitivity of the powder compaction mechanism and the necessity of temperature compensation in predictive models.

Jiang et al. [7] provided an experimental validation of the microstructural consequences of HIP using Inconel 625 alloy consolidated from gas-atomized powder. The alloy was densified at 1140 and 1180 °C under a pressure of 150 MPa for 4 h, producing fully dense billets with uniform grain structures. Electron backscatter diffraction analysis revealed that over 98% of the grain boundaries were high angle type, confirming complete recrystallization after consolidation [7]. The process parameters strongly influenced the development of special coincidence site lattice boundaries, with the $\Sigma 3$ fraction reaching 67% at 1140 °C and 72% at 1180 °C. After solution treatment at 1050 °C, this fraction further increased to approximately 78.5%, indicating enhanced grain boundary engineering through thermally assisted migration and annealing twinning. A high fraction of coincidence site lattice grain boundaries significantly suppressed intergranular defect formation and enhanced mechanical homogeneity, resulting in tensile elongations in the range of 56–60% while maintaining strength levels comparable to those of wrought counterparts.

Similarly, the mechanical homogeneity achieved through PM-HIP processing was further enhanced through optimized post-consolidation heat treatments in Böhler K890 Microclean steel. compared with the conventional quenching and multiple tempering route, the modified treatment incorporating an intermediate cryogenic stage promoted a more homogeneous distribution of V-rich MC carbides and Mo-W-rich M6C carbides within the tempered martensitic matrix, improving hardness, tensile strength, ductility, and fatigue resistance, while preserving the defect-free microstructure characterized by the absence of significant retained austenite, porosity, segregations or carbide clustering [1].

2.2.5 Industrial Applications and Limitations of Traditional Methods

Traditional PM routes such as press and sinter and HIP have long served as the foundation of industrial component fabrication, yet their adaptation to modern applications faces intrinsic limitations in achieving nanoscale control and advanced material performance. As reported by Suryanarayana [30–32], conventional consolidation techniques are capable of producing a variety of metastable and nanostructured phases; however, they often fail to provide the microstructural uniformity and thermodynamic nonequilibrium states necessary for next-generation alloys and composites. Mechanical alloying, developed originally at the International Nickel Company for oxide dispersion strengthened nickel-based superalloys, has emerged as a critical complementary process to overcome these deficiencies. It enables the production of supersaturated solid solutions, amorphous alloys, and nanocrystalline materials directly from elemental or prealloyed powders without the need for melting [33]. This process, based on repeated cold welding and fracturing, extends solubility limits beyond equilibrium predictions, thus facilitating alloy systems that are otherwise unattainable by traditional PM routes.

The integration of mechanical alloying and milling technologies into PM has also broadened industrial applicability (Fig. 2.6). Mechanical milling serves as an advanced nanotechnological approach for fabricating nanocrystalline and nanocomposite materials, providing significant improvements in hardness, wear resistance, and corrosion behavior [31–33]. Through high energy ball milling, powders undergo continuous deformation, fracture, and cold welding, which enhance reactivity and promote alloying between otherwise immiscible elements. This approach has led to the development of high performance materials such as nanocrystalline titanium nitride, tungsten carbide, and oxide dispersion strengthened steels. These materials exhibit superior performance compared to their traditionally sintered counterparts because the milling process generates a high density of defects and fine grain boundaries that act as diffusion pathways during subsequent sintering. Nevertheless, industrial scaling of these nanostructured materials remains constrained by contamination from milling media, limited batch sizes, and high energy consumption, which restrict their feasibility in mass production environments.

The evolution of PM manufacturing has been further expanded by advanced processing routes that integrate both mechanical alloying and sintering technologies into hybrid frameworks. Hybrid consolidation techniques can combine powder blending, mechanical activation, and controlled atmosphere sintering to achieve uniform reinforcement dispersion and enhanced interfacial bonding. These hybridized methods have facilitated the production of composites reinforced with ceramics such as SiC and Al_2O_3, exhibiting increased stiffness and wear resistance suitable for aerospace and automotive applications. However, they also underscore persistent industrial challenges including the control of particle agglomeration, residual porosity, and limited predictability of thermal expansion mismatch [34]. Despite the

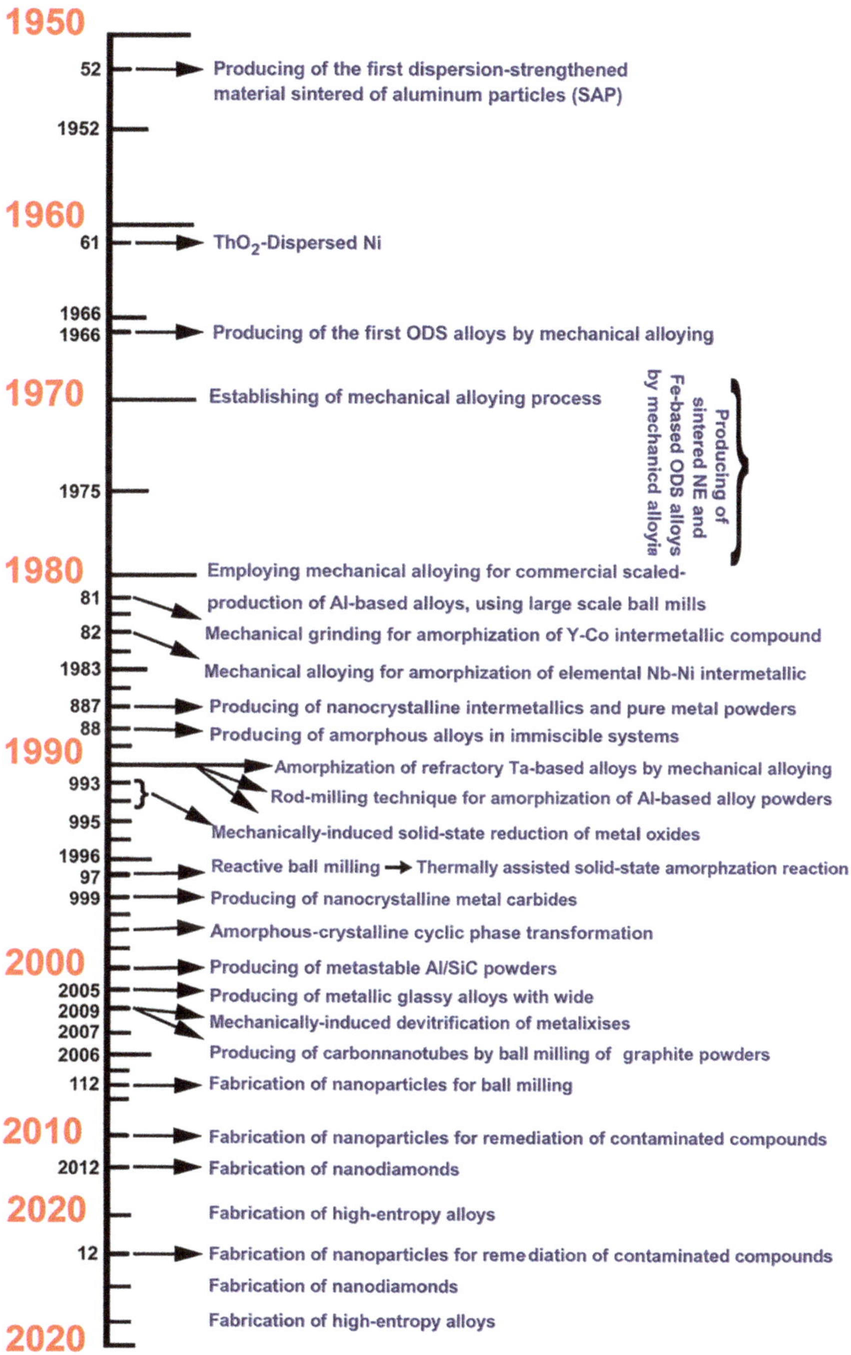

Fig. 2.6 A timeline presentation displaying the evolution of mechanical alloying since 1950, along with some instances of novel materials obtained by this process [33]

improvements in mechanical integrity and microstructural homogeneity, scalability remains limited by the complexity of maintaining uniform temperature and pressure fields during sintering of large components.

In this regard, traditional press and sinter and HIP methods still dominate due to their cost efficiency, established infrastructure, and ability to fabricate near-net-shape components. These techniques are widely used in the automotive and tooling industries for parts such as gears, bearings, and valve seats. However, the major industrial limitations of these traditional processes include incomplete densification, anisotropic mechanical properties, and restricted control over grain growth and phase distribution. To address these limitations, the integration of advanced powder modification techniques such as mechanical activation and high energy premilling has been proposed as a means of enhancing sinterability and achieving finer microstructures without drastically altering existing production lines [31, 32]. Nevertheless, despite these enhancements, the mechanical and microstructural performance attainable through conventional PM routes remains inferior to that achieved via additive manufacturing technologies and advanced sintering techniques.

2.3 Advanced Processes: Metal Injection Molding, Additive Manufacturing, and Binder Jetting

2.3.1 Principles and Workflow of Metal Injection Molding

Metal injection molding is a hybrid manufacturing technique that combines PM and polymer injection molding principles to produce small, complex, and high performance metal parts (Fig. 2.7). The process operates through four sequential stages such as feedstock preparation, molding, debinding, and sintering, each critically influencing final density, geometry, and defect formation. According to Teferi and

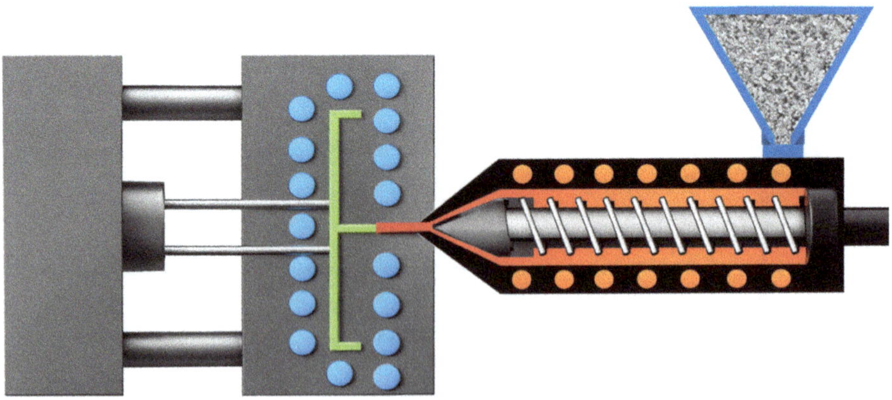

Fig. 2.7 MIM processing flow diagram

Tsegaw [16], the process begins with blending fine metal powders, typically below 20 μm, with thermoplastic binders in proportions of about 60% metal and 40% binder by volume. This mixture is granulated into a uniform feedstock that must display toothpaste-like consistency to ensure homogeneous injection into molds. The choice of powder morphology and binder chemistry directly affects flowability and homogeneity, as agglomeration and poor wetting can cause inhomogeneous microstructures in the final sintered parts. Various atomization and chemical reduction methods, including gas and water atomization, are used to produce spherical powders that achieve high packing density and uniform sintering. The injection stage uses conventional plastic molding machines with cavities scaled up by 20% to compensate for shrinkage during sintering. Subsequent debinding removes the organic binder using thermal, solvent, or catalytic methods, with catalytic debinding being the fastest and most dimensionally stable. Sintering then occurs at approximately 85% of the metal melting temperature, promoting densification to above 95% of theoretical density while preserving fine geometric features.

Teferi and Tsegaw also indicate that defects can arise during feedstock preparation, molding, or debinding, including flash formation, warping, and cracking due to nonuniform temperature and pressure fields [16]. They observed that improper rheology or recycled feedstock increases oxidation and viscosity variations, leading to poor interparticle bonding. Mitigation of these effects was associated with optimization of shear rate, control of particle distribution, and use of pycnometric or rheometric measurements for feedstock quality control. Furthermore, adjustments to injection speed, clamping force, and vent design can eliminate molding defects, while careful control of debinding temperature profiles prevents residual stresses and deformation. The MIM process achieves complex geometries, tight tolerances, and excellent surface finish, but demands strict parameter optimization to ensure defect-free parts suitable for industrial applications in aerospace, biomedical, and telecommunications components.

Kuo et al. [19] demonstrated the fabrication of mold inserts using a composite material composed of aluminum powder and epoxy resin as an alternative to traditional steel molds. The optimal mixture was determined to be 60% aluminum powder and 40% epoxy resin, balancing mechanical strength and processability. Mechanical testing revealed a maximum bending strength of approximately 20 MPa and a Shore D hardness of 81, both sufficient for pilot-scale MIM operations. The aluminum epoxy inserts achieved a service life of about 2100 molding cycles with a cost reduction of 75% compared to conventional mold steels. Experimental results showed that excessive aluminum content reduced flowability and caused air entrapment, whereas too little metallic content decreased thermal conductivity. The investigation also detailed the debinding and sintering of Fe_2Ni feedstock injected through these molds, confirming that the inserts maintained dimensional stability and acceptable surface roughness over repeated use.

Additionally, carbonyl iron powders produced by thermal decomposition provide superior compressibility and surface purity, enabling finer feedstock homogeneity and lower debinding times [17]. However, limitations related to oxidation susceptibility and the necessity for surface coatings or lubricants to prevent premature degradation during high temperature processing are presented. Gas-atomized iron powders produced spherical particles with higher flowability, whereas water-atomized powders exhibited irregular morphology and elevated oxygen content, resulting in reduced sintered density. Experimental data demonstrated that controlling particle size distribution and surface chemistry can reduce defect formation and improve sintering kinetics, directly impacting the mechanical performance of MIM-fabricated iron components [19]. The growing industrial potential of these powders for producing structural parts such as automotive gears and electronic housings is noted, but cost, powder handling, and oxidation control remain key limitations for broader implementation.

2.3.2 *Additive Manufacturing Routes for PM Alloys*

AM routes have emerged as an essential evolution in PM, allowing precise control over microstructure and composition while enabling complex geometries that traditional processes cannot achieve (see a schematic representation of Metal AM processes in Fig. 2.8). Research has highlighted that the adaptation of laser, electron

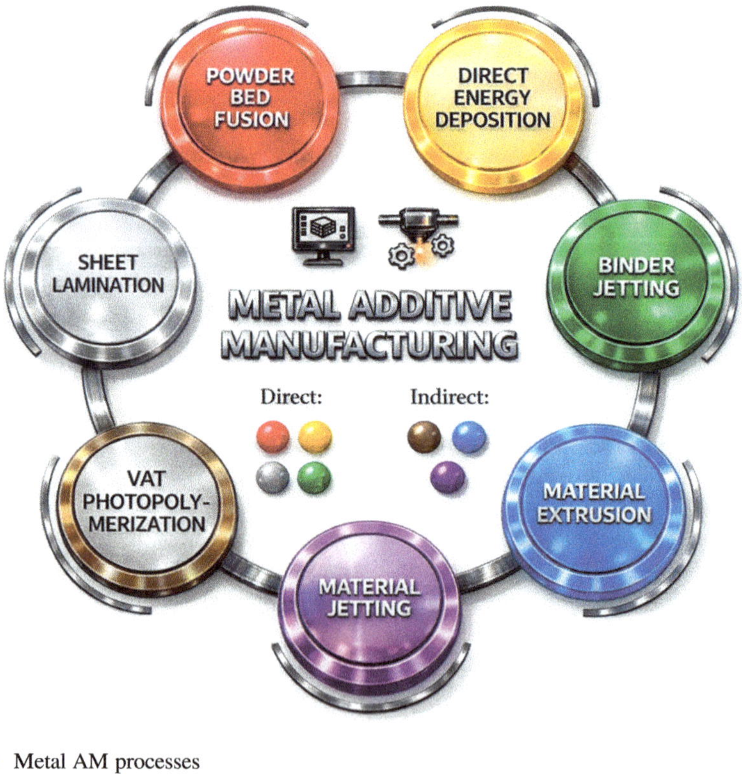

Fig. 2.8 Metal AM processes

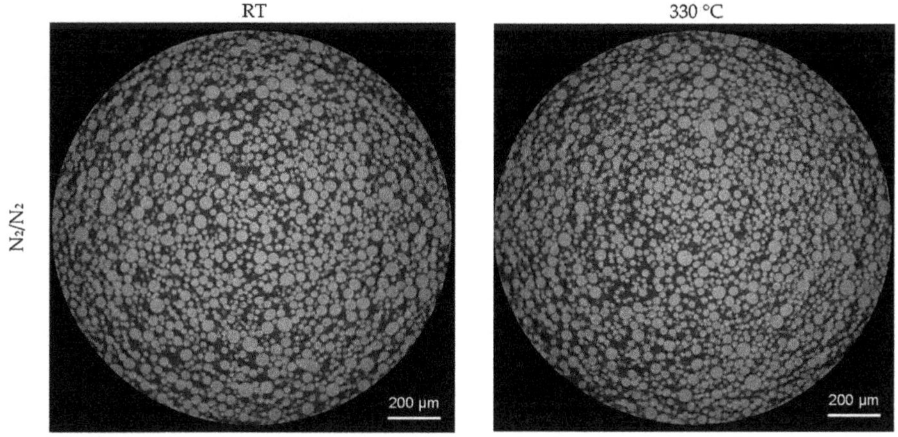

Fig. 2.9 XRM images of gas-atomized X2CrNiMoCuWN25-7-4 powders (particle size 20–63 μm, voxel size 1.56 μm) [35]

beam, and binder-assisted methods to metallic systems enables not only structural optimization but also the development of advanced alloys with tailored functionalities. AM routes including laser powder bed fusion, directed energy deposition, and binder jetting were evaluated for consolidation of WC-based cemented carbides [1]. Compared to conventional PM, these approaches eliminated the need for dies and molds while achieving relative densities exceeding 96% when spherical powders were used. Spherical powders produced by spray drying or plasma spheroidization exhibited increased flowability, reduced internal porosity, and accelerated densification kinetics. It was observed that WC-Co samples fabricated through spray drying followed by L-PBF or binder jetting achieved hardness values above 1200 HV and fracture toughness of approximately 17 MPa·m½, demonstrating the potential of additive routes to match or surpass the mechanical integrity of conventionally sintered counterparts.

It is also described how plasma spheroidization allowed the transformation of irregular carbide particles into spherical feedstock suitable for powder bed fusion. The treatment improved apparent density by nearly 50% and reduced flow time by approximately 4 s per 50 g, both key to forming dense powder beds. However, this technique introduced challenges such as decarburization and the formation of brittle phases like W_2C and W_3Co_3C, which necessitated a secondary heat treatment under carbonaceous atmospheres to restore WC stoichiometry. Despite the complexity of these remedial processes, the resulting powders displayed uniform cobalt distribution and microstructural homogeneity after AM. More generally, X-ray microscopy (XRM) has proven effective for assessing the internal structure and thermal stability of gas-atomized powders, as illustrated by XRM images of X2CrNiMoCuWN25-7-4 particles with sizes between 20 and 63 μm (voxel size 1.56 μm), acquired at room temperature and after heat treatment at 330 °C under N_2/H_2 atmosphere (Fig. 2.9).

Further developments in additive processing have extended beyond homogeneous cemented carbides into the realm of functionally graded metallic materials [36]. AM enables compositional and structural gradients that cannot be achieved by conventional PM or centrifugal casting. The ability to spatially control feedstock composition during layer deposition allows the fabrication of metallic systems with tailored gradients in strength, thermal expansion, or corrosion resistance. Directed energy deposition and powder bed fusion are the dominant technologies for creating functionally graded metallic materials, as they facilitate gradual transitions between metals such as IN718 and SS316L, mitigating issues of solidification cracking common in sharp interfaces. The authors highlighted that through controlled modulation of powder feeding rates and scanning strategies, gradients can be formed continuously in three dimensions, a feature unattainable with traditional compacting or casting. Additively manufactured functionally graded metallic materials exhibited improved stress distribution under thermal cycling and mechanical loading, as the gradient interfaces reduced delamination and concentration of residual stresses.

In addition, it was reported that AM routes allow combinations of metals, ceramics, and intermetallics with previously incompatible thermophysical properties. Graded systems integrating Ti6Al4V with ceramic reinforcements such as Al_2O_3 or TiC demonstrated enhanced wear and oxidation resistance. Similarly, Fe-based graded alloys produced by directed energy deposition exhibited compositional transitions that balanced hardness and ductility while maintaining corrosion stability, considered in nuclear and marine environments. The capacity of AM to control local alloy composition and microstructure enables the design of metallic components with multifunctional zones, such as wear-resistant surfaces over tough cores or thermal barriers within load-bearing sections.

A complementary contribution to this field was provided by Becker and colleagues [37] in their investigation of high nitrogen FeCr(Si)N stainless steels synthesized by combining PM and AM. Their research revealed that the incorporation of nitrogen through additive routes improved mechanical strength and corrosion resistance by promoting the formation of expanded austenite phases and refined grain structures. Nitrogen alloying also mitigated pore formation during laser-based consolidation and enhanced microstructural stability during post-sintering treatments. The results demonstrated that powder-based AM enables controlled nitrogen retention and uniform dispersion, which are challenging to achieve using conventional sintering. This combination of PM precision and additive process flexibility allows the fabrication of stainless steels with customized magnetic, mechanical, and corrosion-resistant properties tailored for advanced engineering applications.

In this regard, research carried out by Zhang and coworkers [28] on metallic multimaterials fabricated by combining AM and PM offered an integrated perspective on hybrid fabrication. The authors described how sequential deposition of dissimilar metals using directed energy deposition or binder jetting enables the creation of complex multimaterial systems with sharp or graded interfaces. By leveraging PM for prealloyed feedstock production and additive manufacturing for spatial deposition, they demonstrated enhanced bonding between layers and minimized interdiffusion defects. Their experiments showed that hybrid AM-PM approaches achieve metallurgical continuity across interfaces and allow localized reinforcement of specific zones. This integration exemplifies how additive manufacturing routes have transformed the conceptual and practical boundaries of PM by introducing spatial material control and digital fabrication precision.

2.3.3 Process Characteristics and Post-processing in Binder Jetting

Binder jetting has emerged as one of the most versatile PM-based additive manufacturing routes, capable of producing complex metallic components at low cost and high throughput (Fig. 2.10). Its process is characterized by the selective deposition of a liquid binder onto a powder bed to form a green part, followed by debinding and sintering to achieve densification. The mechanical integrity of binder-jetted parts is primarily governed by the properties of the powder feedstock, binder chemistry, and post-processing parameters. The group of Fang [38] emphasized that binder jetting enables the fabrication of geometrically complex structures without thermal stresses typically found in fusion-based additive methods, resulting in components with uniform microstructures and reduced residual stresses. The process is adaptable to a wide range of materials, including metals such as stainless steels, nickel-based alloys, and titanium, as well as ceramics and composites. However,

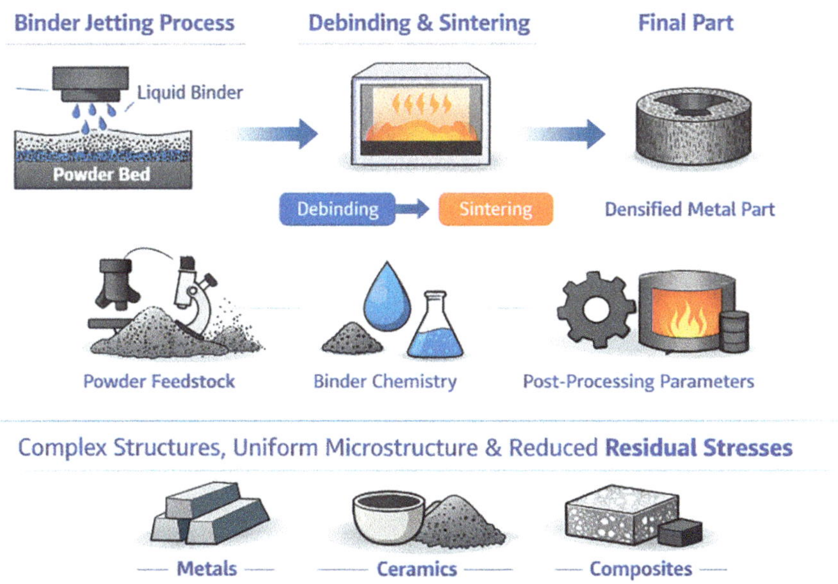

Fig. 2.10 Schematic overview of the binder jetting additive manufacturing process, highlighting powder bed printing, debinding and sintering, and the key factors governing densification and mechanical performance

because the printed parts are inherently porous after printing, post-processing stages such as sintering and infiltration remain essential to achieve full density and mechanical strength comparable to conventionally manufactured components.

One of the key aspects that determine binder jetting performance is the preparation and conditioning of the powder. The morphology, size distribution, and surface condition of metal powders significantly affect the packing density and flowability during recoating. According to Fang and colleagues [38] spherical gas-atomized powders offer superior flow behavior compared to irregular water-atomized powders, resulting in improved bed uniformity and higher sintered densities. They also noted that bimodal or multimodal particle size distributions improve packing density by allowing smaller particles to occupy voids between larger ones, thus minimizing shrinkage and anisotropy during sintering. Additives, including coupling agents and surface modifiers, can further enhance the wettability of powders with binders and mitigate agglomeration. The coupling between binder and powder directly determines the green strength and dimensional accuracy of printed parts, highlighting the importance of both binder selection and droplet deposition dynamics during the printing process.

Recent investigations into titanium alloys have expanded the understanding of material processing interactions in binder jetting. Powder conditioning strategies for Ti-6Al-4V indicated that vacuum and gas-purged drying cycles increased flowability and reduced moisture-induced defects [39]. Through a factorial design of experiments, they determined that drying for 6 h at 200 °C provided an optimal balance between reduced water content and minimized oxidation. This optimized conditioning led to improved dimensional accuracy of printed components, from ±1.5% to 0.3%, and enhanced visual homogeneity of the powder bed. Effective powder conditioning was required to obtain reproducible powder spreading and uniform green density. Powders with fine particle size distributions, typical of MIM feedstocks, exhibited poor flowability unless conditioned, limiting their direct use in binder jetting. Powder conditioning therefore linked conventional PM feedstocks with the requirements of additive manufacturing.

The synergy between powder and binder behavior during the printing and sintering stages also governs the final part quality. Fang et al. detailed that binder infiltration and capillary diffusion play major roles in determining bonding strength and surface finish. Excessive impact velocity of binder droplets can disturb the powder bed, creating uneven layer topographies, while insufficient binder saturation results in weak green parts. The post-processing route, especially sintering temperature and duration, must be optimized to balance densification and dimensional control. In particular, nickel-based alloy 625 fabricated using gas-atomized powder achieved near-full density and superior hardness compared to cast counterparts when sintered under controlled conditions. In titanium systems, as shown by the group of Janzen [39], sintering after optimized drying yielded components with improved structural uniformity and reduced oxidation-related defects.

2.4 Scientific Principles Governing Densification, Sintering Kinetics, and Defect Control

2.4.1 Thermodynamics and Driving Forces for Densification

The thermodynamics and driving forces governing densification in PM systems are determined by the interplay between interfacial energy, curvature-induced chemical potential gradients, and diffusion kinetics. In the case of titanium alloys consolidated by induction sintering, the rapid densification occurs due to the combined effects of electromagnetic field coupling and localized temperature gradients, which enhance atom mobility and diffusion rates at particle contacts. Under induction heating, Ti-6Al-4V powders achieved relative densities above 98% within minutes, a result attributed to the intense thermal gradients that reduced diffusion activation energy and generated a strong capillary pressure between particles [11]. The research group correlated the rapid densification to a significant increase in atomic diffusion flux and to localized Joule heating at interparticle necks, which provided the thermodynamic driving force necessary to overcome conventional sintering rate limitations. Compared with furnace sintering, induction-assisted systems displayed a two to threefold acceleration in densification kinetics due to field-induced thermoelectric and magnetohydrodynamic effects that modified particle surface energies and promoted enhanced mass transport pathways.

In elemental PM-NiTi shape memory alloys, the addition of ultrafine niobium powder generated a eutectic liquid phase with favorable thermodynamics, increasing densification efficiency. Experiments revealed that the addition of fine Nb particles (with a median size of 14 μm) triggered the formation of a NiTi Nb ternary eutectic liquid at approximately 1150 °C, effectively replacing the unstable binary NiTi liquid that typically causes Kirkendall porosity and incomplete densification [40]. The refinement of Nb powder increased the local chemical potential according to the Gibbs-Thomson effect, which raised the reaction driving force and reduced the effective local melting temperature. The thermodynamic analysis showed that decreasing the particle curvature radius amplified the chemical potential difference by nearly threefold, intensifying the capillary pressure gradients that drove liquid migration into pore regions. This capillary-induced liquid infiltration, described by $\Delta P = \gamma sl(1/rl - 1/rp)$, produced pore filling rates up to two orders of magnitude higher than solid-state contact flattening mechanisms, enabling near-full densification with relative densities exceeding 99%. Moreover, kinetic factors such as shortened diffusion paths and increased nucleation site density promoted faster Nb dissolution and uniform eutectic liquid formation, further contributing to pore elimination and compositional homogenization. As a result, alloys produced using fine Nb particles displayed a dramatic reduction in porosity from 10.99% to 0.07% and achieved compressive strengths up to 3051 MPa.

2.4.2 Diffusion Mechanisms in Sintering

Diffusion mechanisms are fundamental to the sintering process, governing densification, grain growth, and the mechanical consolidation of powder particles into a solid structure. In the sintering of metallic systems, diffusion serves as the key atomic scale process that enables neck growth between adjacent particles, elimination of porosity, and development of interparticle bonding (Fig. 2.11). The atomic mobility that drives diffusion is determined by thermodynamic factors such as temperature and chemical potential gradients, as well as by microstructural parameters including particle size, defect density, and interfacial energy. In metallic and composite systems fabricated by PM [3], these mechanisms operate through lattice diffusion, grain boundary diffusion, surface diffusion, and dislocation-mediated transport, each becoming dominant at specific temperature and energy regimes. Molecular dynamics simulations have been instrumental in elucidating these

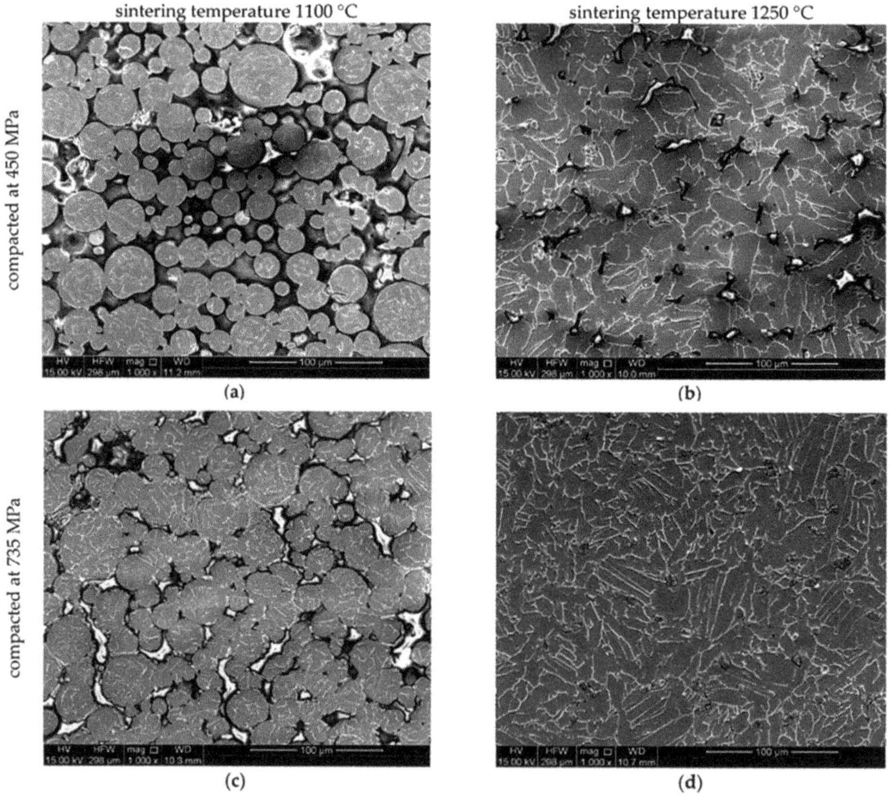

Fig. 2.11 SEM images of PM samples sintered at 1100 °C: (**a**) compacted at 450 MPa and (**c**) 735 MPa, and PM samples sintered at 1250 °C: (**b**) compacted at pressure 450 MPa and (**d**) 735 MPa [24]

mechanisms, providing atomic level insight into how temperature-dependent diffusion transitions occur during sintering.

The atomic scale simulations of Al-Cu nanoparticle sintering by Abedini et al. [41] demonstrated that the dominant diffusion pathway depends strongly on the sintering temperature, with dislocation slip controlling coalescence at lower temperatures and atomic diffusion becoming the prevailing mechanism as temperature increases. Their results revealed that at elevated temperatures, stacking fault density rises, and atoms gain sufficient mobility to diffuse across interfaces, promoting neck growth and enhancing the mechanical strength of the consolidated structure. In systems with high stacking fault energy such as aluminum, dislocation activity interacted with diffusion, enabling a transition from surface controlled to bulk diffusion regimes. Through analysis of mean square displacement and dislocation dynamics, the simulations provided quantitative evidence of how diffusional processes accelerate with increasing temperature, leading to enhanced atomic mixing and densification.

Experimental investigation carried out by Manohar et al. [22] on AA7075/B4C/ZrC hybrid nanocomposites fabricated via SPS illustrated the macroscopic manifestation of diffusion-driven densification. It was observed that the rapid localized heating and pulsed current characteristic of SPS create high energy conditions at particle interfaces, breaking oxide layers and generating plasma discharges that enhance interfacial diffusion. The combination of electric current and temperature gradients promotes both grain boundary and lattice diffusion, enabling rapid consolidation and minimizing the formation of intermetallic compounds or porosity. The densification process is further accelerated by electromigration effects, where the electric field drives charged species through the lattice, effectively augmenting the diffusion coefficient. As a result, the composites exhibited uniform microstructures with strong interfacial bonds and minimal pore content, reflecting the efficiency of diffusion-mediated transport mechanisms under SPS conditions.

Diffusion is also influenced by the intrinsic thermodynamic forces associated with defect mobility and dislocation behavior. In the SPS-processed AA7075-based composites [6], the reduction in dislocation density during sintering was attributed to enhanced diffusion-assisted recovery processes that permit dislocation annihilation and rearrangement. This effect results in increased plasticity and higher strain-to-fracture values due to the balance between diffusion-controlled densification and dislocation motion. Moreover, the pulsed current in SPS was reported to facilitate atomic diffusion through localized Joule heating, which creates transient thermal gradients at the contact points between powder particles. These gradients act as additional driving forces for thermomigration, enhancing the atomic flux along grain boundaries and accelerating neck formation and densification.

At the nanoscale, diffusion is profoundly affected by particle size, as demonstrated by molecular dynamics simulations and experimental results alike. Smaller particles with higher surface-to-volume ratios possess greater surface energy, which provides a strong thermodynamic driving force for diffusion. Abedini et al. found that at high temperatures, atomic diffusion mechanisms dominate, and the mean square displacement of atoms increases significantly, particularly for aluminum atoms due to their lower melting point relative to copper. This leads to faster mass transport across interfaces and improved coalescence of nanoparticles, producing denser and more mechanically robust materials. In contrast, at lower temperatures, diffusion is limited, and coalescence occurs primarily through plastic deformation and dislocation slip, yielding less compact structures.

2.4.3 Kinetics of Neck Formation and Grain Growth

The kinetics of neck formation and grain growth during sintering are governed by diffusion processes and interfacial energy minimization, both of which control the morphological evolution and densification rate of powders (Fig. 2.12). Recent advances in experimental and simulation techniques have enabled a detailed understanding of the relationship between particle interaction, sintering temperature, and grain growth rate in metallic and ceramic systems. The growth of the sintering neck, which forms between adjacent particles at the early stage, follows a power-law relationship with time that reflects the dominant diffusion mechanism, while the subsequent grain growth is controlled by grain boundary migration driven by curvature and stored energy gradients.

Wu et al. [43] carried out molecular dynamics simulations and in situ TEM to analyze neck growth kinetics in silver nanoparticles, revealing that the neck length increases with time following a power law relationship that depends on particle size and temperature. For small silver nanoparticles, the power exponent of neck growth was found to reach values as high as 27 at 250 °C, significantly exceeding the empirical coefficients observed in conventional PM systems. The authors attributed this deviation to the high surface energy and limited grain boundaries in nanoscale systems, which promote enhanced atomic mobility and particle rotation. The sintering process evolved through three distinct stages: initial contact and expansion contraction cycles, directional rotation for lattice alignment, and final coalescence once structural compatibility was achieved. The molecular dynamics results, supported by TEM observations, showed that sintering proceeded without melting, confirming that surface diffusion and atomic rearrangement dominate under these conditions. The in situ analysis demonstrated that as temperature increased, the fitted curves for neck growth kinetics shifted predictably, and the diffusion mechanism coefficients decreased toward values consistent with bulk diffusional control near the melting point.

Microwave sintering of LiH bulk samples provided information on grain growth kinetics under a nonconventional thermal field, with densification accelerated and microstructure refinement observed relative to conventional thermal sintering [44]. Densification in LiH was dominated by grain boundary diffusion with a relatively low activation energy of 73 kJ/mol, indicating efficient atomic transport along interfaces. The grain size increased moderately from 46 to 53.2 μm as sintering temperature rose from 450 to 550 °C, showing that microwave-induced volumetric heating promotes homogeneous temperature distribution and minimizes thermal

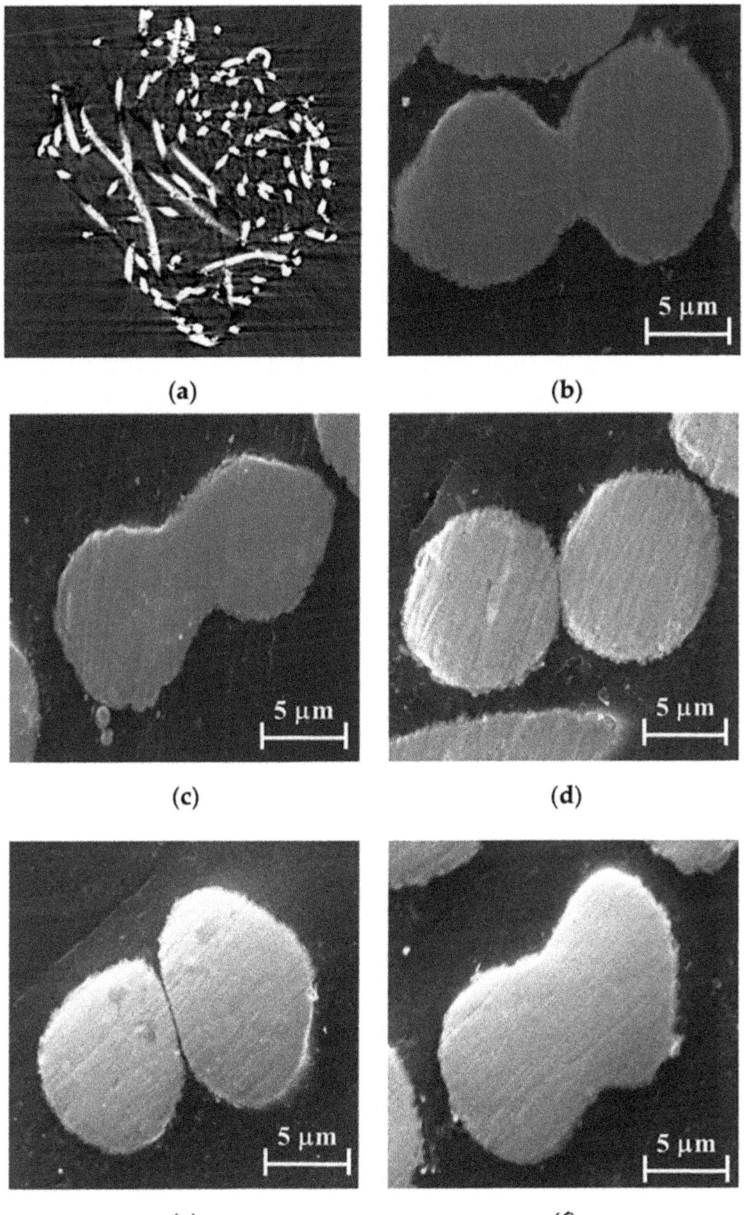

Fig. 2.12 Experimental results of two 8 μm 316L stainless steel fibers with angles of 0–π/2 sintered at 1200 °C for 2 h: (**a**) section morphology of sintering metal fibers; (**b**–**f**) high magnification morphologies [42]

gradients. The activation temperature for densification was determined to be around 475 °C, below which neck formation occurred without significant grain growth. Above this threshold, grains coarsened rapidly, consistent with the classical two-stage model of sintering where densification precedes grain growth.

2.4.4 Origin and Mitigation of Defects in PM Parts

The formation and mitigation of defects in PM parts are fundamentally linked to powder characteristics, processing parameters, and consolidation techniques (see key defects and mitigation strategies in titanium PM in Fig. 2.13). Among titanium-based systems, such as those studied by Fang et al. in their comprehensive review of PM titanium, porosity, oxygen contamination, and microstructural coarsening are the primary defects that limit mechanical performance. Residual porosity originates

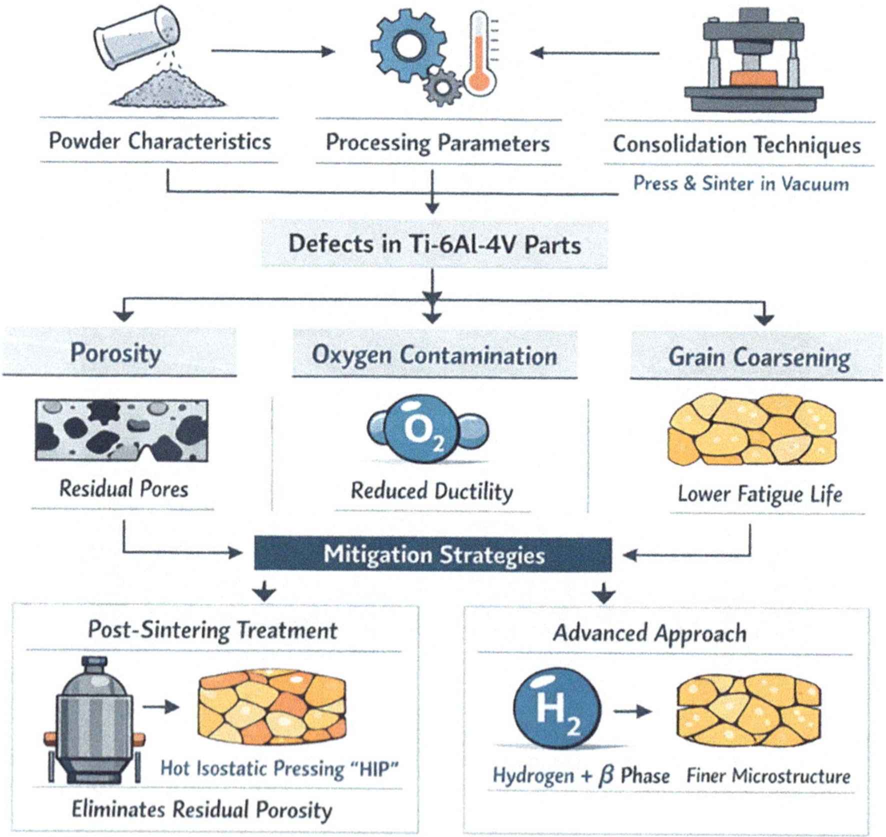

Fig. 2.13 Key defects and mitigation strategies in titanium PM

from incomplete densification during sintering, particularly in the blended elemental route, which involves pressing and sintering elemental powders in vacuum. The fatigue strength of Ti-6Al-4V produced through this method is highly sensitive to pore morphology and distribution, as pores act as preferential sites for crack initiation. Oxygen impurities absorbed during powder handling or sintering exacerbate embrittlement, reducing ductility and fracture toughness [9]. Moreover, grain coarsening during sintering decreases fatigue life and mechanical homogeneity. Post-sintering treatments such as HIP can eliminate residual porosity and improve mechanical integrity although these processes increase cost. To counteract these challenges, Fang et al. proposed hydrogen-enabled phase transformations, where the temporary stabilization of the beta phase during sintering promotes finer microstructures and improved densification without extensive oxidation [45].

Research carried out by Mu et al. [12] demonstrated that reinforcement architecture and dispersion uniformity minimizes internal defects in composite materials. Multilayer graphene (MLG)-reinforced titanium fabricated via flake PM; defect mitigation was achieved by optimizing particle morphology and the sequence of processing steps. The flake ball milling method allowed MLG to embed within titanium flakes, resulting in a homogeneous distribution before consolidation by SPS. This approach avoided the agglomeration of graphene, a common defect in carbon-based composites, which otherwise acts as a stress concentrator and weak interface. The micro laminated structure produced through this process enabled a continuous load transfer pathway and reduced the formation of interfacial voids. After subsequent hot rolling, the interface between MLG and the titanium matrix developed an in situ TiC layer, which enhanced bonding and minimized delamination. The resulting composite exhibited a significant reduction in internal defects and a 280% increase in yield strength compared to pure titanium, evidencing the efficacy of defect control through tailored microstructural design.

2.4.5 Modeling Approaches and Future Trends in Process Optimization

Recent advances in modeling and optimization have transformed PM into a data-driven process where numerical and computational approaches enable the prediction and control of densification, microstructural evolution, and mechanical performance (see Fig. 2.14). Simulation methods such as finite element modeling, discrete element modeling, and molecular dynamics have been employed to describe each stage of the press and sinter route, from powder compaction to sintering shrinkage, capturing the nonlinear relationships between process parameters and part properties. These modeling tools provide a pathway for identifying optimal conditions that minimize defects, enhance density uniformity, and reduce energy consumption [34].

Finite element modeling was applied to the press and sinter process to correlate compaction pressure, interparticle friction, and elastic plastic deformation of powder

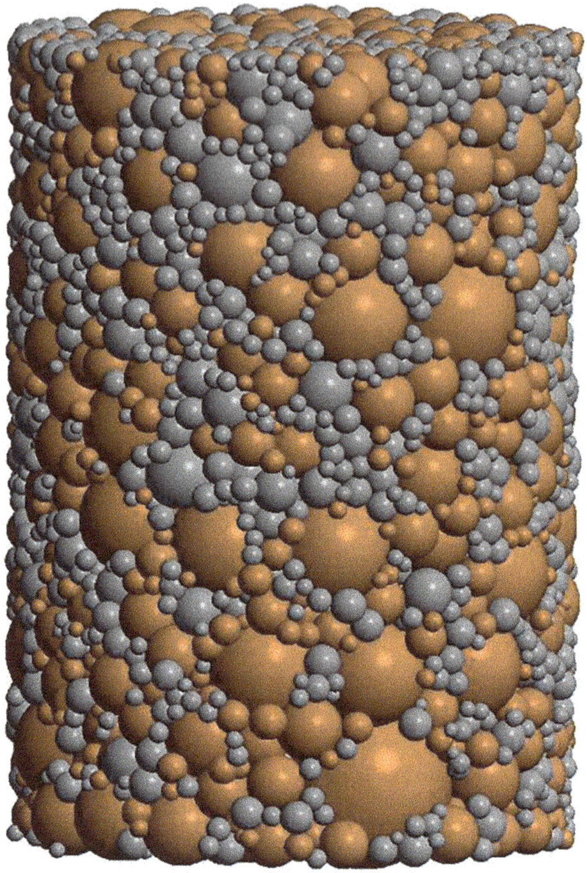

Fig. 2.14 Discrete element specimen of two-phase 80% NiAl 20%Al_2O_3 powder [46]

particles, enabling prediction of green density gradients [47]. The simulation results showed that particle rearrangement and local density heterogeneity depend strongly on the applied load and die wall friction, leading to nonuniform densification during the initial compaction stage. Densification kinetics followed an exponential dependence on applied stress and contact area evolution. Discrete element simulations resolved individual particle motion, with higher coordination numbers associated with reduced residual porosity and stronger interparticle bonding during sintering. In addition, coupled thermomechanical simulations demonstrated that as the sintering temperature increases, the stress relaxation rate accelerates, which affects the pore coalescence mechanism and grain boundary migration.

It is reported that the influence of experimentally controlled process parameters on the resulting characteristics of PM components, while also providing critical insights pertinent to the validation and calibration of computational modeling frameworks [34]. Compaction pressure, sintering temperature, dwell time, atmosphere, and lubrication were evaluated in relation to density, hardness, and microstructure. Final density was primarily governed by compaction pressure and sintering temperature, while ceramic reinforcement and lubrication influenced hardness and wear resistance. Defect-free consolidation required simultaneous control of compaction load, reinforcement fraction, and particle size distribution, indicating strong coupling among processing variables.

2.5 Conclusions

Densification and microstructural stabilization in powder metallurgy are governed by the coupled action of compaction mechanics, thermodynamic driving forces, and diffusion-controlled transport processes operating across multiple length scales. Press and sinter processing establishes the initial density and pore architecture through pressure-dependent particle rearrangement, plastic deformation, and lubricant-mediated friction control, which in turn dictate the evolution of interparticle necks and load-bearing pathways during subsequent sintering. Variations in particle morphology, reinforcement dispersion, and sintering temperature determine whether diffusion promotes uniform bonding or instead drives deleterious interfacial reactions and grain coarsening, as observed in aluminum, titanium, and composite systems. Moreover, hot isostatic pressing extends these mechanisms by introducing isostatic stress that activates temperature-dependent plastic flow and diffusion-assisted pore closure, yielding fully dense materials with engineered grain boundary character and enhanced ductility, as verified in nickel-based alloys and tool steels subjected to optimized HIP and post-treatment schedules. The integration of mechanical alloying, hybrid sintering, and powder conditioning modifies surface energy, defect density, and chemical potential gradients, thereby increasing diffusion rates and enabling metastable, nanocrystalline, and composite microstructures unattainable by conventional consolidation alone.

Advanced powder-based manufacturing routes, including metal injection molding, binder jetting, and laser or binder-assisted additive manufacturing, further reinforce the centrality of feedstock morphology, surface chemistry, and binder powder coupling in determining green strength, shrinkage behavior, and final density. Diffusion-driven neck growth, capillary-induced liquid infiltration, and field-assisted transport processes dictate the kinetics of pore elimination and grain growth, while particle size and curvature control the thermodynamic driving forces that enable near-full densification in both solid-state and liquid phase regimes. Additionally, defect populations in powder metallurgy components arise from incomplete densification, oxidation, and microstructural coarsening, but are shown to be mitigated through tailored powder architecture, controlled atmospheres, hydrogen-assisted phase transformations, and post-sintering treatments such as HIP and hot working. The modeling frameworks reviewed, spanning discrete element, finite element, and molecular dynamics approaches, establish quantitative relationships between processing variables, density gradients, and microstructural evolution, providing predictive capability for multi-parameter optimization. Rigorous control of powder characteristics, compaction conditions, and sintering kinetics provided the basis for producing dense components with limited defects and stable microstructures across conventional and advanced powder metallurgy technologies.

References

1. Morri A et al (2022) Effect of Different Heat Treatments on Tensile Properties and Unnotched and Notched Fatigue Strength of Cold Work Tool Steel Produced by Powder Metallurgy. Metals 12:900
2. Strondl A et al (2015) Characterization and control of powder properties for additive manufacturing. JOM 67(3):549–554
3. Manohar G, Pandey KM, Maity SR (2021) Characterization of boron carbide (B4C) particle reinforced aluminium metal matrix composites fabricated by powder metallurgy techniques—a review. Mater Today Proc 45:6882–6888
4. Baesso I et al (2021) Characterization of powder flow behavior for additive manufacturing. Addit Manuf 47:102250
5. Quan G et al (2025) A characterization of the powder yield behaviors during a hot isostatic pressing process. Metals 15(7):752
6. Manohar G, Pandey KM, Maity SR (2021) Effect of sintering mechanisms on mechanical properties of AA7075/B4C composite fabricated by powder metallurgy techniques. Ceram Int 47(11):15147–15154
7. Jiang Y et al (2024) Optimization of the grain boundary network and tensile properties ofpowder metallurgy-hot isostatic pressed. Mater Sci Eng A 903:176388
8. Susilowati SE, Fudholi A, Sumardiyanto D (2022) Mechanical and microstructural characteristics of Cu–Sn–Zn/Gr metal matrix composites processed by powder metallurgy for bearing materials. Results Eng 14:100377
9. Fang Z et al (2017) Powder metallurgy of titanium—past, present, and future. Int Mater Rev 63: 1–53
10. Tabie VM (2025) Properties of particle-reinforced titanium matrix composites produced by powder metallurgy—current research. J Eng Appl Sci 72(1):53
11. Pan K et al (2024) The rapid densification behavior of powder metallurgy Ti alloys by induction heating sintering. J Mater Sci Technol 181:152–164
12. Mu XN et al (2018) Uniform dispersion of multi-layer graphene reinforced pure titanium matrix composites via flake powder metallurgy. Mater Sci Eng A 725:541–548
13. Pan Y et al (2023) Grain growth kinetics and densification mechanism of Ti/CaB6 composites by powder metallurgy pressureless sintering. J Alloys Compd 939:168686
14. Raynova S, Yang F, Bolzoni L (2021) Mechanical behaviour of induction sintered blended elemental powder metallurgy Ti alloys. Mater Sci Eng A 799:140157
15. Forstner T et al (2025) Influence of molecular structure of POM on processability within metal injection molding. Polymers 17(19):2621
16. Teferi FT, Tsegaw AA (2022) Metal injection molding (MIM) process and potential remedies for its defects: a review. In: Advances of science and technology. Springer International Publishing, Cham
17. Moon AP et al (2021) Opportunity and challenges of iron powders for metal injection molding. ISIJ Int 61(7):2015–2033
18. Lin C-M, Yen P-Y, Tan C-M (2024) Optimization of metal injection molding processing conditions for reducing black lines and meld lines in bone plates. Polymers 16(23):3241
19. Kuo C-C et al (2025) Revolutionizing metal powder injection molding: a cost-effective approach to tooling innovation. J Manuf Process 150:900–913
20. Gobber FS et al (2021) Innovative densification process of a Fe-Cr-C powder metallurgy steel. Metals 11(4):665

21. Hakam A, Taha M (2021) Review on using powder metallurgy method for production of metal-based nanocomposites. Egypt J Chem 64
22. Manohar G, Pandey KM, Maity SR (2022) Effect of spark plasma sintering on microstructure and mechanical properties of AA7075/B4C/ZrC hybrid nanocomposite fabricated by powder metallurgy techniques. Mater Chem Phys 282:126000
23. Tokita M (2021) Progress of spark plasma sintering (SPS) method, systems, ceramics applications and industrialization. Ceramics 4(2):160–198
24. Dareh Baghi A et al (2022) Microstructural development of Ti-6Al-4V alloy via powder metallurgy and laser powder bed fusion. Metals 12(9):1462
25. Hassan SF et al (2024) Tungsten heavy alloy: nano-crystallinity and alloying induced low temperature sintering, microstructure and mechanical properties. Int J Refract Met Hard Mater 120:106627
26. Shakunt NS, Gouthama, Upadhyaya A (2024) Effect of Fe addition in W-Ni-Cu heavy alloy processed through powder metallurgy on microstructure and mechanical properties. J Alloys Compd 970:172578
27. Hu B, Cai G (2022) Effect of hot isostatic pressing process parameters on properties and fracture behavior of tungsten alloy powders and sintered bars. Materials 15(23):8647
28. Zhang D et al (2024) Effect of heat treatment on microstructure and properties of powder metallurgy high-speed steel prepared by hot isostatic pressing. Metals 14(10):1160
29. Jiang Y et al (2024) Optimization of the grain boundary network and tensile properties of powder metallurgy-hot isostatic pressed Ni-based alloy Inconel 625. Mater Sci Eng A 903: 146688
30. Suryanarayana C (2022) Mechanical alloying: a critical review. Mater Res Lett 10:619–647
31. Novák P (2020) Advanced powder metallurgy technologies. Materials (Basel) 13(7):1742
32. Mussatto A et al (2020) Advanced production routes for metal matrix composites. Eng Rep 3
33. El-Eskandarany MS et al (2021) Mechanical milling: a superior nanotechnological tool for fabrication of nanocrystalline and nanocomposite materials. Nanomaterials 11(10):2484
34. Edosa OO, Tekweme FK, Gupta K (2021) A review on the influence of process parameters on powder metallurgy parts. Eng Appl Sci Res 49(3):433–443
35. Cui C et al (2023) Gas atomization of duplex stainless steel powder for laser powder bed fusion. Materials 16(1):435
36. Tian X et al (2023) Progresses on the additive manufacturing of functionally graded metallic materials. J Alloys Compd 960:170687
37. Becker L et al (2024) Powder metallurgy and additive manufacturing of high-nitrogen alloyed FeCr(Si)N stainless steel. Adv Eng Mater 27
38. Fang X et al (2023) State of the art of metal powder bonded binder jetting printing technology. Discov Mater 3(1):15
39. Janzen K et al (2025) Recycling of Ti-6Al-4V powder in metal binder jetting. J Jpn Soc Powder Powder Metall 72:S353
40. Du C et al (2025) Ultrafine niobium powder enabled synergetic liquid-phase sintering toward full densification of elemental powder metallurgy NiTi shape memory alloys. Mater Sci Eng A 946:149090
41. Abedini A et al (2022) Mechanical properties are affected by coalescence mechanisms during sintering of metal powders: case study of Al-Cu nanoparticles by molecular dynamics simulation. Powder Technol 405:117567
42. Dai H, Chen D, Zheng Z (2019) Modelling the sintering neck growth process of metal fibers under the surface diffusion mechanism using the lattice Boltzmann method. Metals 9(5):614
43. Wu P et al (2025) Growth kinetics of sintering neck and particle rotation mechanism of silver nanoparticles by molecular dynamics simulation and in-situ TEM observation. J Mater Res Technol 35:2253–2262
44. Zhao G et al (2025) Microwave sintering of lithium hydride by powder metallurgy: grain growth kinetics and densification mechanism. Adv Powder Technol 36(6):104867

45. Ashebir DA, Mengesha G, Sinha D (2022) An insight into mechanical and metallurgical behavior of hybrid reinforced aluminum metal matrix composite. Adv Mater Sci Eng 2022
46. Nosewicz S et al (2019) Discrete element modeling of intermetallic matrix composite manufacturing by powder metallurgy. Materials 12(2):281
47. Chung S-H et al (2010) Modeling and simulation of press and sinter powder metallurgy. ASM handbook, vol 22

Chapter 3
Additive Manufacturing with Metal Powders

Abstract Additive manufacturing routes that rely on particulate feedstocks integrate thermophysical and metallurgical phenomena in which powder characteristics, energy input, and transient melt pool behavior jointly govern microstructural development, defect formation, and overall performance. Processes such as laser powder bed fusion, electron beam melting, and directed energy deposition operate under steep thermal gradients and rapid solidification conditions that promote fine cellular, columnar, or martensitic microstructures, enabling high strength-to-weight ratios, localized property control, and near-net-shape fabrication. Feedstock morphology, particle size distribution, flowability, and surface chemistry regulate powder bed uniformity, energy absorption, and melt stability, with spherical and narrowly distributed particles consistently yielding higher density and reduced porosity. Processing parameters related to energy density, scan strategy, and deposition kinematics determine solidification behavior, anisotropy, and residual stress evolution, while in situ monitoring and data-informed control increasingly support defect mitigation. Hybrid and extrusion-based approaches broaden process accessibility and sustainability although challenges associated with debinding-induced shrinkage and incomplete densification remain. Across aerospace, biomedical, and automotive applications, topology optimization and functional grading establish these technologies as a key route for producing structurally efficient, application-specific metallic components.

Keywords Additive manufacturing · Particulate feedstocks · Rapid solidification microstructures · Process-induced defects

3.1 Introduction

The selective melting or bonding of metallic feedstocks underpins an additive manufacturing (AM) approach that extends the principles of powder metallurgy to the fabrication of near-net-shape components with controlled geometry and microstructure. Techniques such as laser powder bed fusion, electron beam melting, and directed energy deposition involve strongly coupled thermal, fluid, and metallurgical

R. Pérez-Bustamante, *Fundamental, Practical and Disruptive Applications of Powder Metals*, Topics in Mining, Metallurgy and Materials Engineering,
https://doi.org/10.1007/978-3-032-27130-3_3

interactions that control melt pool stability, solidification kinetics, and defect formation across a broad range of alloy systems [1–10]. In current L-PBF practice, the process is characterized through multiphysics modeling approaches that couple heat transfer, phase transformation, and porosity evolution in order to relate laser parameters and powder layer characteristics to the resulting metallurgical outcomes; however, fully unified predictive models are still lacking. Aluminum, titanium, cobalt-chromium, and nickel-based alloys exhibit process-specific solidification structures, from fine cellular dendrites in Al-Si systems to columnar epitaxial grains and martensitic α' in Electron Beam Melting (EBM)-processed Ti-6Al-4V, reflecting the dominant role of thermal gradients and cooling rates in microstructural selection [11–14].

The performance envelope of these technologies is constrained by the properties of the powder feedstock [3, 15–18]. Particle size distribution, morphology, surface chemistry, and flowability control powder bed uniformity, melt pool absorption, and pore nucleation, while atomization routes and post-processing such as spheroidization and classification determine packing density and chemical stability. Deviations in powder morphology or spreading dynamics generate local energy imbalances that manifest as lack-of-fusion pores, balling, or surface roughness, particularly under the rapid thermal cycling intrinsic to LPBF and DED. These sensitivities are amplified by part geometry and scale, as demonstrated by size-dependent thermal histories that alter grain size, density, and defect distribution even under nominally optimized processing conditions [19–22].

The chapter further establishes that defect formation is a multiscale consequence of melt pool instability, powder contamination, and energy density misalignment, motivating the adoption of in situ sensing and data-driven control to detect porosity, cracking, and denudation in real time [9]. In parallel, filament-based systems using powder-loaded thermoplastic binders extend additive manufacturing toward metal injection molding-like feedstocks, where rheology, binder chemistry, and debinding kinetics govern green density and final sintered integrity [3, 23–25]. Deterministic control of powder attributes, thermal input, and post-processing sequences enables fabrication of dense and dimensionally accurate components tailored to specific applications. The chapter analyzes quantitative relations among feedstock engineering, processing conditions, microstructural evolution, and defect suppression that govern the viability of powder-based metal additive manufacturing in aerospace, medical, and automotive sectors.

3.2 Laser Powder Bed Fusion, Electron Beam Melting, and Directed Energy Deposition

3.2.1 Fundamentals of Powder-Based Additive Manufacturing

Powder-based AM encompasses several processes: Laser Powder Bed Fusion (LPBF), Electron Beam Melting (EBM), and Directed Energy Deposition (DED) that share the fundamental principle of melting and solidifying metallic powders to produce near-net-shape components layer by layer (Fig. 3.1). Each process involves complex physical and metallurgical phenomena, where heat transfer, phase transformation, and microstructural evolution define the resulting properties of the fabricated parts. Recent research has focused on modeling these mechanisms, optimizing material systems, and mitigating defects, providing a deeper understanding of their process-structure-property relationships.

Laser Powder Bed Fusion has been extensively analyzed due to its precision and versatility in producing intricate geometries. Modeling of laser powder bed fusion addressed metallurgical response, temperature fields, and defect formation, with thermal simulations required to capture nonlinear heat transfer and solidification kinetics during laser scanning [9]. Finite element and finite volume simulations were used to evaluate temperature gradients, which strongly influence solidification modes and grain morphology. Metallurgical models discussed in the review emphasized the prediction of grain size, phase transformation, and texture evolution under varying laser power, scanning speed, and cooling rates. Porosity, cracking, and lack of fusion were associated with unstable melt pool dynamics and incomplete powder melting. Numerical simulations incorporating coupled physical interactions

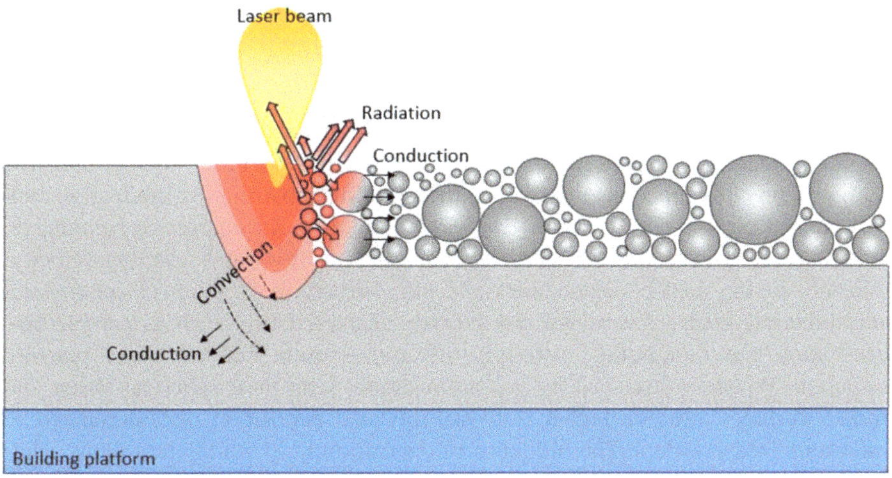

Fig. 3.1 Physical phenomena involved in the PBF LB/M process [6]

indicated that porosity could be reduced through adjustment of laser parameters and powder layer thickness. Nabavi and colleagues concluded that integrating temperature, metallurgical, and defect models enables predictive control of part quality, though current efforts still lack unified frameworks linking these domains.

LPBF aluminum alloys, particularly those in the Al-Si system, exhibit fine cellular dendritic microstructures resulting from high solidification rates, leading to tensile strengths ranging from 200 to 600 MPa [14]. Alloy composition governed printability, with AlSi alloys exhibiting improved weldability through silicon-induced reduction of thermal expansion, whereas 2XXX and 7XXX series alloys showed increased susceptibility to hot cracking due to copper and zinc additions. To improve printability, researchers have developed aluminum matrix composites reinforced with nanoceramic particles and microalloyed powders specifically tailored for LPBF. Experimental observations demonstrated that optimizing laser parameters can control grain refinement and minimize thermal cracking by altering the melt pool morphology and solidification rate. It was observed that manufacturing costs remain high, between 2 and 120 times higher than conventional methods, mainly due to the cost of atomized powders and the need for post-processing to remove surface defects.

EBM represents another key powder-based process distinguished by its use of an electron beam as the energy source. In 2014, Murr investigated the metallurgy of EBM across several alloy systems, including Ti-6Al-4V, Co-Cr-Mo, and nickel-based superalloys [26]. Prealloyed powders were preheated and selectively melted layer by layer under vacuum, with electron beam scanning generating Gaussian temperature profiles governed by beam current, scan speed, and layer thickness. The resulting heat input and cooling rate produced columnar grain architectures aligned with the build direction through epitaxial growth during successive layer melting. Microstructural analysis using scanning and transmission electron microscopy revealed that slower scan speeds and higher beam currents increase melt pool stability and part density, whereas rapid cooling in thin sections produces martensitic α' phases in Ti-6Al-4V. This phase transformation increases hardness up to 4.8 GPa compared to 3.5 GPa for coarser structures, underscoring the importance of thermal control in microstructure tailoring.

DED extends powder-based AM to larger components and repair applications. Piscopo et al. [10] analyzed the mechanisms governing the Laser Powder Directed Energy Deposition process, emphasizing the coupling between laser irradiation, melt pool generation, and solidification. They identified up to 14 interacting parameters, including laser power, powder feed rate, shielding gas flow, and scanning velocity, each influencing melt pool dynamics and microstructural evolution. Experimental and numerical analyses revealed that powder characteristics, such as particle size distribution and morphology, strongly affect deposition efficiency and porosity formation. Powders produced by gas atomization, with their spherical shape and smooth surfaces, provide higher flow stability and deposition rates compared to water-atomized powders. The influence of deposition head configuration was evaluated, revealing that coaxial nozzle systems achieve higher powder capture efficiency and more isotropic deposition than lateral feed arrangements. Observations of melt pool behavior indicated that Marangoni convection, buoyancy, and recoil pressure jointly determine pool shape and dilution rate. These fluid flow phenomena directly impact cooling rate and solidification microstructure.

3.2.2 Powder Characteristics and Material Considerations

The performance and reliability of metal AM processes are highly dependent on the properties of the powders used as feedstock (Fig. 3.2). Powder characteristics, including particle size distribution, morphology, flowability, and chemical composition, govern powder bed quality, melt and solidification behavior, and the resulting component microstructure. Understanding these powder attributes and their influence on AM techniques such as LPBF, EBM, and DED is essential for ensuring process stability and part integrity.

The production of metal powders involves several atomization techniques that aim to generate uniform and spherical particles with controlled size distributions (Fig. 3.3). Priyadarshi et al. [12] utilized ultrafast synchrotron X-ray and high speed optical imaging to investigate the cavitation and capillary dynamics during atomization. The results demonstrated that ultrasonic atomization combines the effects of capillary wave formation and acoustic cavitation, where bubble implosions disrupt

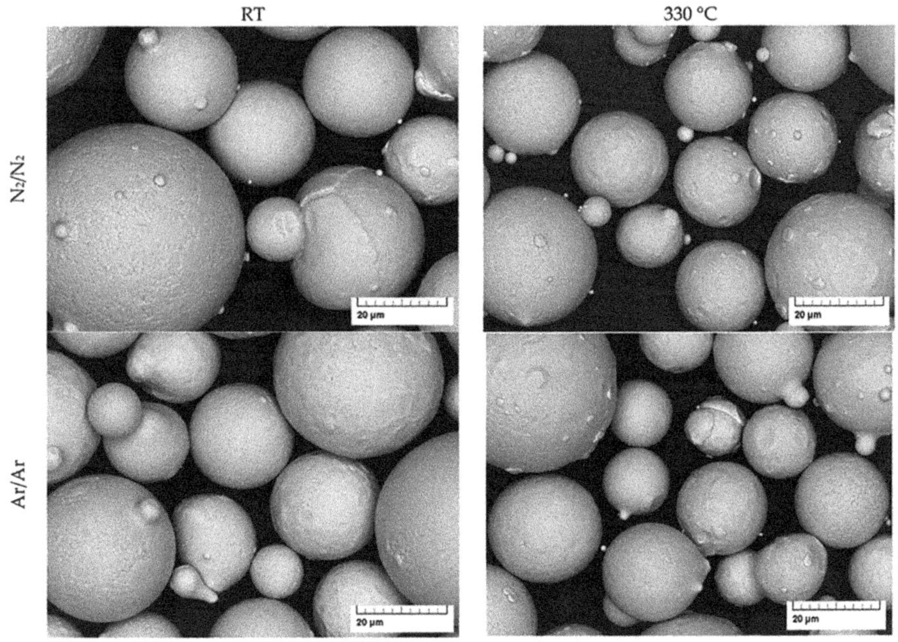

Fig. 3.2 SEM images of gas-atomized X2CrNiMoCuWN25 7 4 powders (particle size 20–63 µm, at high magnification) [4]

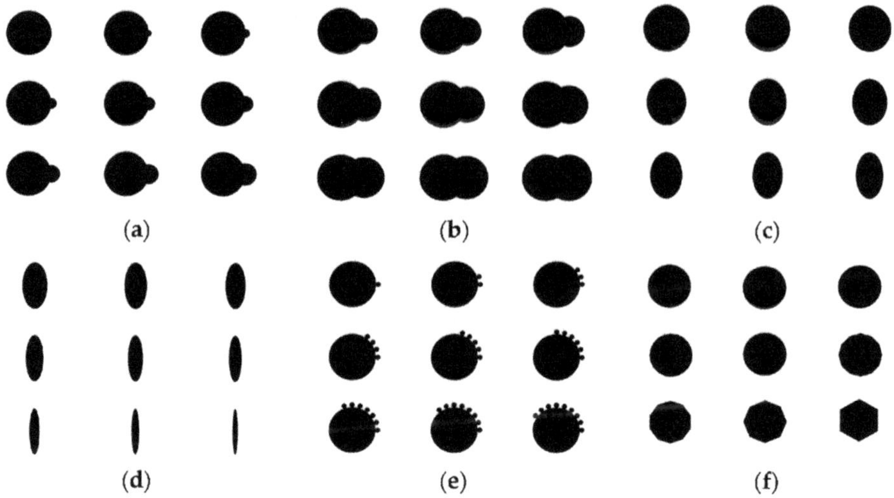

Fig. 3.3 Collection of schematized particles used for the evaluation of basic measurements and shape descriptors for a single satellite (**a**, **b**), elongated (**c**, **d**), multiple satellites (**e**), and faceted (**f**) particles [28]

the liquid surface to form fine droplets. Experiments using aluminum melts revealed that ultrasonic atomization can yield nearly perfect spherical powders with narrow size distributions, and the average particle size decreases as vibration amplitude is reduced. Powder morphology significantly affects the packing uniformity and surface topology of the powder bed in layer-based manufacturing. In this regard, spherical powders produce smoother and denser powder beds compared to irregularly shaped particles, which tend to create voids and uneven layers [19, 27]. By using laser profilometry and statistical surface analysis, the researchers quantified how particle shape irregularity and size distribution heterogeneity increased surface roughness and local variation in layer thickness. They also investigated spreading speed and recoater blade velocity, finding that slower speeds promote higher uniformity at the expense of process time.

Flowability is another fundamental property influencing powder deposition and consolidation [29]. The flow behavior of various metal powders has been reported using Hall flow rate, angle of repose, and shear testing methods. Spherical powders exhibit superior flowability due to lower interparticle friction and cohesive forces, while irregular particles display significant resistance to flow, especially under low stress conditions. It was also established that correlations exist between particle size, shape factor, and dynamic flow indices, noting that fine particles below 20 μm are prone to agglomeration, reducing their spreadability. It was observed that a quantitative flow function exists that relates the cohesive index to apparent density and moisture content, enabling prediction of powder flow under different storage and processing conditions. Such characterization is particularly important for DED, where stable powder feeding and consistent mass flow are essential for controlling

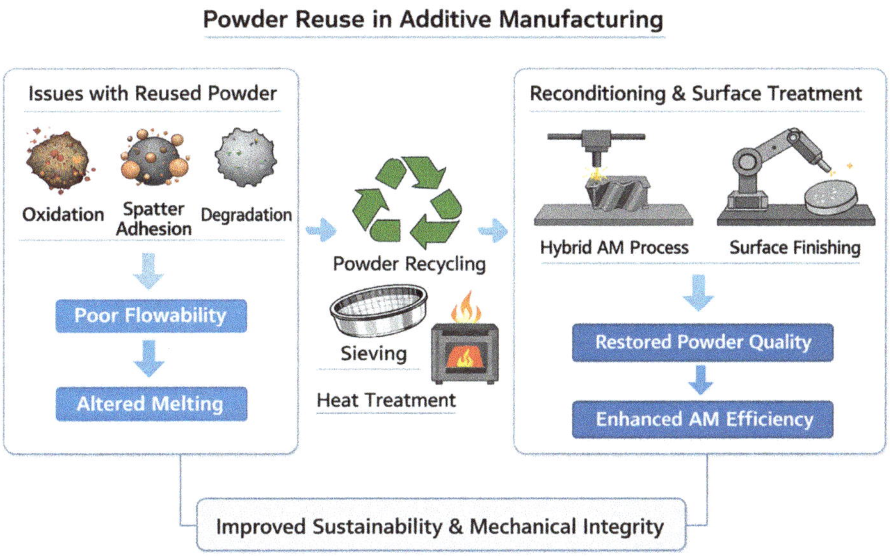

Fig. 3.4 Schematic overview of powder reuse in metal AM, illustrating degradation mechanisms, reconditioning strategies, and their role in improving sustainability and process efficiency

melt pool geometry and layer adhesion. The contamination from oxygen and moisture during atomization or storage alters powder reactivity, leading to oxidation and gas porosity during melting. In LPBF and EBM, oxygen-rich powders promote oxide inclusion formation, while in DED, this contamination affects bonding and reduces mechanical strength.

The quality of powders is also closely related to their production method and post-processing treatments. Gas atomization generally produces spherical particles with smooth surfaces, whereas water atomization leads to irregular morphologies with higher internal porosity. Plasma spheroidization and mechanical classification can enhance powder sphericity and narrow the particle size distribution. Optimizing atomization gas pressure and cooling rates directly impacts particle solidification behavior and microstructure uniformity, which in turn determine the laser absorptivity and melt pool stability during fabrication [8, 12, 27].

Material considerations in AM extend to powder reuse and sustainability aspects (Fig. 3.4). Kahhal et al. reported remanufacturing technologies using metal AM and surface treatment, highlighting that powder recycling plays a vital role in reducing material waste and cost [19]. However, it was observed that repeated reuse alters particle morphology due to oxidation, spatter adhesion, and thermal degradation. The accumulation of satellite particles and changes in surface chemistry degrade powder flowability and modify melting behavior. It was proposed that combining powder reconditioning techniques, such as sieving and heat treatment, with surface finishing could recover powder quality for subsequent AM cycles.

3.2.3 Process Parameters and Their Influence on Microstructure

The control of microstructure in metal AM is highly dependent on process parameters such as energy input, scan speed, layer thickness, and feedstock geometry (Figs. 3.5 and 3.6). These parameters govern the thermal history, solidification rate, and defect formation during LPBF, EBM, and DED. In LPBF, the thermal gradient and solidification velocity define the resulting grain morphology and texture, while in solid-state processes such as Friction Rolling Additive Manufacturing (FRAM), the mechanical interaction between the tool and the feed material dictates dynamic recrystallization and grain refinement.

Part size and processing parameters were examined in Fe–Ni alloys produced by laser powder bed fusion, with a laser power of 200 W, scanning speed of 700 mm s^{-1}, and hatch spacing of 120 μm, producing nearly defect-free microstructures in 10 mm cubes, while larger components exhibited altered thermal histories and increased defect variability [1]. Finite element simulations revealed that larger parts experienced lower peak temperatures and slower cooling rates, leading to

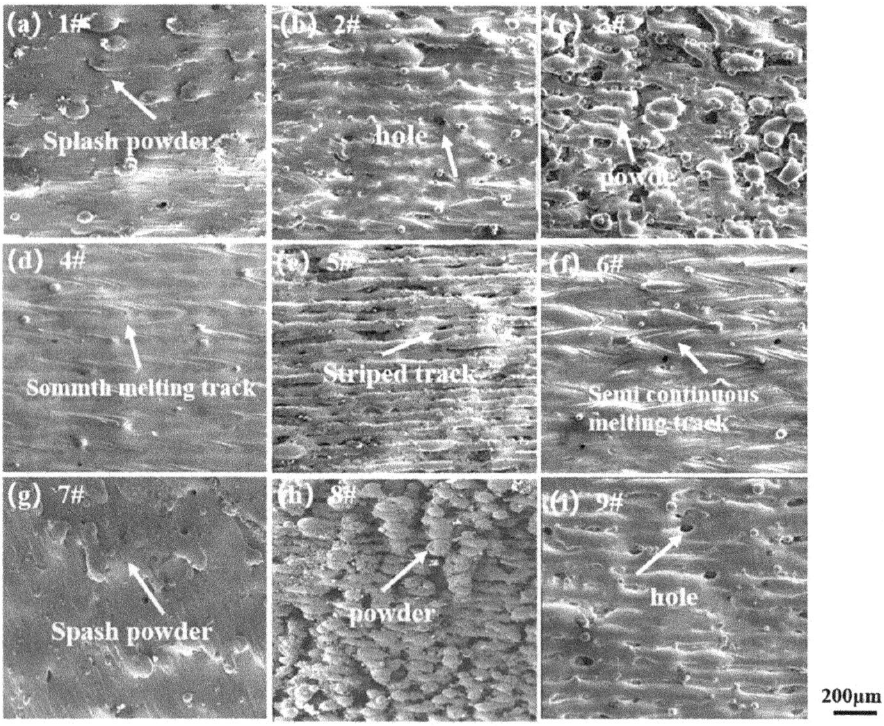

Fig. 3.5 SEM micrographs showing top surface topography of as-received samples fabricated at the energy density **(a)** 95.00 J/mm^3, **(b)** 35.19 J/mm^3, **(c)** 18.63 J/mm^3, **(d)** 65 J/mm^3, **(e)** 32.1 J/mm^3, **(f)** 76.47 J/mm^3, **(g)** 55.00 J/mm^3, **(h)** 122.22 J/mm^3, and **(i)** 48.53 J/mm^3 [30]

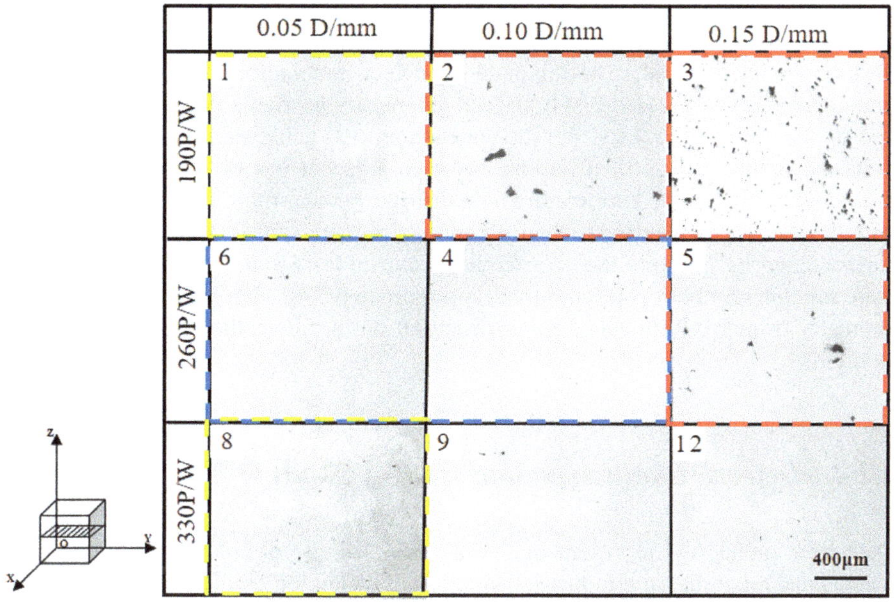

Fig. 3.6 Porosity distribution of 316L stainless steel samples fabricated via LPBF at various laser powers and scan speeds [30]

increased lack-of-fusion defects and crack propagation along the scan direction. Experimental data indicated a density decrease from 8.11 g cm^{-3} in small parts to 8.02 g cm^{-3} in larger ones, with grain size reducing from 41.88 to 23.07 μm in the scanning plane due to constrained grain growth and higher thermal gradients. Thermal accumulation and cooling rate variations, driven by part geometry, alter the melt pool morphology and crystallographic orientation, affecting both mechanical strength and defect sensitivity.

Microstructural control through powder size has also been identified as a decisive factor in AM. Finer powders enhance melt pool homogeneity due to better packing density and absorption efficiency, while coarse particles can lead to heterogeneous melting and pore formation. A balance between powder size and process parameters must therefore be achieved to maintain consistent microstructural refinement throughout the build height. These effects are particularly relevant in processes such as LPBF, where localized laser powder interactions and repeated thermal cycling promote epitaxial grain growth and texture alignment along the build direction.

Feedstock geometry and deposition kinematics were evaluated in friction rolling AM of 6061-T6 aluminum, with wire and strip feedstocks processed under identical conditions of 1000 rpm rotation speed, 0.8 mm lift height, and 100 mm min^{-1} traverse speed, isolating the effect of feed geometry on microstructure and mechanical response [31]. Results showed that wire feedstock promoted greater plastic deformation and grain refinement than strip feedstock, with grain size reductions of 95% and 81%, respectively. The wire-fed process facilitated improved heat generation and dynamic recrystallization due to enhanced frictional contact between the tool and the material, resulting in dense, defect-free microstructures. Microstructural observations revealed that both feed geometries formed a thin plasticized layer approximately 112 μm thick, but the wire-fed process achieved more homogeneous flow and minimized unfilled regions between adjacent passes. Numerical simulations confirmed that the temperature field distribution was more uniform when using wire feedstock, contributing to the formation of refined equiaxed grains and superior surface integrity. Despite these differences, tensile tests indicated comparable ultimate strengths for both configurations, approximately 144 MPa, suggesting that feed geometry primarily influenced microstructural uniformity rather than macroscopic strength.

3.2.4 Defects, Challenges, and Quality Control

Defects in metal AM represent one of the most persistent challenges limiting the mechanical reliability and industrial adoption of technologies such as L-PBF, EBM, and DED. These processes involve complex thermal, mechanical, and material interactions that can generate porosity, cracking, balling, and surface oxidation, each influenced by processing conditions and material behavior. Understanding the formation mechanisms of these defects and developing detection and mitigation strategies are critical to achieving defect-free parts with predictable performance (see formation of the powder building layer with a narrow and uneven particle size distribution in Fig. 3.7).

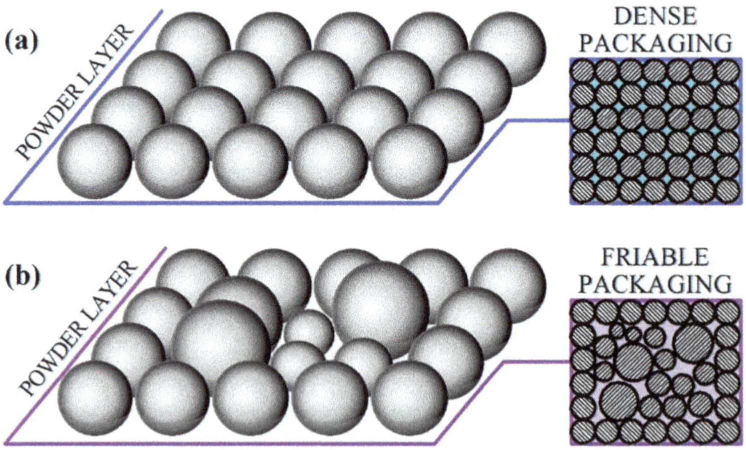

Fig. 3.7 Formation of the powder building layer: (**a**) powder with a narrow particle size distribution; (**b**) powder with uneven particle size distribution [13]

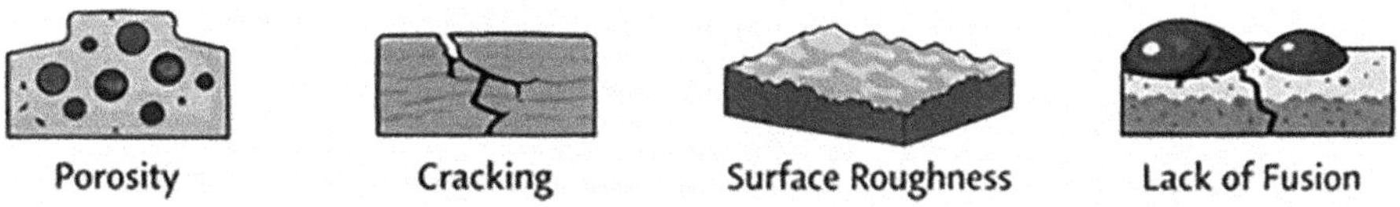

Fig. 3.8 Schematic representation of defect types and formation mechanisms in LPBF

Defect formation mechanisms across AM routes included porosity, cracking, surface roughness, and lack of fusion, grouped into geometrical, surface-related, microstructural, and mechanical categories. Laser power, scan speed, hatch spacing, and layer thickness governed defect formation, with low energy density producing lack of fusion porosity and excessive energy causing keyhole porosity through vapor depression instability [32]. Oxidation and contamination of powders during storage or handling can also lead to microstructural defects that propagate during solidification, particularly in aluminum and titanium alloys. Post-processing strategies, including hot isostatic pressing and surface machining, were reported to reduce the severity of these defects although their complete elimination remains difficult due to the stochastic nature of melt pool dynamics.

An interesting research related to the understanding of surface and geometrical defects in LPBF was presented by Guillen et al. [33], who categorized them as balling, surface roughness, oxidation, and denudation. It is reported that balling is caused by inadequate wettability of the molten pool and excessive surface tension, often triggered by low laser power or high scanning speeds. Experimental results indicated that increasing laser power within the appropriate energy density window decreases melt viscosity, improving wetting behavior and reducing spherical bead formation. Surface oxidation, even at oxygen levels as low as 0.1–0.2%, was observed to severely affect scan stability in alloys containing magnesium and silicon, producing rough and irregular surfaces. Denudation effects, characterized by the removal of surrounding powder due to vapor plume dynamics, were linked to gas flow patterns and scan velocity (Fig. 3.8). These defects can accumulate and degrade interlayer bonding, leading to porosity and structural anisotropy. It was also emphasized that combinations of sensing technologies, optical, thermal, and acoustic are essential to capture these transient defect formation events and enable real-time correction during fabrication.

Advanced process monitoring combined with machine learning was applied to defect detection and quality control in laser powder bed fusion, using photodiode signals, infrared thermography, and acoustic emission to identify anomalies during processing. Convolutional neural networks and random forest classifiers resolved spatter, keyhole porosity, and layer delamination from high speed imaging data [5, 32, 33]. By correlating sensor data with post-build microstructural analysis, the researchers established predictive models capable of identifying defective layers in real time. It was demonstrated that the potential for autonomous feedback control systems reduce the need for post-process inspection. It was noted that combining multiple sensor modalities enhances detection reliability, particularly when fusing thermal and optical data streams for spatiotemporal feature extraction (Fig. 3.9).

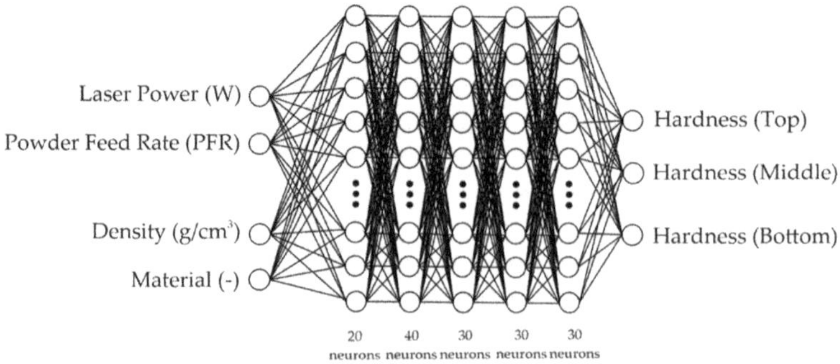

Fig. 3.9 Representation of the network architecture of the three sets [34]

Intrinsic reduction of defects required control of material and thermal parameters during processing. A normalized model-based processing diagram was applied to laser powder bed fusion of a TiZrNbTa refractory high entropy alloy, with in situ alloying of elemental powders producing nearly defect-free specimens in which unmelted particles, keyhole porosity, and thermal cracking were suppressed [5, 7, 32, 33]. The processing diagram normalized laser power, scan speed, and hatch spacing against the thermophysical properties of the alloy, allowing identification of a processing window that balances energy input and solidification behavior. Experiments with 28 different parameter combinations confirmed that this method produced dense, isotropic microstructures with minimal porosity. The high cooling rates achieved, on the order of 10^6 K s^{-1}, promoted homogeneous solidification and suppressed segregation-driven cracking [7]. Compared to conventional refractory alloys, the printed TiZrNbTa demonstrated enhanced elastic isotropy and mechanical strength, validating the framework as a quantitative tool for minimizing defects in complex multicomponent systems.

3.2.5 Industrial Applications and Emerging Innovations

Industrial applications of metal AM have rapidly expanded, supported by improvements in powder bed fusion-based technologies and their integration into aerospace, automotive, and energy sectors. The development of LPBF, EBM, and DED processes has enabled the production of components with optimized geometries and mechanical performance suitable for demanding operational environments. A key advancement enabling this expansion is the enhanced control over powder spreading and layer uniformity, which directly impacts the mechanical and microstructural consistency of printed parts. Shaheen et al. [35] demonstrated that the spreading behavior of metal powders in LPBF depends on both material characteristics and process parameters, such as interparticle cohesion, particle shape, and spreading tool design. Irregular particle geometries, rough surfaces, and higher cohesive forces degrade powder layer uniformity, whereas counter-rotating rollers improve layer packing density compared to blades.

AM combined with alloy design enabled rapid prototyping of compositions with tailored strength, corrosion resistance, and high temperature performance. Laser powder bed fusion and directed energy deposition were applied to refractory and high entropy alloys, producing dense, fine-grained microstructures not accessible by conventional processing routes [2, 18]. The flexibility to process custom alloy compositions has allowed industries such as aerospace and biomedical engineering to tailor material performance to specific environments. AM not only allows complex part fabrication but also enables site-specific compositional gradients, allowing components to combine different mechanical properties within a single structure. However, challenges remain, including managing residual stress, preventing microcracking, and standardizing quality assurance procedures for multicomponent systems.

From a sustainability standpoint, Kahhal et al. [19] highlighted the increasing role of AM in remanufacturing technologies. Traditional remanufacturing often faces limitations in restoring complex geometries or ensuring consistent material properties, but metal AM overcomes these barriers by enabling near-net-shape reconstruction of worn or damaged parts. Techniques such as DED and Laser Cladding allow for the restoration of high value components such as turbine blades, engine blocks, and hydraulic cylinders. The integration of post-processing and surface treatments, including thermal spraying and electroplating, improves surface hardness and corrosion resistance of remanufactured parts.

The transition of AM from prototyping to large-scale industrial production is also influenced by the optimization of powder handling and spreading systems. Uniform powder deposition remains a crucial factor determining build quality and mechanical properties. Material flowability, particle size distribution, and spreading velocity govern powder bed uniformity and density, with quantitative relations established between powder morphology and layer homogeneity to guide spreader design in laser powder bed fusion systems [35].

Industrial deployment was accelerated through the combination of process modeling and machine learning for predictive control of quality parameters. Data-driven monitoring of melt pool stability, defect formation, and real-time adjustment of processing conditions improved batch-to-batch consistency and supported certification requirements in aerospace production. In hybrid AM remanufacturing systems, adaptive control of deposition and surface finishing optimized processing of complex geometries and variable material properties [36]. These developments reflected the integration of digital manufacturing with materials engineering in contemporary industrial AM.

3.3 Use of Filaments with Embedded Powders in Fused Deposition Modeling-Like Processes

3.3.1 Concept and Evolution of Powder-Loaded Filaments

In the production of metal AM components (Fig. 3.10), the concept of powder-loaded filaments emerged as a hybrid solution combining the principles of metal injection molding (MIM) and fused filament fabrication (FFF) to enable metal and ceramic AM through extrusion-based systems. This evolution represents a significant advancement in AM, offering a cost-effective and accessible pathway for producing dense metallic and ceramic parts using standard Fused Deposition Modeling (FDM)-like printers. Ramkumar and Rijwani [37] described this process as hybrid fused filament fabrication (hybrid FFF), emphasizing its potential to merge the precision of MIM feedstock processing with the flexibility of FFF. The idea involves the extrusion of a composite filament containing a mixture of metal or ceramic powders suspended in polymer binders. During printing, only the binder phase melts, forming the part layer by layer, while the metal or ceramic particles remain suspended until post-processing steps such as debinding and sintering transform the printed green body into a dense metallic or ceramic structure.

The origins of powder-loaded filaments can be traced back to the multiphase jet solidification (MJS) process proposed in the 1990s, which later evolved into modern hybrid FFF systems. Ramkumar and Rijwani [37] detailed how this hybridization allows traditional polymer FFF printers to be adapted for metal and ceramic production without extensive hardware modifications. The success of this approach depends on achieving a filament with the appropriate mechanical and rheological

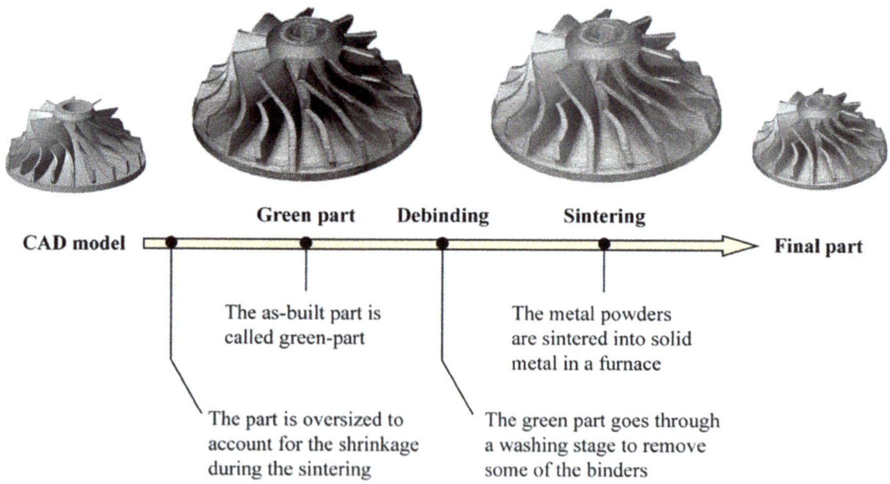

Fig. 3.10 Workflow of the Bound Metal Deposition (BMD) and the Atomic Diffusion Additive Manufacturing (ADAM) [22]

properties, sufficient flexibility to be wound on spools and fed through the printer, but enough stiffness to avoid buckling under compression. The feedstock preparation thus becomes critical, requiring careful selection of powder size distribution, binder composition, and additives to achieve optimal viscosity and homogeneity. Early studies established that a typical mixture consists of 55–60 vol.% of powder and 40–45 vol.% of binders, incorporating thermoplastic polymers, waxes, and elastomers to balance flexibility and strength. Additives such as oleyl alcohol or tackifiers are often included to enhance filament flexibility and adhesion between printed layers, enabling continuous feeding during printing.

Powder-loaded filaments required control of diameter and surface quality to maintain stable extrusion flow and uniform printing. Single-screw and twin-screw extrusion were used to produce filaments with circular cross sections and low porosity, with inline laser systems providing real-time diameter control and torque rheometry verifying homogeneous dispersion of powders within the binder [3, 21]. Mixing of silicon nitride with binder exhibited stable torque when uniform dispersion was achieved, and similar feedstock design principles applied to stainless steel, titanium, and neodymium-based alloys, for which identical binders produced filaments with markedly different flexibility and tensile response depending on powder chemistry.

Once prepared, these powder-loaded filaments undergo AM via standard FFF equipment, where process parameters such as layer height, nozzle temperature, extrusion multiplier, and chamber conditions must be optimized to ensure uniform deposition and minimize defects. Direct extrusion systems were favored over Bowden configurations because metal-loaded filaments exhibited higher brittleness and surface roughness, increasing the probability of fracture under tensile loading [21, 37]. Extrusion temperatures must be carefully calibrated to exceed the binder melting point without causing excessive flow or nozzle clogging. Likewise, layer height and cooling rates strongly influence green part density and interlayer bonding. Over extrusion of approximately 5% increased green density by filling interline voids, resulting in improved densified microstructure and mechanical response. Subsequent debinding and sintering, analogous to metal injection molding, removed the polymer binder and fused the metal or ceramic particles into a dense structure, with relative densities reaching approximately 97%.

3.3.2 Filament Manufacturing and Composition Optimization

The manufacturing and optimization of powder-loaded filaments are the steps that define the performance and reliability of metal fused filament fabrication (MF3) and other FDM-like AM processes. The development of filaments with well-balanced mechanical flexibility, uniform powder dispersion, and consistent dimensional accuracy is essential for achieving homogeneous extrusion and high density sintered

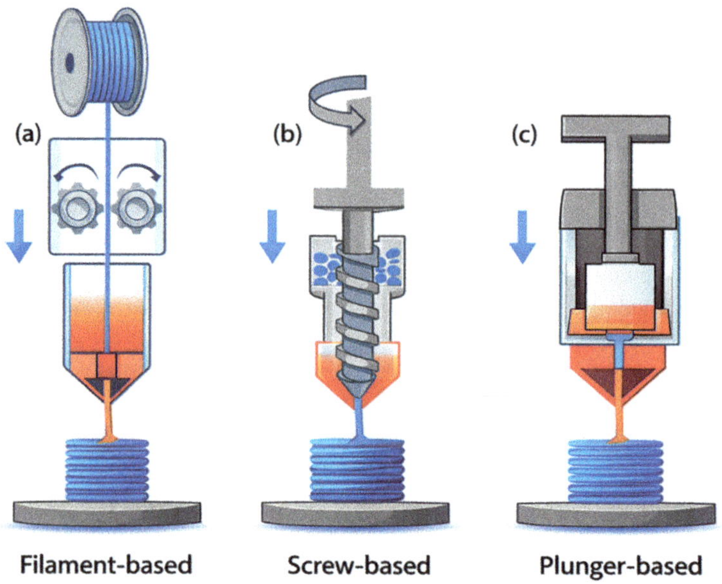

Fig. 3.11 Different approaches for material extrusion (MEX) AM that can be applied to producing metal parts (metal MEX): (**a**) filament-based, where the feedstock is a filament pushed into the heating zone by a mechanism consisting of gears or a roller; (**b**) screw-based, where the feedstock is made of composite pellets driven to the nozzle by a screw system; and (**c**) plunger-based, where the feedstock is loaded in a cartridge and pushed out by a plunger working like a piston

parts (Fig. 3.11). The rheological, morphological, and thermal characteristics of the metallic powders used directly influence feedstock behavior, filament stability, and the final quality of printed components. Metallic powder characteristics were varied to examine their influence on extrusion behavior and feedstock performance in indirect AM, with particle size, morphology, and surface roughness evaluated for stainless steel powders. Spherical particles with narrow size distributions promoted uniform dispersion within the binder and reduced internal porosity during extrusion [25]. Conversely, irregularly shaped or coarser powders increased shear stress and torque during compounding, resulting in inhomogeneous feedstock and premature filament breakage. It was noted that achieving a powder loading of around 60 vol.% represents an optimal balance between flexibility and stiffness, ensuring smooth feeding while maintaining sufficient green part density after printing.

It was also reported that binder composition and the powder-binder interaction significantly affect filament manufacturability. Thermoplastic polymers such as polyethylene, polypropylene, and polyoxymethylene combined with low-molecular-weight waxes and surfactants increased wetting of metallic particles and reduced interparticle friction during mixing and extrusion. Addition of stearic acid produced a lubricating surface film that improved powder dispersion and lowered extrusion torque, whereas excessive surfactant promoted phase separation and pore formation within the filament [25, 38].

Three-dimensional printing of metallic components by fused filament fabrication was performed using optimized metal polymer composites, with the rheological behavior of stainless steel and copper-based filaments governed by extrusion temperature, shear rate, and binder viscosity [39]. Experimental results showed that higher extrusion temperatures reduce viscosity but risk binder degradation, whereas lower temperatures increase pressure and cause filament buckling. In this regard, feed rate and nozzle diameter control the shear-induced alignment of metallic particles within the polymer matrix, affecting the mechanical anisotropy of the printed part [40]. By combining rheological modeling with empirical testing, the researchers proposed a processing window that minimizes shear instability while maintaining extrusion continuity. Furthermore, it was demonstrated that maintaining a constant filament diameter within ±0.05 mm is crucial for consistent material flow, as minor deviations lead to volumetric fluctuations and print defects.

Feedstock and processing optimization was applied to Ti-6Al-4V, with metal fused filament fabrication producing sintered components comparable to those obtained by metal injection molding. A binder system based on polyamide and paraffin wax enabled powder loadings up to 65 vol.% while retaining filament flexibility. Powder sizes below 25 μm increased sintered density to approximately 98% of theoretical, whereas coarser particles produced incomplete sintering and intergranular porosity [7, 40, 41]. The thermal debinding process was optimized to gradually remove the binder without structural collapse, followed by sintering under high vacuum to prevent oxidation. Microstructural analysis revealed a homogeneous $\alpha + \beta$ lamellar structure with tensile strength exceeding 900 MPa, closely matching that of MIM produced Ti-6Al-4V.

3.3.3 Printing Parameters and Post-processing Requirements

The optimization of printing parameters and post-processing conditions in MF3 and FFF has proven to be fundamental for achieving dense, defect-free metallic parts (Fig. 3.12). Finite element simulations were used to evaluate the sensitivity of metal fused filament fabrication of Ti-6Al-4V feedstock containing 59 vol.% solids to extrusion temperature, layer thickness, printing speed, and toolpath. Geometric and thermal variables governed print time, warpage, and residual stress, with reduced layer thickness improving surface quality at the expense of longer build times, and variations in bead width modifying dimensional accuracy and thermal gradients [42]. Results also indicated that extrusion temperature and build plate temperature play critical roles in ensuring adequate interlayer bonding and minimizing thermal stresses, with an optimal range identified between 220 and 260 °C. The simulations further demonstrated that printing speed must be kept below 15 mm/s for Ti-6Al-4V filaments to prevent delamination or poor layer adhesion, given the high viscosity of the metal polymer composite. The identification of these relationships enabled the definition of an optimized window for MF3 printing, highlighting how coupled thermal and mechanical effects determine structural fidelity during deposition.

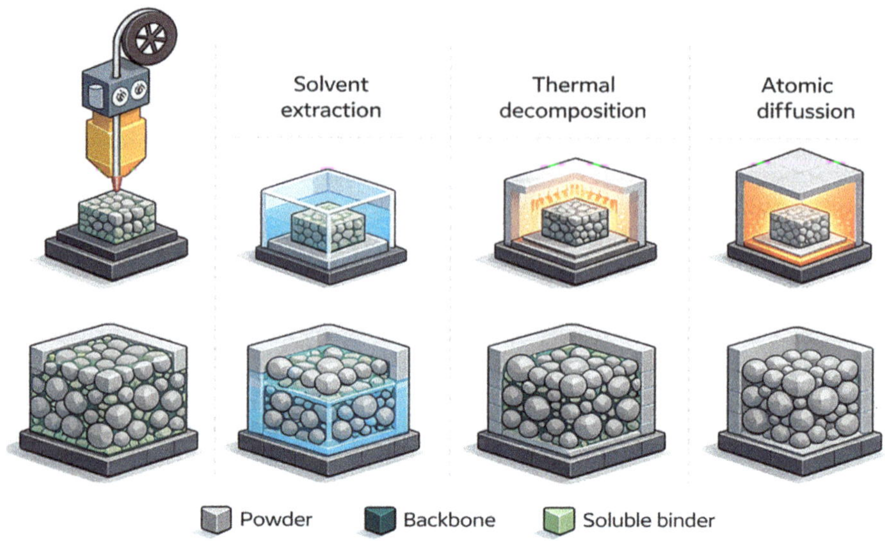

Fig. 3.12 Schematic of the FFF process showing shaping, debinding, and sintering stages

Wagner et al. [25] extended this understanding by experimentally characterizing stainless steel 316L feedstocks in the context of FFF, focusing on binder formulation and its implications for post-processing. The binder, composed of polyethylene (PE) backbones, thermoplastic elastomers (TPE), and stearic acid, was optimized to achieve balance between filament stiffness, flexibility, and rheological stability during extrusion. Through mechanical and rheological testing, it was observed that increasing the PE70 fraction in the binder reduced viscosity and improved printability at lower nozzle diameters without filament breakage. The optimized binder system allowed extrusion through 0.25 mm nozzles at 230 °C, demonstrating precise control over melt flow behavior. This mechanical stability directly influenced post-printing performance, as uniform material deposition reduced the formation of voids that could propagate during sintering. It was also noted that solvent and thermal debinding steps must be carefully synchronized to prevent distortion. A two-step debinding strategy, solvent removal followed by thermal decomposition, successfully avoided cracking and blistering, enabling sintering at 1360 °C under hydrogen to yield nearly fully dense 316L structures. The resulting parts exhibited isotropic mechanical properties and enhanced ductility, confirming that binder selection and debinding control are as critical as the printing parameters themselves.

Optimization of thermal and mechanical inputs during printing influenced residual stress development and densification during subsequent post-processing [25, 38, 40, 42]. For Ti-6Al-4V filaments, build plate temperature and toolpath orientation were found to affect the anisotropy of thermal gradients during deposition, which in turn determined the uniformity of shrinkage during sintering. Layer orientation between 0° and 90° provided better heat distribution compared to alternating angles of 45–135°, reducing warpage and improving the final part geometry. The post-processing step required a controlled debinding and sintering schedule to eliminate polymer residues while maintaining structural integrity. In particular, the thermal debinding phase, where the binder backbone decomposes, demanded precise temperature ramp rates to prevent collapse of the partially sintered green parts. Results confirmed that the sequence of printing parameters and post-processing conditions forms a continuous process chain where any deviation can cascade into major structural defects.

3.3.4 Advantages, Limitations, and Technical Barriers

The development of FFF as an alternative to conventional powder metallurgy and MIM has generated growing interest due to its capability to produce functional metal components at reduced costs and with greater geometric flexibility. However, the transition of this technology from laboratory-scale experiments to industrial applications faces technical barriers associated with the extrusion of powder-loaded filaments, the debinding and sintering stages, and the mechanical limitations of the final parts. Metal fused filament fabrication improved material utilization and customization, while limitations remained in achieving fully dense components with mechanical properties comparable to those produced by metal injection molding. The extrusion process imposes inherent constraints due to the thermoplastic binder system, which governs flow behavior, dimensional stability, and layer adhesion. Improper control of extrusion temperature or feed rate results in poor layer fusion, interfacial voids, and anisotropic mechanical performance. These drawbacks limit the structural applications of metal FFF components, especially where fatigue resistance and high temperature stability are required [14, 20, 24].

Metal filament-based processing was evaluated for replacement of metal injection molding in structural component fabrication, with elimination of mold tooling enabling cost-effective production of low-volume and customized parts. Furthermore, the environmental impact of FFF-based AM is lower than that of conventional sintering routes because of its reduced material waste and energy consumption. Several barriers limited industrial implementation, with powder loading in polymer matrices typically restricted to about 65 vol.%, reducing achievable sintered density and increasing shrinkage during debinding. Residual carbon from incomplete binder decomposition also degraded metallurgical bonding and introduced microstructural defects [14, 20, 24].

The manufacturing of functional components, such as metallic cutting inserts using FFF, further exemplifies both the promise and limitations of this method. Martín Fernández et al. [20] successfully produced H13 tool steel cutting inserts by adapting fused filament fabrication for metal powders and comparing the results with commercial carbide inserts. Their experiments demonstrated that the inserts manufactured via additive methods achieved similar chip formation behavior to that of conventionally manufactured tools, particularly in dry machining of aluminum alloys. However, the FFF inserts exhibited increased builtup edge formation and slightly higher wear rates due to microstructural heterogeneity and residual porosity. Thermal analysis revealed that cutting temperatures were approximately 70 °C higher than those of conventional carbide inserts, and surface roughness values were modestly inferior although within acceptable limits for semi-finishing operations.

3.4 Additive Manufacturing in Aerospace, Medical, and Automotive Sectors

3.4.1 Aerospace Applications: Lightweight and High Performance Components

AM has emerged as a transformative technology in the aerospace industry, as can be observed in Fig. 3.13, providing pathways for lightweight and high performance

Fig. 3.13 This rocket engine fuel pump has hundreds of parts including a turbine that spins at over 90,000 rpm. This turbopump was made with AM and had 45% fewer parts than pumps made with traditional manufacturing. (Courtesy of NASA)

components that significantly reduce production costs and environmental impact. According to Altıparmak and Xiao [43], the aerospace sector has been one of the largest adopters of metal AM technologies, primarily because of its ability to produce geometrically complex and lightweight structures with superior strength-to-weight ratios. The capability of AM to reduce raw material waste by approximately 40% compared to subtractive methods, and to recycle up to 98% of unused material, has further strengthened its industrial appeal. This sustainability advantage is particularly relevant for aerospace manufacturers, where titanium and nickel-based alloys are commonly used and where scrap reduction directly translates to economic and environmental gains. Moreover, the process flexibility of AM enables the fabrication of consolidated components with internal lattice structures and integrated functionalities, which are unachievable through conventional machining or casting techniques.

Balakrishnan Subeshan and coauthors detailed how AM processes such as selective laser melting (SLM), direct metal laser sintering (DMLS), and wire and arc additive manufacturing (WAAM) are being widely employed for aerospace and space exploration applications [44]. These processes allow for the creation of near-net-shape parts that meet the stringent mechanical and thermal requirements of flight environments while minimizing the buy-to-fly ratio to nearly one-to-one. The capacity of AM to manufacture lightweight parts without compromising structural integrity has been exploited by major aerospace organizations such as Boeing, SpaceX, and NASA. For example, the production of titanium brackets, turbine blades, and engine housings using powder bed fusion techniques has resulted in substantial mass savings, directly improving fuel efficiency. A particularly significant application has been the adoption of AM by NASA for the fabrication of rocket engine injectors and combustion chambers. This approach has enabled the implementation of optimized cooling geometries, resulting in enhanced thermal management and overall engine performance. Furthermore, the launch of the Metal3D printer to the International Space Station by Airbus Defense and Space in 2024 represented a major milestone, as it tested the viability of metal printing in microgravity conditions, potentially enabling in orbit manufacturing of structural components and repair parts, thereby reducing logistical dependence on Earth-based resupply missions.

In terms of market evolution, the assessment by Altıparmak and Xiao underscored the rapid expansion of the global aerospace AM market, forecasting continuous growth as materials science and process reliability improve [43]. Aerospace manufacturing incorporated additive routes to reduce supply chain complexity, shorten lead times, and enable performance-driven design. Part consolidation replaced assemblies of conventionally machined components with single printed units, decreasing interfaces and associated failure sites. General Electric reported a reduction in component count from 885 to 12 in a jet engine module through additive consolidation, resulting in lower mass and improved operational efficiency. In parallel, Airbus applied direct metal laser sintering to fabricate nacelle hinge brackets for the A320, achieving a 64% mass reduction while maintaining load-bearing capability [16, 43].

3.4.2 Medical Sector, Patient-Specific Implants, and Biocompatibility

AM has become a transformative tool in the medical sector, enabling the development of patient-specific implants with precise geometries and high biocompatibility, as presented in Fig. 3.14. Additive manufacturing enabled fabrication of patient-specific devices directly from computed tomography and magnetic resonance imaging data, reducing lead times and improving surgical outcomes [10]. Titanium and its alloys, particularly Ti-6Al-4V, are among the most commonly employed materials because of their excellent biocompatibility and mechanical strength, which are essential for long-term integration with human tissue. Powder bed fusion and directed energy deposition techniques are especially suited for producing implants with porous structures that facilitate osseointegration, a critical factor in bone replacement and regeneration. It was also observed that AM enables the production of complex geometries that mimic trabecular bone structures, resulting in improved mechanical compatibility and reduced stress shielding. This precision contributes to the growing clinical success of 3D printed implants in craniofacial, orthopedic, and dental applications.

Amaya Rivas and colleagues [45] highlighted how AM technologies have evolved to address the increasing demand for personalized implants through processes such as SLM, EBM, and FDM, each selected according to the required accuracy, material, and biocompatibility. Powder bed fusion has demonstrated exceptional results in producing implants from Ti-6Al-4V and Co-Cr-Mo alloys, as well as in creating ceramic composites such as zirconia alumina, which exhibit enhanced biofunctionality for load-bearing applications. These processes allow precise control of pore geometry, which directly influences the cellular response, particularly osteoblast differentiation and bone ingrowth. Directed energy deposition has also been applied to fabricate bone implants with controlled porosity, enabling the creation of structures that replicate the anisotropic mechanical properties of bone. Amaya Rivas et al. [45] emphasized that such developments are paving the way for cementless joint implants that rely on optimized surface textures to enhance biological fixation.

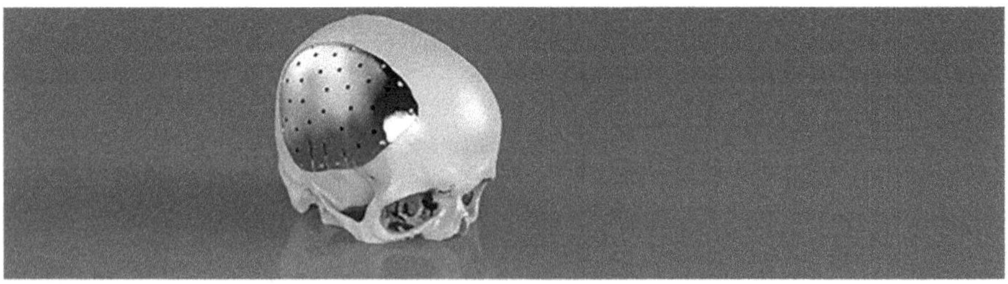

Fig. 3.14 3D printed cranial plate implant. (Courtesy of Renishaw)

Another key contribution of AM to the medical field lies in its capacity for rapid prototyping and pre-surgical planning [46]. The production of anatomical models from patient imaging data facilitates the preparation of complex surgical procedures by allowing surgeons to analyze individual anatomies and simulate interventions before actual operations. This practice has been shown to reduce operation times, improve precision, and lower the risk of complications. Furthermore, AM has been applied in fabricating temporary scaffolds for tissue regeneration. By utilizing biodegradable polymers and bioceramics, scaffolds can be engineered to degrade gradually while supporting cell growth, thus promoting natural tissue recovery. The inclusion of bioactive materials and surface coatings further enhances the integration between artificial structures and living tissue.

3.4.3 Automotive Sector: Rapid Prototyping to Serial Production

The automotive industry has evolved from utilizing AM primarily for rapid prototyping to increasingly adopting it for the serial production of end-use metal components (Fig. 3.15). This transition is driven by the need for lightweighting, part consolidation, and supply chain simplification. Binder jetting, combined with generative design, was used to fabricate a metallic radar mounting bracket for an automotive intelligent cruise control system [47]. The original bracket, made of hot rolled steel and manufactured through die forming and arc welding, weighed

Fig. 3.15 The turbo intake manifold, printed with the Meltio M600 in dual material, is capable of withstanding temperatures above 800 °C, while manufacturing costs and time were reduced by 70% compared to traditional methods. (Courtesy of Meltio)

0.283 kg. Through a topology optimization workflow, which involved defining the design space and iterating through several concepts to maximize stiffness under displacement and vibration constraints, a final design was achieved. This redesigned bracket, fabricated from stainless steel 316L via binder jetting, achieved a 42% weight reduction to 0.163 kg while meeting all structural performance requirements, including a maximum displacement of less than 0.01 mm and a natural frequency exceeding the 90 Hz minimum. A production cost analysis for manufacturing 30,000 brackets annually was conducted, comparing the Desktop Metal P50 and ExOne X160Pro systems. It was reported that while both systems yielded competitively priced final parts, the cost of metal powder was the predominant factor, constituting 63–85% of the total part cost, and remains a significant limiting factor for wider adoption in high volume automotive production [47].

Further elaborating on the application of AM technologies within the automotive sector, Yang et al. [48] classified AM processes into seven categories, including material extrusion, vat photopolymerization, binder jetting, and powder bed fusion. They further analyzed the adoption and implementation of these technologies by major automotive manufacturers. For instance, BMW employs multiple processes, including fused deposition modeling and SLM, while Ford utilizes binder jetting for aluminum components. Additive manufacturing in the automotive sector enabled design flexibility, lightweight complex geometries, and part consolidation, with eight separate components integrated into a single seat bracket by General Motors.

The environmental implications of this shift toward AM for serial production are investigated by Priarone et al. [18] through a life cycle assessment of a redesigned automotive bracket. The case study involved substituting a conventionally manufactured cast iron differential mounting bracket with a topologically optimized version made of AlSi10Mg alloy using direct metal laser sintering. The redesign process, which involved iterative topology optimization and finite element analysis, resulted in a substantial 69% mass reduction. The Life Cycle Assessment (LCA), conducted under cradle-to-grave boundaries, compared the energy demand and CO_2 emissions of both manufacturing routes. The results indicated that although the AlSi10Mg powder has a significantly higher embodied carbon footprint than cast iron, the high material usage efficiency of AM and the extensive mass reduction can offset this initial environmental cost.

3.4.4 *Economic and Sustainability Considerations*

Economic viability governed the adoption of metal additive manufacturing, with cost comparisons against conventional metal processing determining implementation thresholds [49]. In a detailed cost comparison for fabricating a titanium part, it was found that the PBF process has the most significant cost driver at the printing stage, while DED and Binder Jetting (BJP) processes spend more money on the raw material preparation process. This highlights the importance of understanding cost structures across different AM technologies when making economic decisions.

Complementing this, a life cycle costing (LCC) analysis of the SLM process reveals a more granular cost distribution. The dominant cost driver is machining, accounting for 41.1% of the total production cost, followed closely by machine-related costs at 39%, which include expenses for purchasing, maintaining, and tooling machine tools. Labor costs represent another significant contributor at 29%, while post-processing accounts for 26%. In contrast, consumables like electricity and shielding gas are a modest 2.1%, and raw material costs, such as powders, are minimal at 1.5% though this can be higher for specialized alloys [50].

From a sustainability perspective, LCA serves as an effective tool for evaluating how metal AM contributes to global sustainability by quantifying every elementary step of a product service life from cradle to grave. Research consistently show that life cycle energy consumption of metal AM is only 30–50% of that required by traditional methods [49]. A comparative LCA of an industrial hydraulic valve produced via SLM versus conventional manufacturing (casting and CNC machining) demonstrated that SLM reduced environmental impacts by 37.42%, with further reductions of 10–23% achievable through process optimization [50]. For a rocker arm component, conventional manufacturing produced 14.53% lower environmental impact than SLM in the absence of topology optimization, whereas application of topology optimization reduced the environmental burden of SLM to 21.31% below that of the conventional route. The environmental benefits of AM are multifaceted, including significant material efficiency and waste reduction. AM produces complex geometries with minimal material waste, as it is an additive process that layers material precisely, unlike subtractive methods that generate substantial scrap [51]. This near-net-shape characteristic minimizes waste and enables the recycling and reuse of unmelted powder in processes like LPBF, which is particularly beneficial for high value materials like titanium alloys.

A major environmental advantage of metal AM, particularly in sectors like aerospace and automotive, is its ability to produce lightweight, topology-optimized components. This capability directly translates to improved fuel efficiency and reduced emissions during the product use phase. In particular, adopting lightweight additively manufactured parts in aircraft could reduce annual US air fleet fuel consumption by 6.4% by 2050 [50]. Similarly, every kilogram of weight reduction in an aerospace component can save around 90,000 L of fuel annually and prevent nearly 230 tons of CO_2 emissions. Furthermore, SLM-enabled conformal cooling channels in tooling can reduce injection molding cycle times and electricity requirements by up to 30%, offering significant energy savings during production cycles. Beyond performance and design advantages, additive manufacturing also plays a fundamental role in reshaping supply chain structures toward greater sustainability. AM allows for distributed and localized manufacturing, which reduces transportation costs, material consumption, and the carbon footprint associated with long-distance shipping and extensive logistics. This localized, on-demand production model simplifies the production flow, prolongs product service life, and minimizes the need for physical inventory and warehousing, thereby enhancing overall supply chain sustainability.

3.4.5 Cross Sectoral Trends and Future Outlook

Additive manufacturing showed cross sector adoption across aerospace, medical, and automotive industries, driven by capabilities in customization, lightweight structural design, and supply chain adaptability [52]. It was reported that laser-based additive manufacturing (LBAM) processes, such as LPBF and DED, enable the production of complex, high performance metal components that are difficult or impossible to manufacture using conventional methods. In the aerospace sector, the integration of digital twins and AI facilitates real-time monitoring and optimization of AM processes, enhancing part quality and reducing defects. Specifically, the results highlight advancements in LPBF for metals, emphasizing improved material innovation and process control to decrease porosity and enhance mechanical properties, which are critical for aerospace applications. The medical sector leverages these technologies for bioprinting and the creation of patient-specific implants, utilizing materials like titanium and cobalt-chromium alloys to produce biocompatible structures with complex geometries, such as spinal fusion scaffolds and acetabular cups, which promote osteogenesis and improve patient outcomes. Similarly, the automotive industry adopts LBAM for manufacturing high stress components, such as optimized pistons and electric drive housings, which exhibit reduced weight and enhanced performance compared to traditionally forged parts, as demonstrated by case studies from Porsche.

Nyamekye et al. further elaborate on these cross sectoral trends by providing specific case studies that align with Industry 5.0 goals of sustainability, human centricity, and resilience [11]. Their research underscores the role of LBAM in producing high performance metals like nickel alloys (e.g., IN718) and titanium alloys, which are essential for demanding applications across these sectors. In aerospace, the ADDIFLAP project utilized wire-based laser DED to manufacture a flap track support system, resulting in a 60% reduction in material waste and a 26% decrease in carbon dioxide emissions, showcasing the environmental benefits of AM. The medical field benefits from the ability of LBAM to create highly porous titanium structures, such as architectures inspired by Stryker's Tritanium material, as illustrated in Fig. 3.16. These structures mimic cancellous bone and promote tissue integration, thereby advancing personalized medicine. In the automotive sector, the application of LBAM for producing lightweight pistons and drive housings not only improves fuel efficiency but also demonstrates the potential of the technology for mass customization and agile production, which are key aspects of a resilient supply chain.

Fig. 3.16 Illustration of a porous metal implant with a graded lattice architecture, representative of additively manufactured designs aimed at enhancing osseointegration and mechanical fixation

3.5 Conclusions

Metal additive manufacturing based on particulate feedstocks is governed by the coupled interaction of feedstock properties, energy input, and solidification behavior, which together control microstructural development, defect formation, and mechanical performance. In processes such as laser powder bed fusion, electron beam melting, and directed energy deposition, melt pool stability and solidification pathways are dictated by energy density, scan strategy, powder layer uniformity, and part geometry, leading to characteristic cellular, columnar, or martensitic microstructures and spatial variations in grain size and density. Powder morphology, size distribution, and surface chemistry play a central role in packing efficiency, energy absorption, and flow stability, thereby influencing the reproducibility of melting and the occurrence of porosity, lack of fusion, and related defects. Hybrid extrusion-based routes extend these principles through additional dependencies on binder chemistry, rheology, and debinding behavior, while in situ monitoring and data-assisted control increasingly support defect identification and process optimization. Across aerospace, biomedical, and automotive applications, the ability to regulate feedstock condition, thermal history, and post-processing enables the production of dense metallic components with controlled defect populations and application-specific performance, despite ongoing challenges related to cost, residual stress, and large-scale manufacturing.

References

1. Ahn SY et al (2024) Size matters: exploring part size effects on microstructure, defects, and mechanical property in optimized laser powder bed fusion (L-PBF) additive manufacturing. Mater Sci Eng A 902:146616
2. Bandyopadhyay A et al (2022) Alloy design via additive manufacturing: advantages, challenges, applications and perspectives. Mater Today 52:207–224
3. Cerejo F, Gatões D, Vieira MT (2021) Optimization of metallic powder filaments for additive manufacturing extrusion (MEX). Int J Adv Manuf Technol 115(7):2449–2464
4. Cui C et al (2023) Gas atomization of duplex stainless steel powder for laser powder bed fusion. Materials 16(1):435
5. Herzog T et al (2023) Process monitoring and machine learning for defect detection in laser-based metal additive manufacturing. J Intell Manuf 35(4):1407–1437
6. Martucci A, Aversa A, Lombardi M (2023) Ongoing challenges of laser-based powder bed fusion processing of Al alloys and potential solutions from the literature—a review. Materials 16(3):1084
7. Mooraj S et al (2024) Additive manufacturing of defect-free TiZrNbTa refractory high-entropy alloy with enhanced elastic isotropy via in-situ alloying of elemental powders. Commun Mater 5(1):14
8. Mussatto A et al (2021) Influences of powder morphology and spreading parameters on the powder bed topography uniformity in powder bed fusion metal additive manufacturing. Addit Manuf 38:101807
9. Nabavi SF, Dalir H, Farshidianfar A (2024) A comprehensive review of recent advances in laser powder bed fusion characteristics modeling: metallurgical and defects. Int J Adv Manuf Technol 132(5):2233–2269
10. Piscopo G et al (2023) An overview of the process mechanisms in the laser powder directed energy deposition. Appl Sci 13(1):117
11. Nyamekye P, Westman S, Tepponen V (2023) Enhancing industry 5.0 goals through laser based additively manufactured high-performance metals. IOP Conf Ser Mater Sci Eng 1296(1): 012001
12. Priyadarshi A et al (2024) New insights into the mechanism of ultrasonic atomization for the production of metal powders in additive manufacturing. Addit Manuf 83:104033
13. Shinkaryov AS et al (2021) Aluminum powder preparation for additive manufacturing using electrostatic classification. Coatings 11(6):629
14. Wang X et al (2024) A review on traditional processes and laser powder bed fusion of aluminum alloy microstructures, mechanical properties, costs, and applications. Materials 17(11):2553
15. Chandra S et al (2024) Powder-size driven facile microstructure control in powder-fusion metal additive manufacturing processes. Nat Commun 15(1):3094
16. Gradl P et al (2022) Robust metal additive manufacturing process selection and development for aerospace components. J Mater Eng Perform 31(8):6013–6044
17. Ledniczky G et al (2025) Impact of recycling on polymer binder integrity in metal injection molding. Sci Rep 15(1):21600
18. Priarone PC, Catalano AR, Settineri L (2023) Additive manufacturing for the automotive industry: on the life-cycle environmental implications of material substitution and lightweighting through re-design. Prog Addit Manuf 8(6):1229–1240
19. Kahhal P, Jo Y-K, Park S-H (2024) Recent progress in remanufacturing technologies using metal additive manufacturing processes and surface treatment. Int J Precis Eng Manuf-Green Technol 11(2):625–658
20. Martín-Fernández F et al (2024) Metallic cutting inserts fabrication by means of additive manufacturing with fused filament fabrication technology. Results Eng 24:103194
21. Sadaf M, Bragaglia M, Nanni F (2021) A simple route for additive manufacturing of 316L stainless steel via fused filament fabrication. J Manuf Process 67:141–150

22. Galati M, Minetola P (2019) Analysis of density, roughness, and accuracy of the atomic diffusion additive manufacturing (ADAM) process for metal parts. Materials 12(24):4122
23. Jacob J et al (2024) A review of fused filament fabrication of metal parts (metal FFF): current developments and future challenges. Technologies 12(12):267
24. Quarto M, Giardini C (2022) Additive manufacturing of metal filament: when it can replace metal injection moulding. Prog Addit Manuf 8:1–10
25. Wagner MA et al (2022) Fused filament fabrication of stainless steel structures—from binder development to sintered properties. Addit Manuf 49:102472
26. Murr L (2014) Metallurgy of additive manufacturing: examples from electron beam melting. Addit Manuf 5:40–53
27. Slotwinski JA et al (2014) Characterization of metal powders used for additive manufacturing. J Res Natl Inst Stand Technol 119:460–493
28. Charbonneau C et al (2025) Classification of metallic powder morphology using traditional and automated static image analysis: a comparative study. Powders 4(2):15
29. Baesso I et al (2021) Characterization of powder flow behavior for additive manufacturing. Addit Manuf 47:102250
30. Zheng Z, Peng L, Wang D (2021) Defect analysis of 316 L stainless steel prepared by LPBF additive manufacturing processes. Coatings 11(12):1562
31. Xie R et al (2024) Effect of feeding material shape on microstructures and mechanical properties in friction rolling additive manufacturing. Mater Des 241:112952
32. Raja H, Kumar S (2023) Defects in metal additive manufacturing: formation, process parameters, postprocessing, challenges, economic aspects, and future research directions. 3D Print Addit Manuf 11:e1629–e1655
33. Guillen D, Wahlquist S, Ali A (2024) Critical review of LPBF metal print defects detection: roles of selective sensing technology. Appl Sci 14(15):6718
34. Wong V, Aversa A, Rodrigues AR (2024) A deep learning model for estimating the quality of bimetallic tracks obtained by laser powder-directed energy deposition. Materials 17(22):5653
35. Shaheen MY et al (2021) The influence of material and process parameters on powder spreading in additive manufacturing. Powder Technol 383:564–583
36. Dejene ND, Lemu HG (2023) Current status and challenges of powder bed fusion-based metal additive manufacturing: literature review. Metals 13(2):424
37. Ramkumar PL, Rijwani T (2022) Additive manufacturing of metals and ceramics using hybrid fused filament fabrication. J Braz Soc Mech Sci Eng 44(10):455
38. Singh P et al (2021) Factors affecting properties of Ti-6Al-4V alloy additive manufactured by metal fused filament fabrication. Powder Technol 386:9–19
39. Buj-Corral I, Fenollosa-Artés F, Minguella-Canela J (2024) Three-dimensional printing of metallic parts by means of fused filament fabrication (FFF). Metals 14(11):1291
40. Singh P et al (2021) Additive manufacturing of Ti-6Al-4V alloy by metal fused filament fabrication (MF3): producing parts comparable to that of metal injection molding. Prog Addit Manuf 6(4):593–606
41. Santos C et al (2021) Influence of metallic powder characteristics on extruded feedstock performance for indirect additive manufacturing. Materials (Basel) 14(23):7136
42. Shaikh Q et al (2021) Process sensitivity and significant parameters investigation in metal fused filament fabrication of Ti-6Al-4V. J Mater Eng Perform 30:5118
43. Altıparmak SC, Xiao B (2021) A market assessment of additive manufacturing potential for the aerospace industry. J Manuf Process 68:728–738
44. Subeshan B, Ali Z, Asmatulu E (2024) Metal additive manufacturing in space and aerospace exploration: current processes, material selection and challenges. J Eng Appl Sci 11:35
45. Amaya-Rivas JL et al (2024) Future trends of additive manufacturing in medical applications: an overview. Heliyon 10(5):e26641
46. Li Y et al (2025) Revolutionizing medical implant fabrication: advances in additive manufacturing of biomedical metals. Int J Extreme Manuf 7(2):022002

47. Zhao N et al (2023) Direct additive manufacturing of metal parts for automotive applications. J Manuf Syst 68:368–375
48. Yang J et al (2024) Application of additive manufacturing in the automobile industry: a mini review. Processes 12(6):1101
49. Gao C, Wolff S, Wang S (2021) Eco-friendly additive manufacturing of metals: energy efficiency and life cycle analysis. J Manuf Syst 60:459–472
50. Rahmani R et al (2025) Sustainable additive manufacturing: an overview on life cycle impacts and cost efficiency of laser powder bed fusion. J Manuf Mater Process 9(1):18
51. Javaid M et al (2021) Role of additive manufacturing applications towards environmental sustainability. Adv Ind Eng Polym Res 4(4):312–322
52. Bănică C-F, Sover A, Anghel D-C (2024) Printing the future layer by layer: a comprehensive exploration of additive manufacturing in the era of industry 4.0. Appl Sci 14(21):9919

Chapter 4
Metal Injection Molding and Emerging Variants

Abstract The production of high precision components with complex geometries is achieved through a manufacturing route in which feedstock rheology, binder chemistry, debinding kinetics, and sintering behavior are tightly linked to control densification, dimensional accuracy, and microstructural evolution. Fine, high purity powders with tailored particle size distributions enable high solid loading and near-net-shape capability, while multicomponent binder systems balance flowability, green strength, and contamination control. Debinding remains a defect-sensitive stage that requires careful matching of removal routes to binder architecture in order to generate interconnected porosity without inducing internal stresses. Final densification and shrinkage during sintering are governed by particle size, oxide stability, alloy chemistry, and furnace atmosphere, with optimized processing achieving relative densities above 97% and mechanical properties approaching those of wrought materials. Emerging variants such as microscale injection molding and hybrid injection molding–additive routes extend geometric freedom but introduce narrower processing windows and increased sensitivity to powder binder interactions. The growing use of in situ monitoring and predictive digital tools continues to support scalable, high precision component production across medical, electronic, and automotive applications.

Keywords Feedstock rheology · Debinding kinetics · Sintering shrinkage · High precision components

4.1 Introduction

The combination of fine metal powders with thermoplastic binder systems enables a manufacturing route that merges the design freedom of polymer injection molding with the microstructural control and mechanical performance of metallic materials. Through molding, debinding, and subsequent sintering, this approach produces high density, near-net-shape components with well-controlled microstructures and properties, reflecting the maturity of metal injection molding as a reliable powder-based forming technology [1–3]. Powder characteristics including particle size,

R. Pérez-Bustamante, *Fundamental, Practical and Disruptive Applications of Powder Metals*, Topics in Mining, Metallurgy and Materials Engineering,
https://doi.org/10.1007/978-3-032-27130-3_4

morphology, and surface chemistry govern feedstock rheology, packing density, and sintering behavior, defining dimensional stability and final mechanical performance. Gas-atomized spherical powders with narrow size distributions are emphasized for alloys such as stainless steels and titanium, where high purity and controlled oxygen content are decisive for densification and fatigue resistance [4–9].

Densities in the range of 94–98% of the theoretical value are obtained by metal injection molding in stainless steels, titanium alloys, aluminum alloys, and composite systems when powder characteristics, binder composition, and thermal processing schedules are appropriately coordinated. Experimental data for 316L stainless steel, Ti-6Al-4V, aluminum alloys, and W-Cu pseudo alloys indicate that microstructure, residual porosity, and impurity content directly determine tensile strength, fatigue behavior, and functional properties such as thermal conductivity and corrosion resistance. However, intrinsic limitations are presented. Reactive powders are prone to oxygen and carbon uptake during debinding; ultrafine powders generate high viscosity and instability in feedstocks, and sintering shrinkage remains sensitive to powder size, loading, and thermal gradients [10–16].

Binder systems composed of waxes, polyolefins, polyoxymethylene, and water-soluble polymers are analyzed as functional architectures that create transient pore networks and maintain green strength, yet their decomposition pathways and compatibility with solvents or catalysts impose constraints on impurity control and dimensional fidelity. Feedstock design emerges as the unifying element that links powder loading, viscosity, shear thinning, and thermal sensitivity to mold filling and defect suppression [6, 11, 17, 18].

This chapter establishes quantitative relations between powder morphology, binder chemistry, and processing parameters across metal injection molding, micro-metal injection molding, and hybrid MIM additive routes, and shows how these relations enable fabrication of high precision, high performance components for medical, electronic, and lightweight engineering applications [19]. The integration of tailored powders, multistage debinding, and controlled sintering, supported by digital twins, surface functionalization, and hybrid additive processing, provides a reproducible route toward metallic components with controlled defect populations and application-specific properties [12, 15, 16, 20–22].

4.2 Overview of Metal Injection Molding: Material Feedstock, Binder Removal, and Sintering

4.2.1 Fundamentals of Powder Selection and Particle Characteristics

Metal Injection Molding (MIM, Fig. 4.1) represents a well-established manufacturing technology that combines the design flexibility of plastic injection molding with the material performance of metals. The process involves mixing fine metal

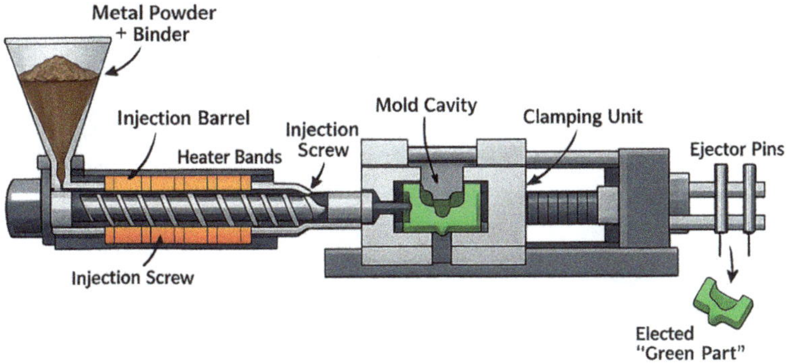

Fig. 4.1 Scheme of an injection molding machine

powders, typically below 45 μm, with a multicomponent thermoplastic binder system to form a feedstock [23]. This feedstock is then injection molded into a desired shape, after which the binder is removed through a process called debinding. The final step is sintering, where the brown part is heated to a high temperature to densify the metal particles into a solid, high strength component. The ideal MIM powder is characterized by a small particle size, often with a D90 of less than 22 μm for standard alloys, a spherical shape for high packing density, and high surface purity to promote sintering. Powders for refractory metals and cemented carbides are typically even finer, with a D90 below 5 μm. The particle size distribution is critical, as it influences the feedstock rheology, solids loading, and final dimensional control. A broad distribution can enhance packing density, but for reactive metals like titanium, a narrower, coarser distribution is sometimes preferred to minimize surface area and reduce oxygen pickup [13].

The selection of powder fabrication technique is fundamental to achieving the desired powder characteristics and final part properties. Common methods include gas atomization, water atomization, thermal decomposition, and chemical reduction [23]. Gas-atomized powders are spherical, exhibit high purity and packing density, and are commonly used for stainless steels and superalloys. Water-atomized powders are more irregular in shape and have a higher oxide content, but they are produced at a lower cost. For titanium MIM (Ti-MIM), the powder quality is paramount due to the material's sensitivity to interstitial impurities like oxygen and carbon. Successful titanium metal injection molding required high purity spherical powders, typically produced by gas or plasma atomization, with median particle sizes between 30 and 60 μm and oxygen contents below approximately 0.15 wt.%. Impurities introduced during processing were difficult to remove, making initial powder purity a controlling factor for final material quality.

The mechanical properties of MIM components are primarily governed by the density and microstructure obtained after sintering. Microstructural features observed after debinding and sintering, such as sintering neck formation and secondary phases, provide useful context for understanding this relationship (see Fig. 4.2). Research on stainless steel 316L produced via MIM by Hwang et al.

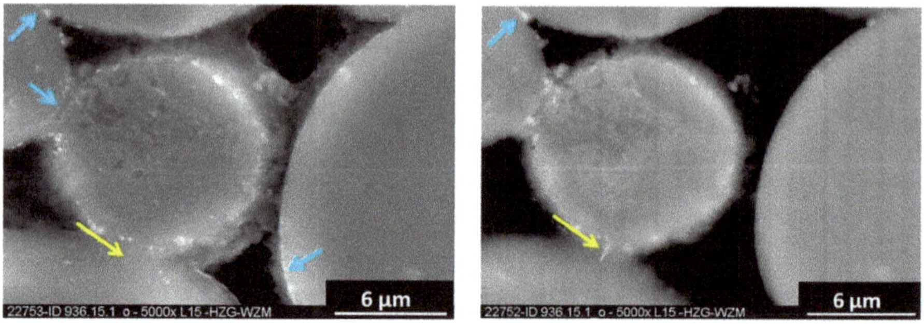

(a) (b)

Fig. 4.2 (**a**) SEM SE micrographs of thermal debound and sintered MIM specimens after 64 h sintering time using PE-VA copolymer showing bright layers on the particles' surface and sintering necks (see blue arrows), and a needle-shaped crystal slightly underneath the polished surface using 18 kV; (**b**) the needle-shaped structure inside the sintering neck can be seen more clearly using backscattered electrons mode [24]

[16] demonstrated a relative density of 94.62% of theoretical, which resulted in a tensile strength of approximately 539 MPa and a very high elongation to fracture of 92%. This high ductility contrasts with the lower elongation typically seen in cold rolled 316L, which the authors attribute to the specific sintering temperature of 1320 °C. However, the presence of porosity led to a lower tensile strength compared to wrought material. The fatigue limit of the MIM 316L was found to be 297 MPa, which is 55% of its tensile strength, positioning it between selective laser melting (SLM)-fabricated and cold rolled 316L in terms of performance. Microstructural analysis using Electron Backscatter Diffraction (EBSD) revealed an average grain size of 88.51 μm. Results showed that with increasing tensile deformation, the austenite grains became fragmented due to increased misorientation, twin boundary density, and deformation bands, which act to suppress dislocation movement and enhance mechanical properties.

For titanium, the properties are even more sensitive to density and interstitial content. Density, purity, alloying, and microstructure were identified as governing variables for Ti metal injection molding. Fatigue strength in Ti-6Al-4V increased by approximately 400% as density rose from 94% to full theoretical density, reflecting the strong sensitivity of cyclic performance to residual porosity. Near-full density was obtained through post-sintering hot isostatic pressing. A baseline Ti metal injection molding route used spherical prealloyed powder, a multicomponent wax polymer binder, sequential solvent and thermal debinding, high temperature vacuum sintering, and a final HIP cycle. This process can yield tensile strengths around 900 MPa with 12% elongation for Ti-6Al-4V, provided the oxygen content is controlled to between 0.20% and 0.22%. The challenges in Ti-MIM, such as controlling contamination during debinding and sintering and managing the cost of high quality powders, continue to drive research and process refinement for its application in demanding aerospace and medical fields.

4.2.2 Binder Systems: Composition, Function, and Design Criteria

In both MIM and its additive manufacturing counterpart, material extrusion (MEX), the binder system acts as a temporary medium that allows metal powders to be shaped into complex geometries (Fig. 4.3). An ideal binder must fulfill a multifaceted set of criteria to ensure successful processing. According to Wen et al. [10], these requirements include a low melting temperature for easy molding, sufficient strength at room temperature to retain the shape of the green part, and a low viscosity with good fluidity at the molding temperature to facilitate mold filling. Furthermore, the binder must be easily removable without leaving contaminating residues, which is particularly crucial for reactive metals like titanium. Given the difficulty for a single material to meet all these demands, binder formulations are typically multicomponent systems. They generally consist of a major component, such as a wax, which wets the powder particles and provides flowability, and a polymeric backbone component that confers strength to the molded part after the major component is removed. A surfactant is often included as a third component to improve compatibility between the hydrophobic polymers and the metallic powder surface.

The design and composition of a binder system are heavily influenced by the chosen debinding method, which is the process of removing the binder prior to sintering. Several binder system classifications based on their debinding mechanisms are discussed. Wax-based systems, one of the most common, often utilize a two-step debinding process involving solvent extraction followed by thermal pyrolysis to

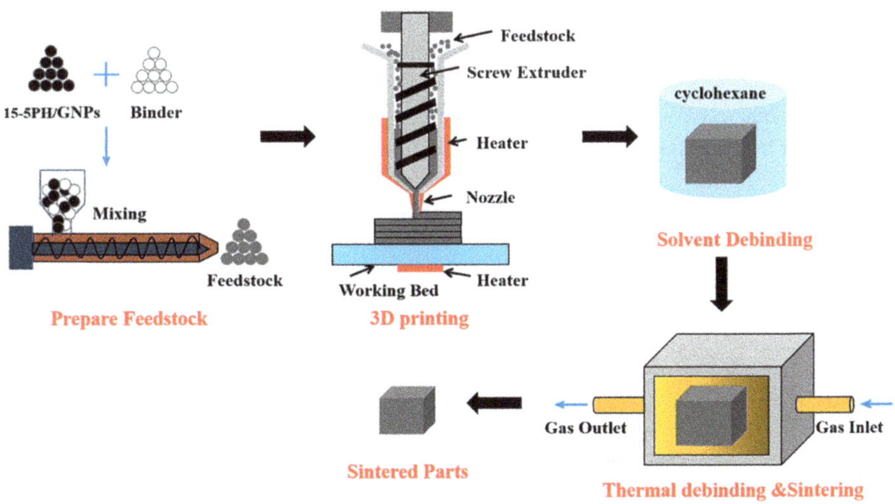

Fig. 4.3 Schematic illustration of the MIM-based additive manufacturing process, including feedstock preparation, extrusion-based 3D printing, solvent debinding, and subsequent thermal debinding and sintering to obtain densified parts [25]

minimize contamination. Guo et al. [26] demonstrated that a binder composed of paraffin wax, polyethylene glycol (PEG), polyethylene (PE), polypropylene (PP), and stearic acid could achieve low oxygen and carbon levels in sintered Ti-6Al-4V parts after solvent and thermal debinding. Another prominent system is the polyoxymethylene (POM)-based binder, which is removed via catalytic debinding in a gaseous acid atmosphere at temperatures below the polymer's softening point, allowing for rapid debinding and good shape retention, though it requires significant capital investment [10]. Alternatively, water-soluble binder systems have gained interest for their environmental benefits. Binder systems use either non-gelation routes based on water-soluble components such as polyethylene glycol or gelation routes employing natural polymers such as agar. A PEG PMMA stearic acid binder applied to Ti-6Al-4V enabled complete removal of polyethylene glycol by water leaching followed by thermal debinding of the PMMA backbone [4, 10].

The selection of specific binder components directly dictates the rheological behavior of the feedstock, which is vital for shaping. Mixing behavior in aluminum feedstocks was examined through the relation between powder loading, shear rate, and viscosity using two binder systems composed of HDPE with paraffin wax and stearic acid or HDPE with stearic acid alone. Viscosity decreased with increasing shear rate, indicating pseudoplastic flow, and increased with higher powder loading. It was indicated that Type A binder with 62% powder loading was most suitable, as it provided a lower flow behavior index and an acceptable viscosity for molding, a conclusion validated by a fuzzy logic expert system model [27]. The role of binder composition in determining the success of the entire process is further emphasized in a comparative study on copper feedstocks for MEX and MIM by Sadaf et al. [6]. They evaluated three feedstocks: F1_ST and F2_ST, designed for solvent and thermal debinding, and F3_T, intended for thermal debinding only. Their results showed that the binder composition was paramount for defect-free debinding and sintering. Components shaped with F1_ST and F2_ST survived the thermal and sintering steps, whereas those from F3_T, which lacked a soluble component to create a porous network during initial debinding, experienced nearly complete shape loss during thermal debinding due to trapped decomposition gases [6]. This underscores that a binder must not only provide good processability but also be designed for a specific, often multistage, debinding sequence to ensure the integrity of the part is maintained throughout the process.

4.2.3 Feedstock Preparation and Rheological Behavior in Injection Molding

The preparation of metal injection molding feedstock governed molding, debinding, and sintering behavior through the rheology of a homogeneous mixture of metal powder and a multicomponent binder. Binder composition and powder loading were varied to evaluate their effects on the rheological and magnetic response of Nd-Fe-B feedstocks [28]. The research group formulated several binder systems using linear low-density polyethylene (LLDPE) as a backbone, alongside paraffin wax and stearic acid. It was found that the apparent viscosity of the feedstock was highly dependent on the binder formulation; Feedstock A, with the highest LLDPE content (52.5 vol.%), exhibited the highest viscosity, while reducing the backbone content lowered the viscosity. However, formulations with excessively high volumes of low viscosity components like PW and SA (Feedstocks C and D) led to processing challenges such as leakage during compounding, indicating poor homogeneity. The study concluded that a binder system with 45 vol.% LLDPE (Feedstock B) offered an optimal balance, providing acceptable viscosity below 1000 Pa·s, stable pressure profiles during capillary rheometry, and favorable shear thinning behavior, which is essential for complete mold filling without powder binder separation (see relative viscosity as a function of powder loading for PP- and POM-based feedstocks at elevated temperatures, showing model fittings at different shear rates on a logarithmic scale in Fig. 4.4).

The rheological evaluation of feedstock extends beyond viscosity measurements to include parameters such as the flow behavior index (n) and flow activation energy (E), which quantify sensitivity to shear rate and temperature, respectively. Ahmad et al. [14] performed a detailed rheological analysis on titanium alloy (Ti-6Al-4V) feedstocks using binder systems incorporating palm stearin (PS), paraffin wax, and stearic acid. Their results demonstrated that all tested feedstocks exhibited pseudoplastic behavior ($n < 1$), as required by MPIF 35 standards. However, the composition significantly influenced stability. The feedstock designated PE/PW/PS, which contained palm stearin and paraffin wax, showed fluctuating viscosity and unstable pseudoplastic behavior at temperatures above 120 °C, attributed to the low molecular weight of its components increasing the likelihood of powder binder separation. In contrast, the PE/PS 4060 feedstock, with a 40/60 vol.% ratio of polyethylene to palm stearin, demonstrated moderate values for both the flow behavior index and flow activation energy. This combination indicates a viscosity that is not excessively sensitive to sudden changes in shear rate or temperature, thereby reducing the risk of defects like jetting or stress-induced cracking during molding. The superior performance of the PE/PS 4060 formulation was further confirmed by its high moldability index (α), a composite parameter derived from both n and E, identifying it as the most suitable feedstock for the titanium MIM process.

The interaction between the powder and binder is another crucial factor dictating feedstock properties, which can be enhanced through powder surface modification. Ma et al. [8] explored the effect of coating 17 4PH stainless steel powder with aluminum stearate (AlSA) on feedstock characteristics. Their findings revealed that the surfactant coating successfully broke up powder agglomerations, as confirmed by scanning electron microscopy and zeta potential measurements, which showed improved dispersion stability. This enhanced compatibility between the powder and a polymethyl methacrylate/polyethylene glycol (PMMA/PEG) binder system led to a significant reduction in feedstock viscosity. The most favorable results were achieved with a 0.6 wt.% AlSA coating, which resulted in the lowest shear viscosity

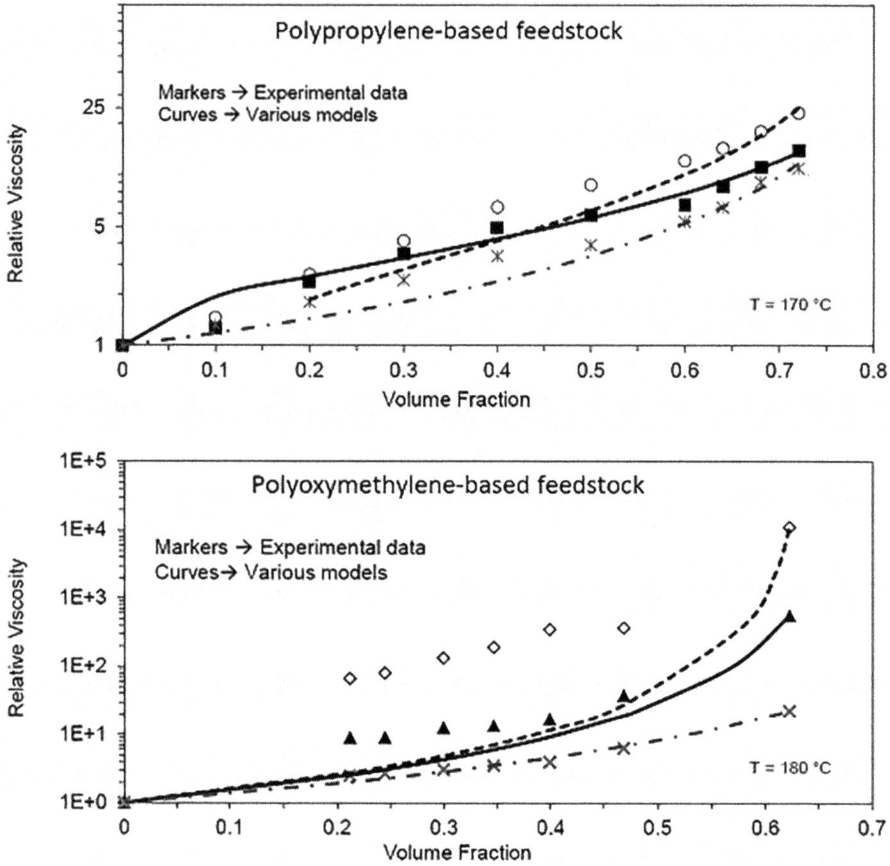

Fig. 4.4 (**a**) Relative viscosity as a function of powder loading for PP-based feedstock at 170 °C. Lines represent the fittings done by the model proposed by Frankel et al. Please note that the vertical axis is in logarithmic scale. (**b**) Relative viscosity as a function of powder loading for POM-based feedstock at 180 °C. Lines represent the fittings done by the model proposed by Mendoza and Santamaria Holek at different shear rates. Please note that the vertical axis is in logarithmic scale [29]

and the highest melt flow index of 81.5 g/10 min at 200 °C. This optimal coating amount, slightly above the theoretical monolayer calculation of 0.26 wt.%, ensured the powder particles were uniformly wetted by the binder, facilitating smoother flow. Furthermore, the improved powder binder interaction yielded green parts with higher density and flexural modulus, and significantly enhanced shape retention during the subsequent thermal debinding stage, reducing the length change rate from 1.71% for uncoated powder to 0.82%.

4.2.4 *Binder Removal Strategies: Thermal, Solvent, and Catalytic Debinding*

Polymer binder removal in the MIM process is carried out using thermal, solvent, or catalytic debinding strategies. Each method relies on distinct mechanisms and offers specific advantages, while also introducing potential defect pathways that must be carefully controlled to preserve the structural integrity of the component (Fig. 4.5). Thermal debinding operates through the thermal degradation of binder components into gaseous byproducts. A significant risk associated with this method is defect formation, such as cracking and blistering, if the heating rate is excessively rapid. The pressure from evolving gases can build up within the part if adequate pore channels have not yet formed, leading to internal stresses and part failure [31]. Consequently, thermal debinding often requires very slow heating rates, sometimes extending over several days, to allow for controlled gas escape. To mitigate issues like distortion, components may require support from ceramic setters during the process. The efficacy of thermal debinding is highly dependent on the binder composition, as demonstrated by Ledniczky et al. [32], who used thermogravimetric analysis (TGA) to identify distinct decomposition stages for wax and polypropylene components in their feedstock over multiple recycling cycles. Furthermore, research into cleaner binder systems for reactive metals like titanium emphasizes the use of polymers with lower decomposition temperatures. Lim et al. [11] developed a binder system based on polypropylene carbonate (PPC) and polyoxymethylene (POM), which decompose fully at 275 and 370 °C, respectively, thereby minimizing the risk of carbon and oxygen contamination during thermal debinding.

In contrast, solvent debinding removes the binder by dissolving it in a chemical or water, a process that introduces different challenges. The primary defects in this method are swelling, cracking, and distortion, often caused by the solvent causing the polymer to expand within the part. The compatibility between the solvent and the binder system is paramount; an unsuitable combination can lead to excessive swelling and internal stress [31]. Solvent debinding using water was evaluated in a polyethylene glycol-based binder system, with increasing PEG content from 30% to

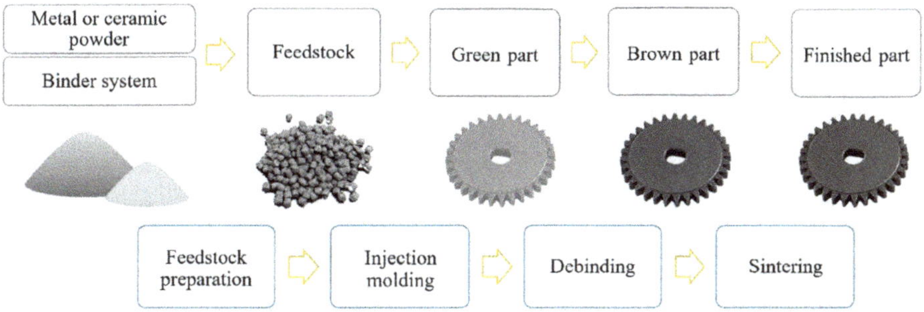

Fig. 4.5 Material flow and processing sequence in low pressure injection molding (LPIM) process [30]

50% accelerating extraction while increasing swelling and blistering due to reduced backbone support. A composition of 40% PEG, 48% PPC, and 10% POM provided a balance between PEG removal rate and shape retention, with the POM backbone supplying green strength [11]. The creation of interconnected open pore channels during solvent debinding is essential for the subsequent efficient removal of the residual backbone binders in the thermal debinding stage.

4.2.5 Sintering Dynamics: Shrinkage, Densification, and Microstructural Evolution

In MIM, the sintering stage is a thermal process that governs the final part density, dimensional accuracy, and mechanical properties, with its evolution strongly influenced by the initial material characteristics and the selected process parameters. Particle size effects on sintering shrinkage (Figs. 4.6, 4.7, and 4.8) and porosity in stainless steel metal injection molding were evaluated by multiphysics simulations for 17-4PH, 316L, and 304 alloys. The 17-4PH powder with a D50 of 6.03 μm exhibited a higher sintering driving force owing to its larger specific surface area, resulting in enhanced densification behavior [26]. This led to greater dimensional shrinkage rates and improved densification. The ANSYS Workbench simulations were instrumental in visualizing the total deformation and stress distribution during sintering, revealing that regions farther from the injection gate exhibited larger gaps between particles, higher porosity, and significantly increased deformation, with the maximum total deformation for 304 material reaching 0.0879 mm. These simulation results were directly validated by SEM observations, which showed a higher concentration of pores in areas distant from the gate. The relationship between particle size and shrinkage was mathematically expressed, demonstrating an inverse

Fig. 4.6 Green part (left) and "guidepost" sintered part made of MIM 4140 steel (right) [17]

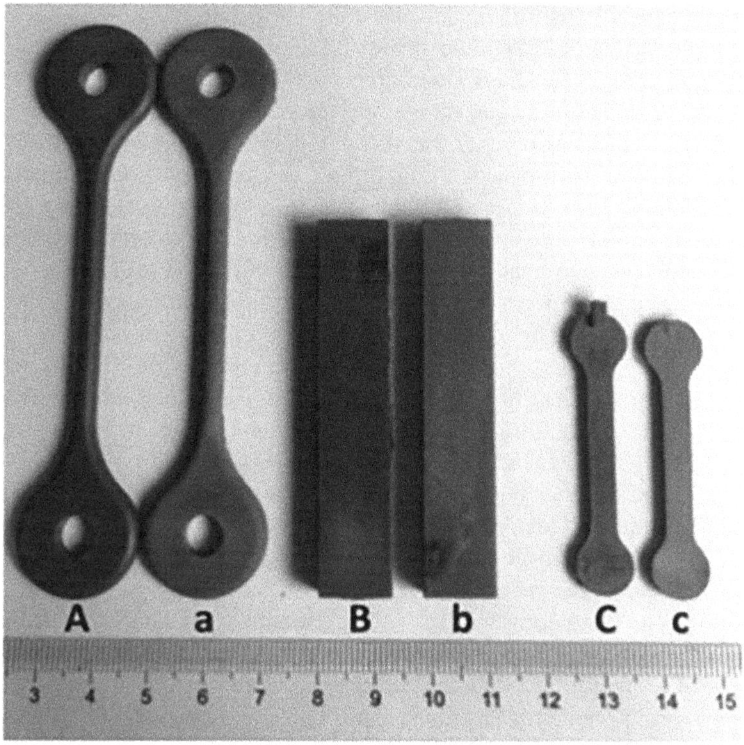

Fig. 4.7 Demonstration of green parts (samples A–C) and solvent debinded parts (samples a–c) [1]

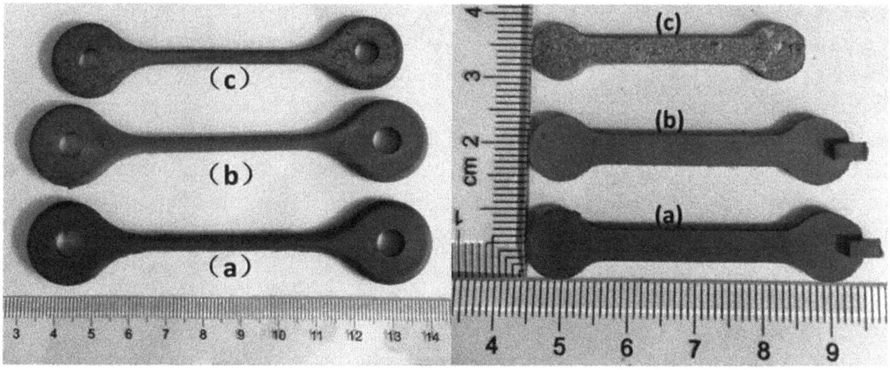

Fig. 4.8 Shrinkage variation of samples A and C (a) green part; (b) solvent debinded part; and (c) sintered part [1]

proportionality where finer powders promote more uniform shrinkage and a lower final porosity, ultimately leading to superior mechanical properties and hardness distribution in the sintered component.

The challenges of sintering densification are particularly acute for active metals like aluminum, primarily due to the stable surface oxide layer that impedes atomic diffusion. Zou et al. [4] address aluminum alloy MIM systems in which alloying elements such as magnesium and tin are introduced to disrupt the continuous oxide film and promote sintering activity. In Al-Mg-Si-Cu compositions, the densification process during sintering proceeds through three distinct stages. The initial stage is associated with the formation of a low-melting-point Al-Mg eutectic liquid phase, which generates initial porosity while initiating the breakdown of the surface oxide layer. The second stage is a rapid densification phase where additional liquid phases, such as Al-Cu and Al-Si, form and fill particle gaps through capillary action, leading to a significant increase in relative density. The final stage achieves near-full densification through grain growth and further pore elimination. The study emphasized that a nitrogen sintering atmosphere is highly beneficial as it promotes the formation of aluminum nitride (AlN). This AlN formation not only assists in disrupting the oxide layer but also reduces the pressure within pores, facilitating their collapse and leading to sintered densities that can exceed 98% of theoretical density under optimized temperature and time profiles.

In the context of composite materials, the work by Ouyang et al. [33] on tungsten copper (W 20 wt.% Cu) pseudo alloys offers profound insights into how powder morphology dictates sintering dynamics. The research group addressed the problem of agglomerated ultrafine powders, which are typically difficult to process via powder injection molding (PIM) due to poor flowability and low powder loading. They employed a bonding treatment and ball milling (BTBM) process that transformed the loose, agglomerated particles into dense, pebble-shaped larger particles. This transformation resulted in a dramatic increase in tap density from 3.25 to 7.22 g/cm^3 and a wider particle size distribution, characterized by a decrease in the slope parameter Sw from 4.22 to 2.84. Modification of feedstock composition increased the critical powder loading from 45 to 56 vol.% and improved rheological behavior. During sintering, the higher solids content reduced shrinkage and improved dimensional stability, with the oversizing factor decreasing from 1.297 to 1.216. Furthermore, although the powder particles were larger after BTBM, the intrinsic tungsten particle size within them remained submicron, preserving the high sintering activity. This combination led to a sintered composite with a higher relative density of 98.6%, compared to 97.8% for the untreated powder, and a corresponding increase in thermal conductivity from 218 to 233 W/(m·K), demonstrating a direct link between powder processing, sintering shrinkage control, and final performance.

Microstructural evolution during and after sintering can be further tailored using post-sintering heat treatments to enhance specific properties (Fig. 4.9). The research by Liang et al. [12] on titanium-coated 316L stainless steel fabricated via MIM illustrates this principle effectively. A dip coating process using a titanium slurry was applied, followed by vacuum sintering at temperatures between 1100 and 1200 °C. During sintering, titanium diffused into the stainless steel substrate, leading to the formation of a surface layer enriched with intermetallic compounds such as Fe_2Ti, Ti_2Ni, and NiTi, as confirmed by X-ray diffraction analysis. The thickness of this

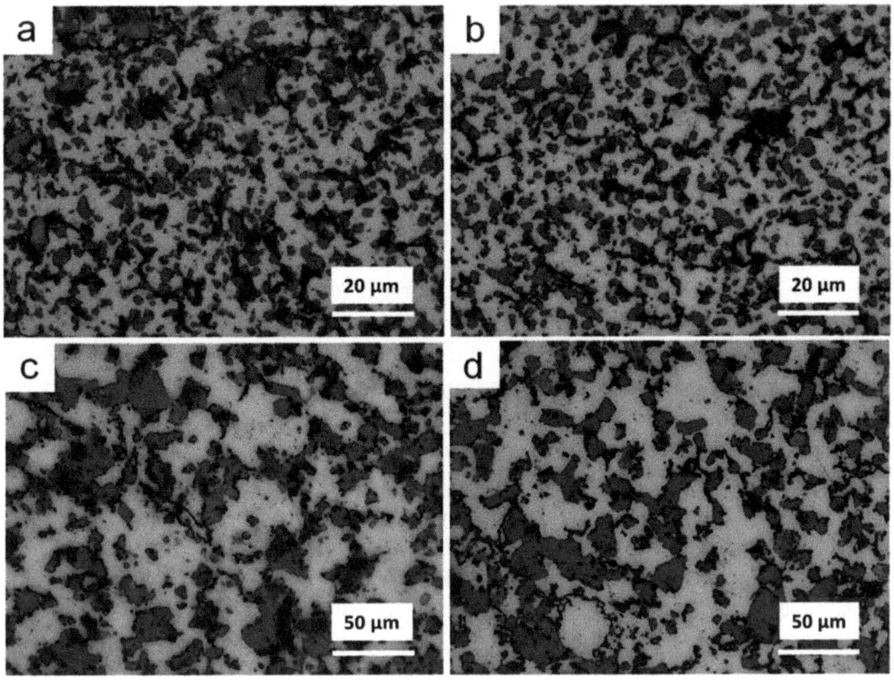

Fig. 4.9 Optical microstructures of parts sintered at (**a**) 530 °C; (**b**) 550 °C; (**c**) 570 °C; and (**d**) 590 °C for 1 h [1]

diffusion layer was observed to increase with sintering temperature, reaching up to 120 μm at 1200 °C, which contributed to a surface hardness of 762.8 HV. Subsequent heat treatment, involving annealing at 750 °C for 4 h followed by oil quenching and aging at 550 °C for 3 h, induced precipitation strengthening phases. This thermal cycle promoted the formation of numerous fine carbides and further homogenized the microstructure, which resulted in an enhanced yield strength of 344.48 MPa and a significant increase in polarization impedance to 1243.3 $\Omega \cdot cm^2$, demonstrating that strategic post-sintering thermal management can optimize both the mechanical and corrosion-resistant properties of MIM components without compromising their structural integrity.

4.3 New Technologies in Injection Molding, Including Micro MIM and Hybrid Techniques

4.3.1 Advances in Micro MIM: Scaling Challenges and Precision Control

MIM has been successfully adapted for the fabrication of microscale components, a technology referred to as micro MIM (Fig. 4.10). This adaptation, however, introduces significant scaling challenges, particularly concerning powder characteristics, feedstock rheology, and dimensional control during sintering. The primary distinction from conventional MIM lies in the requirement for finer metal powders to replicate microfeatures accurately. As noted by Liao et al. [9], the powder particle size must be sufficiently fine, with a D50 typically less than 5 μm, to facilitate the filling of minute mold cavities and achieve acceptable surface roughness. For Fe-4Ni-0.8Mo-0.6C steel small-module gears, spherical powder with a D50 of 4.79 μm was employed at a powder loading of 58 vol% within a polyformaldehyde-based binder system. The large specific surface area associated with such fine powders requires binders that combine low viscosity with sufficient mechanical integrity to avoid damage during demolding of the fragile green parts. Rheological characterization confirmed the feedstock exhibited pseudoplastic behavior and a low flow activation energy, which is crucial for complete mold filling without defects in complex micro-geometries.

The sintering process in Micro MIM demands precise control to manage substantial shrinkage and achieve near-full density while preserving fine features. Liao et al. [9] systematically investigated the effect of sintering temperature on Fe-4Ni-0.8Mo-0.6C gears, finding that relative density and mechanical properties increased with temperature up to an optimum of 1320 °C, yielding a relative density of 97.6%, a tensile strength of 2034 MPa, and a hardness of 709 HV. They observed isotropic shrinkage averaging 17.3%, which is critical for maintaining dimensional accuracy in microparts. Microstructural analysis revealed that higher sintering

Fig. 4.10 Scheme of micro MIM [34]

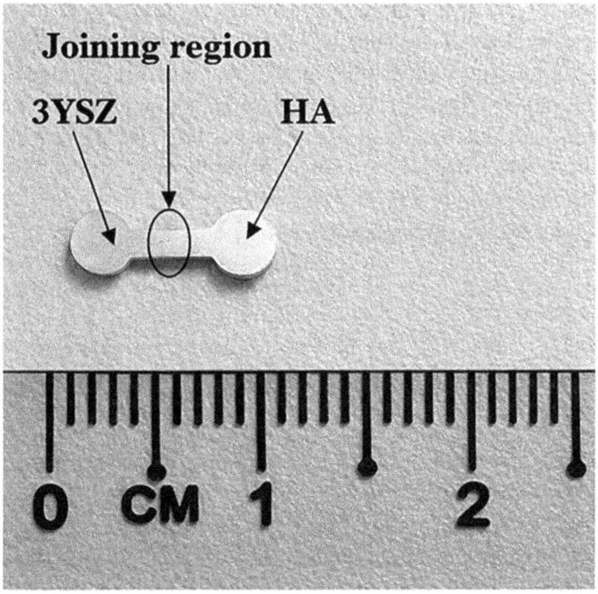

Fig. 4.11 Green micro-sized bi-material component of HA and 3YSZ [2]

temperatures promoted the formation of martensite at the expense of ferrite and pearlite, contributing to enhanced strength. However, temperatures exceeding 1320 °C led to abnormal grain growth and a decrease in mechanical properties, underscoring the narrow processing window for optimal performance in micro MIM. This sensitivity to thermal profile is a central challenge in scaling down MIM processes.

Beyond material and sintering refinements, alternative powder metallurgy approaches are being explored to overcome the inherent limitations of Micro MIM, particularly for part dimensions below 1 mm. Kieback et al. [35] proposed a method combining three-dimensional screen printing with subsequent sintering and optional sizing or compaction. This technique uses fine metal powders with a D90 smaller than 10 μm to create green parts with resolutions down to 100 μm, a feat difficult to achieve with traditional die compaction due to powder flowability issues. Complex 316L stainless steel gear wheels and microtubes were produced, illustrating the potential for mass production of intricate microparts. Two processing routes were compared: printing-sintering-sizing and printing-compaction-sintering. The latter approach, which includes a slight compaction step of the green printed part prior to sintering, showed improved performance for 316L, reducing porosity to 2.3% and enhancing surface roughness to Rz = 5.0 μm relative to parts subjected only to sintering. This hybrid approach offers a potential pathway for manufacturing high density, high precision metallic microparts where Micro MIM might face economic or technical hurdles (see Fig. 4.11).

Tool wear and flash control, which are well-known challenges in conventional hardmetal MIM, become more pronounced at the microscale. The abrasive character of fine, irregular powders such as tungsten carbide leads to increased wear of mixing equipment and tooling, as reported by Johnson et al. [36]. Furthermore, the small particle size allows feedstock to enter minute gaps in the tooling, leading to flash that is particularly detrimental to features like cutting edges on micro tools. They discuss the development of low pressure injection molding for carbides to mitigate these issues, which requires customized binder and solids loading formulations to achieve a low viscosity slurry. This is directly relevant to Micro MIM, where controlling injection parameters is essential for preventing defects such as weld lines and jetting, which can be catastrophic in micro-sized components.

4.3.2 Hybrid MIM Processes: Combining Additive Manufacturing and Injection Molding

The exploration of hybrid manufacturing processes that merge additive manufacturing (AM) with MIM represents a significant advancement in the production of complex metal parts (Fig. 4.12). Novák et al. [37] investigated the integration of MEX-based additive manufacturing and powder injection molding for 17-4PH stainless steel components. Surface characteristics and mechanical performance were compared for parts produced by MEX, PIM, and direct metal laser sintering (DMLS). A key finding was the statistical proximity between the surfaces created via PIM and MEX, with a cluster analysis revealing approximately 58% similarity between PIM parts made from PIM quality powder (PIMPIM) and MEX parts. This similarity suggests a potential pathway for using MEX for rapid prototyping and PIM for mass production of identical part geometries.

It is reported that an environmentally benign binder system for the PIM feedstocks was used, and advanced statistical tools, including principal component analysis, were employed to objectively quantify surface roughness parameters (Ra, Rz, RSm). Mechanically, while DMLS samples exhibited the highest ultimate tensile strength (1140 MPa) and ductility (19% strain at break), PIMPIM parts demonstrated a high yield strength (800 MPa), indicating that PIM parts from such

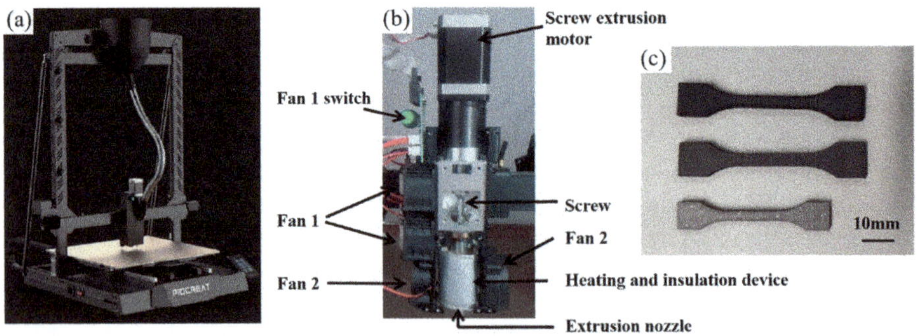

Fig. 4.12 (**a**) G5Pro printer, (**b**) the structure of G5Pro printer, and (**c**) the parts after being printed (up), debinded (middle), and sintered (down) [25]

feedstocks can successfully compete with certain AM routes. The research also highlighted the influence of powder characteristics, as using DMLS grade powder in the PIM process (PIMDMLS) led to different rheological behavior and inferior mechanical properties, underscoring the importance of tailored feedstock compositions.

Beyond process level integration, the broader concept of hybrid manufacturing encompasses the combination of additive manufacturing with post-processing technologies within single- or multi-setup production routes, as described by Jiménez et al. [19]. Although the combination of laser-based additive manufacturing processes, such as laser powder bed fusion (LPBF) and direct laser deposition (DLD), with machining is addressed in detail, the underlying concept of process synergy is equally applicable to hybrid MIM-AM systems. In this context, a key motivation for hybrid manufacturing lies in the inherent limitations of standalone AM technologies, including limited surface quality, dimensional deviations, and the reliance on support structures. These challenges are similarly addressed when merging AM with the high volume, net-shape capabilities of PIM. Hybrid processing combines the geometric freedom and low material waste of additive routes with the surface integrity and dimensional accuracy of subtractive or formative operations. This rationale supports the hybrid MIM-AM approach, where the "green" part from either process undergoes a unified debinding and sintering cycle, as demonstrated in the experimental work by Novák et al. [37]. Hybrid techniques were applied to fabrication of complex end-use geometries and component repair, consistent with the use of combined metal injection molding and additive processing to produce intricate prototypes and transition them to mass production. Process parameter control and material machinability remain limiting factors for broader industrial adoption, requiring standardization to obtain consistent mechanical performance and surface quality in hybrid MIM-AM workflows [19].

4.3.3 Process Monitoring and Digital Twins for Quality Assurance

Based on the research by Wang et al. [38], the implementation of a digital twin system within the injection molding industry provides a comprehensive framework for process monitoring and quality assurance. A smart factory architecture has been proposed based on a five-dimensional digital twin framework that integrates physical entities, virtual models, data, services, and connectivity (Fig. 4.13). In this regard, real-time data acquisition from the production floor, including personnel, materials, molds, machines, and environmental conditions, supports system level integration and process coordination. To address the challenge of heterogeneous equipment with varied interface protocols, the system employs a unified communication architecture based on the open platform communications unified architecture (OPC UA) protocol. This protocol enables the aggregation of field level data from sources such

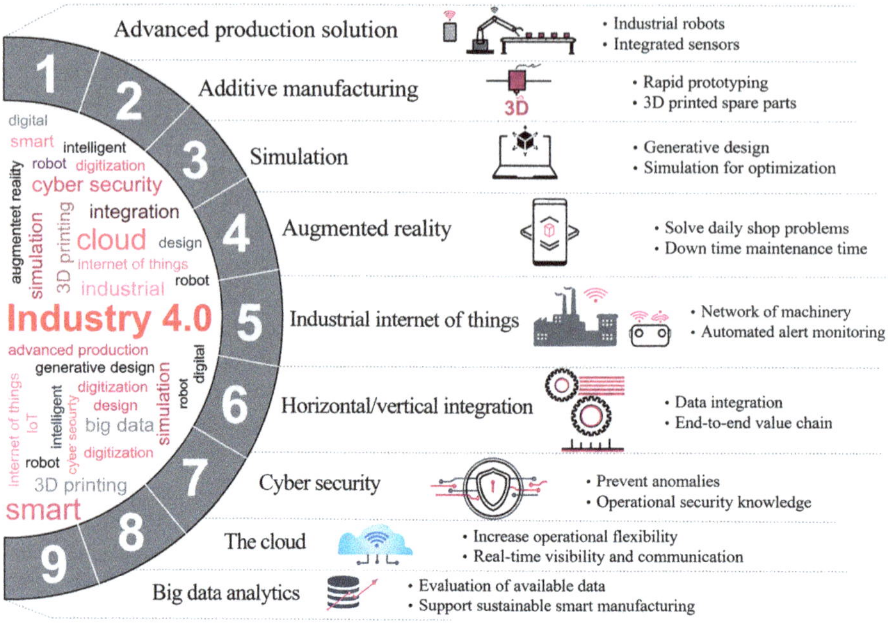

Fig. 4.13 Main pillars of Industry 4.0 in the production process [39]

as injection molding machine controllers and radio frequency identification systems into a centralized data server, allowing real-time interactive mapping between the physical and virtual factory environments. The resulting bidirectional data exchange permits continuous monitoring of the physical production process within the virtual factory and supports iterative optimization activities, including production scheduling simulations and fault prediction, thereby contributing to stable product quality and efficient operation.

The digital twin concept was described as a cyber physical representation of multistage manufacturing processes, implemented through data-driven models or physics-based simulations, with physical modeling applied to complex operations such as metal forming [40]. A significant contribution of their research is the FlexMM (Flexible Material Modeling) approach, which provides a generalized formulation for constitutive material behavior. This method is designed to solve interoperability problems when different Finite Element Method (FEM) solvers are used across a production chain, ensuring that material state variables remain consistent and transferable between processes. For a digital twin to be effective in real-time control and quality assurance, the underlying physical models must be robust, validated, and mature. However, the computational intensity of high fidelity FEM models often precludes their direct use in real-time applications. To overcome this, the study proposes the creation of meta models using techniques like Design and Analyses of Computer Experiments (DACE), which are fast, statistical representations of the complex physical models, enabling real-time simulation and process control for quality prediction. This approach was validated in a case study on the sintering stage of MIM, where the digital twin model was calibrated against physical samples to predict shape changes and defects accurately.

Wang et al. [38] detail specific technologies that feed critical process data into the digital twin. The injection molding integrated intelligent control system is a key enabler, consisting of a high precision injection molding machine control system and an in-mold online quality monitoring system. The high precision control system, often based on an EtherCAT architecture, utilizes advanced algorithms like two-dimensional, time-iterative predictive control to achieve precise regulation of key parameters such as barrel temperature, injection speed, and holding pressure. This enhances the repeatability and stability of the process, which is fundamental for quality assurance. Simultaneously, the in-mold online quality monitoring system uses capacitance measurement to sense the state of the plastic melt within the mold cavity. As the dielectric constant of the melt changes during injection, packing, and cooling stages, the capacitance signal variations are used to establish a detection model for online monitoring of flow state, product weight, and optimal cooling time, effectively creating a closed-loop perception of the entire process. The integration of these two systems allows for the establishment of a data-driven quality model, enabling fault diagnosis and automatic optimization of process parameters, moving the process toward what the authors term "intelligent injection."

The synergy between advanced process monitoring and the digital twin creates a powerful quality assurance ecosystem. The real-time data from the intelligent control systems and heterogeneous sensors populate the digital twin, which then acts as a virtual replica of the physical process. The factory service system uses this data for real-time prediction and optimization. For instance, during production, if the virtual factory detects an anomaly through the real-time data stream, it can immediately run simulations to evaluate the situation, predict potential outcomes, and generate optimized production scheduling or control instructions to correct the issue [38]. This capability for iterative, real-time optimization based on a continuously updated digital twin ensures that quality is not merely inspected at the end but is assured throughout the manufacturing process, significantly reducing defects and enhancing overall production efficiency.

4.3.4 Surface Functionalization and Post-processing Techniques

Surface functionalization and post-processing techniques are critical for enhancing the performance and expanding the application range of components produced via advanced injection molding methods. In the context of LPIM, the surface energy of the mold material itself is a pivotal factor influencing feedstock behavior. Research by Kurusu et al. [22] investigated the surface energy of various mold materials, including steel and polytetrafluoroethylene (PTFE), and correlated these properties with feedstock moldability and mold adhesion. The research demonstrated that while PTFE molds exhibited superior moldability due to their low thermal conductivity, which delays feedstock solidification, they also resulted in high adhesion, rendering them impractical for use without modification. In contrast, steel molds, owing to their higher thermal conductivity, exhibited lower moldability while markedly reducing adhesion. Adhesion behavior was found to be primarily governed by the interfacial energy between the mold surface and the binder, rather than by the solidification rate. This relationship was demonstrated by applying a thin gold coating to a polytetrafluoroethylene mold, which combined the high moldability typically associated with polymeric molds with the low adhesion characteristic of metallic molds, arising from the modified interfacial energy. Such surface coating strategies illustrate how functionalization can be used to decouple and optimize competing process requirements.

Beyond mold surfaces, functionalization is equally critical for enhancing the bond strength in polymer metal hybrid structures created through processes like injection overmolding. Wang et al. [41] investigated the functionalization of an aluminum alloy (Al) surface to create ultra-high strength joints with polyamide-6 (PA6). Their method involved a synergistic treatment of sandblasting to create microstructures for mechanical interlocking, followed by the application of a reactive epoxide silane (γ-epoxysilane). The hydrolyzed silane formed chemical bonds with the aluminum oxide layer, while its epoxy groups subsequently reacted with the amide groups of the PA6 during injection molding. This dual mechanism of mechanical interlocking and chemical bonding resulted in a dramatic 182% increase in joint strength, reaching 37.8 MPa, compared to samples that were only sandblasted. X-ray photoelectron spectroscopy (XPS) analysis confirmed the formation of Si-O-Al and Si-O-Si bonds, validating the successful functionalization of the metal surface. The study further demonstrated that the joint strength could be elevated to 46.6 MPa through annealing, which increased the crystallinity of the PA6, and that the joints retained significant strength at elevated temperatures [41].

In addition to chemical treatments, mechanical post-processing techniques are employed to significantly alter the surface and subsurface properties of molded metallic parts. Wiratkapun et al. [42] explored the effects of severe shot peening (SSP) on 316L stainless steel manufactured by MIM. The SSP treatment induced severe plastic deformation, resulting in a surface layer with a nanocrystalline structure, the formation of deformation-induced martensite (α' martensite), and a work-hardened subsurface with high dislocation density. This microstructural evolution led to a substantial 126% increase in surface microhardness. In tribological tests, the SSP-treated specimens exhibited significantly improved wear resistance under loads of 10 and 20 N, which was attributed to the enhanced surface strength imparted by the nanocrystalline layer and strain hardening. However, under a higher load of 30 N, the strengthening effect was surpassed, and the wear resistance became comparable to that of the as-sintered specimens, with severe adhesive and oxidative wear dominating in both cases. It is noteworthy that the SSP treatment also altered surface topography, increasing the arithmetic average roughness (Ra) but decreasing the maximum height of the profile (Rz) by eliminating surface porosity, a common feature in MIM parts.

4.3.5 Sustainability-Oriented Variants: Energy Efficiency and Recyclability

The pursuit of sustainability within injection molding has led to significant innovations focused on energy efficiency and the integration of recyclable materials. A primary strategy for enhancing energy efficiency involves the optimization of the cooling phase, which constitutes up to 80% of the total injection molding cycle time. Research by Clark et al. [43] explores this through the implementation of additively manufactured conformal cooling channels (CCCs) and the use of alternative mold insert materials with high thermal conductivity. Their simulation-based study compared conventional straight-drilled cooling channels with various CCC designs, such as U-shaped and zigzag patterns, manufactured via LPBF. The molds were fabricated from different materials, including 420 stainless steel (420SS), H13 tool steel, AlSi10Mg aluminum alloy, and CuNi2SiCr copper alloy. The results demonstrated that for conventional cooling channel designs, higher thermal conductivity materials like the AlSi10Mg alloy substantially reduced cooling times by up to 47%, which directly translated to lower energy consumption. The specific energy consumption (SEC) per part decreased by a maximum of 15% when using the aluminum alloy compared to the 420SS control. However, the study noted that the benefits of high thermal conductivity materials diminished with increasingly complex CCC designs, where convective heat transfer became more dominant. In such cases, the AlSi10Mg alloy provided only a 10% cooling time reduction in the most complex zigzag CCC design, indicating that material selection must be tailored to the cooling channel geometry.

Beyond energy efficiency, the principles of a circular economy are advanced through material recyclability and the use of sustainable composites. Mechanical behavior of acrylonitrile butadiene styrene composites reinforced with glass fiber and granite powder derived from industrial waste was evaluated [21]. The researchers prepared four composite batches via injection molding: pure ABS (Batch A), ABS with 10% GF (Batch B), ABS with 10% GP (Batch C), and a hybrid of ABS with 10% GF and 10% GP (Batch D). Mechanical testing revealed that Batch B exhibited the highest tensile strength of 45.76 MPa, a significant improvement over pure ABS at 41.6 MPa. In contrast, the incorporation of granite powder alone in Batch C reduced the tensile strength to 36.9 MPa. The hybrid composite, Batch D, showed a moderate tensile strength of 42.54 MPa. Furthermore, Batch D demonstrated the highest compressive strength at 53.6 MPa and the greatest hardness, indicating that the synergistic use of both fillers can enhance specific mechanical properties suitable for demanding applications. However, the impact strength was highest for pure ABS, and all reinforced composites showed reduced toughness, with the hybrid composite performing the worst in this category. Morphological analysis using Scanning Electron Microscopy (SEM) confirmed good dispersion of the fillers but also highlighted the presence of porosity, which could affect overall durability.

The adoption of additive manufacturing for producing mold inserts themselves presents another avenue for sustainability, particularly for low volume production. Coronado et al. [44] evaluated the environmental impact and cost-effectiveness of 3D printed mold inserts (MIs) using Vat Photopolymerization (VPP) with High Temp and Rigid 10k resins. Their life cycle assessment (LCA) found that the carbon footprint of these MIs ranged from 27.466 to 28.579 kg CO_2 equivalent per insert, with propylene production for resin and isopropyl alcohol for post-processing being the largest contributors, accounting for 15.7–16.3% of emissions. Functionally, the Rigid 10k resin with a solid design (MI-C) demonstrated superior durability, withstanding 250 injection molding shots, whereas inserts made from High Temp resin failed after only 4–5 shots. From an economic perspective, the study revealed that 3D printing MIs for production batches of 250 units or fewer reduced fabrication costs by 43.19% compared to traditional subtractive manufacturing methods. This combination of adequate functional performance, reduced carbon footprint relative to traditional manufacturing waste, and significant cost savings positions 3D printed MIs as a viable, sustainability-oriented alternative that aligns with circular economy goals by minimizing material waste through layer-by-layer fabrication.

4.4 Industrial Applications: Small, High Precision Components in Medical and Electronics

4.4.1 Medical Applications: Orthopedic Implants, Dental Devices, and Surgical Tools

Among advanced powder-based manufacturing routes, MIM plays a central role in enabling the fabrication of small, high precision medical components, ranging from orthopedic implants to dental devices and surgical tools (Fig. 4.14). This process combines the design flexibility of plastic injection molding with the material integrity of powder metallurgy, enabling the cost-effective mass production of complex-shaped parts with excellent mechanical properties. A significant application is in the orthopedic sector, where the demand for affordable and customized implants is rising. For instance, research conducted by Wiranegara et al. [46] focused on optimizing MIM process parameters for manufacturing green part orthopedic implants from the facial and jaw region using 316L stainless steel feedstock. Sigmasoft simulations were used to examine the influence of filling temperature and filling time on incomplete filling, with a melt temperature of 180 °C, a filling time of 1 s, and a filling pressure of 700 bar producing complete cavity filling and thereby preserving structural integrity of the sintered implant.

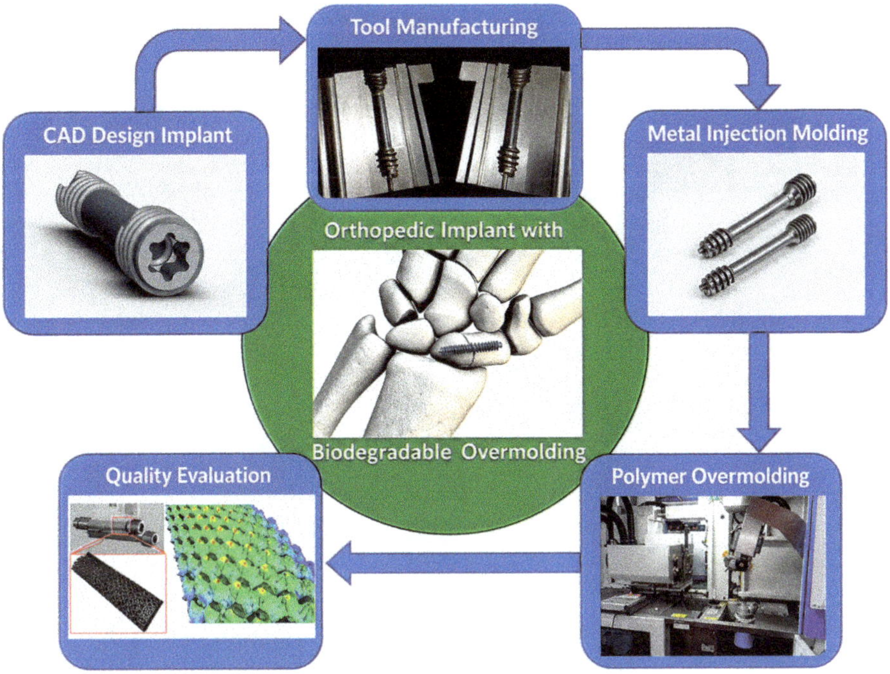

Fig. 4.14 Schematic representation of MIM orthopedic implants with a functionalized thin layer of biodegradable polymer [45]

Beyond traditional biomaterials like stainless steel, MIM is also being explored for next-generation biodegradable implants (Fig. 4.15). Wermuth et al. [5] investigated the use of pure iron for this purpose, manufacturing samples via MIM with a novel, eco-friendly feedstock based on natural rubber from *Hevea brasiliensis*. This binder system offered advantageous elastic properties for green part handling. The physical, mechanical, and biological properties of the sintered iron samples were evaluated, yielding a sintered density of 6.61 g/cm^3 and an average porosity of approximately 16%. This level of porosity is considered advantageous for promoting implant degradation and supporting drug adhesion. Mechanically, the samples exhibited a yield strength between 65 and 96 MPa and high elongation between 35% and 53%; properties deemed suitable for biomedical applications such as withstanding balloon expansion in stents. In vitro biocompatibility tests with adipose-derived mesenchymal stromal cells (ADSCs) showed no significant cytotoxicity, with cell viability remaining high after exposure to iron extracts. Furthermore, hemocompatibility tests revealed a hemolysis rate of less than 5%, categorizing the material as non-hemolytic according to standard ASTM F756 08. In vivo subcutaneous implantation in a Wistar rat model confirmed the material biocompatibility, with no significant adverse effects on animal health, organ

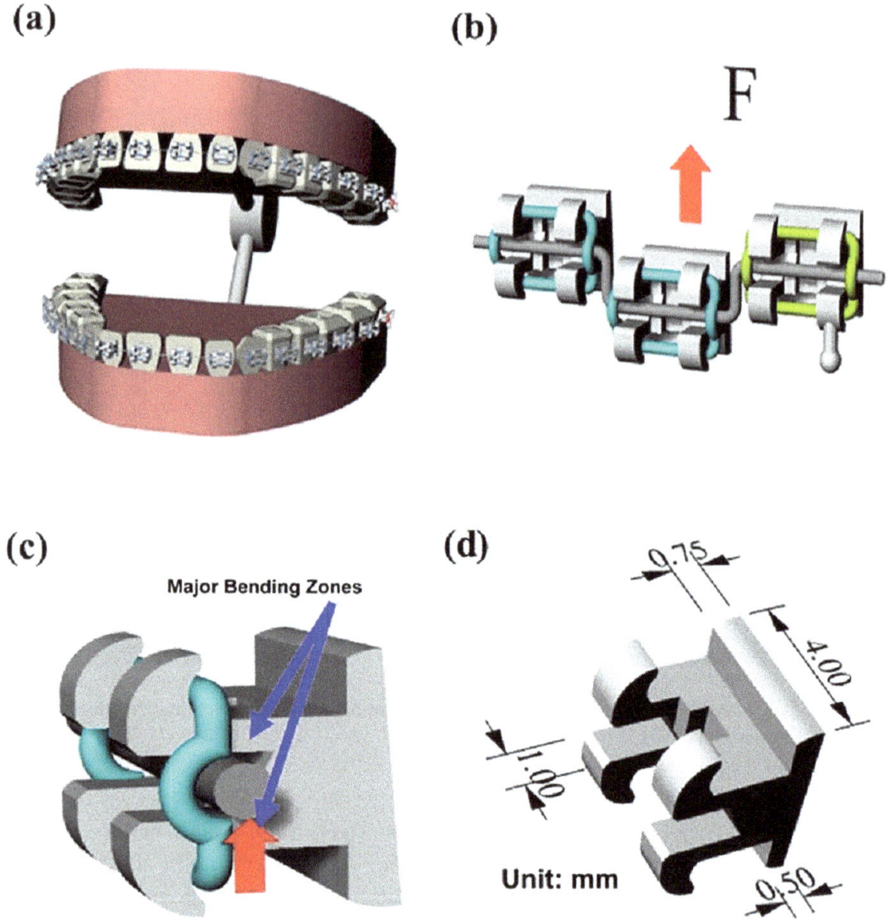

Fig. 4.15 Orthodontic brace for oral correction: (**a**) general arrangement; (**b**) lateral forces acting on cantilever regions of orthodontic brace; (**c**) bending forces acting on cantilever regions of orthodontic brace; and (**d**) geometric dimensions of orthodontic brace [47]

histology, or hematological parameters observed over a 6-month period, supporting its potential for clinical use.

The application of MIM extends to vascular medical devices, as demonstrated by Shu et al. [48], who evaluated vascular stents manufactured from 316L stainless steel. The performance of 316L stainless steel during metal injection molding was shown to be sensitive to variations in carbon content. While moderate increases in carbon content enhanced tensile strength, levels exceeding 0.30 wt.% resulted in pronounced reductions in both tensile strength and elongation, together with diminished corrosion resistance, particularly in sodium chloride environments. Through precise control of carbon content, stents were fabricated and successfully implanted in canine models. The stents demonstrated no significant cytotoxicity in vitro and exhibited favorable in vivo performance, including fast endothelialization and long-term patency without complications such as thrombosis or restenosis, underscoring the biocompatibility and structural stability of MIM-fabricated vascular implants.

4.4.2 Electronics Industry: Connectors, Heat Dissipation Parts, and Micro-components

MIM has established itself as a pivotal manufacturing technology for producing small, high precision components within the electronics industry, including connectors, heat dissipation parts, and various micro-components, as illustrated in Fig. 4.16. This process is particularly advantageous for fabricating intricate geometries that are challenging to achieve through conventional manufacturing methods. The versatility of MIM allows for the use of a wide range of materials, including stainless steels and copper alloys, which are commonly selected for electronic applications due to their favorable electrical and thermal properties [50]. A significant application highlighted is the production of fiber optic parts and heat sinks, which are critical for managing thermal loads in electronic devices. The design flexibility of MIM enables the creation of advanced thermal management solutions, such as components with aerodynamic fins, heat pipes, and microchannels, which can be mass produced

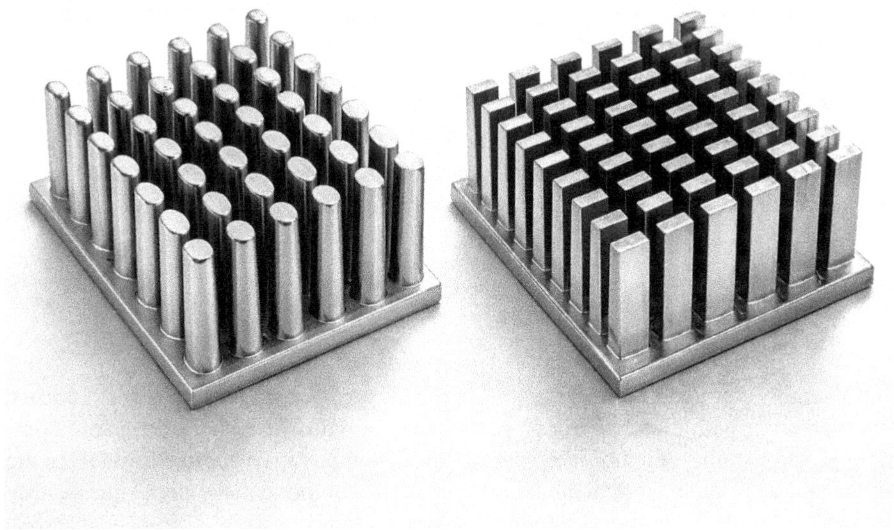

Fig. 4.16 Examples of heat sink geometries produced with aluminum MIM. Heat sinks up to $50 \times 50 \times 50$ mm^3 are most amenable to processing by MIM [49] (based on the original reference image)

with greater geometrical freedom compared to processes like extrusion or die casting [51]. This capability is essential for meeting the high power demands and miniaturization trends in modern electronics, where efficient heat dissipation is paramount.

The drive toward miniaturization in the electronics sector has further amplified the relevance of MIM, particularly through its derivative, micro MIM. This technology is adept at manufacturing extremely small and complex parts, such as micro-gears and micro-optical benches, which are integral to consumer electronics, precision instruments, and aerospace applications [51]. In micro MIM, the use of nano- and microscale iron powder feedstocks has been investigated as a means to improve surface finish and dimensional accuracy. Combining microscale iron powder with nanoscale iron powder has been shown to alleviate issues such as powder binder separation and to enhance sintering behavior, resulting in improved quality of microscale components. The replication of fine features, such as the 1500 pyramidal bumps on a dental endodontic tool, demonstrates the process's capability for high detail resolution, which is directly transferable to manufacturing intricate micro-components for the electronics industry.

Furthermore, the integration of knowledge-driven engineering and intelligent design systems is revolutionizing the MIM process for electronic components. Kong and colleagues [52] developed an automated technological design system on the NX platform specifically for cylindrical spur gears, a common element in precision electronic devices. Their system utilizes feature recognition algorithms and intelligent retrieval from a comprehensive knowledge base to automate the design of molds and process parameters. This approach significantly reduces the reliance on designer experience and shortens development cycles. The system demonstrated remarkable efficiency improvements, with recognition and retrieval times enhanced by over 98% and the number of required trial moldings reduced by up to 75%. This method of intelligent analogy design ensures that the manufacturing process for new, complex electronic components can be rapidly and accurately derived from proven similar parts, thereby accelerating production and improving consistency in the electronics sector. The application of such systems underscores a broader trend toward digitization and smart manufacturing within MIM, facilitating its expanded use in producing high value, precision-driven electronic parts.

4.4.3 Lightweight Components

Lightweight components represent a critical area of application for additive manufacturing with metal powders, particularly in industries where precision, complexity, and weight reduction are paramount. One innovative approach involves the use of lattice structures to reduce the mass of metal components without significantly compromising their structural integrity. A lightweight injection mold was designed and fabricated using Ti-6Al-4V lattice structures produced by additive manufacturing [7]. Compression behavior was evaluated for two lattice topologies, octet truss, characterized by stretch-dominated deformation, and diamond, associated with bending-dominated behavior, across different relative densities. The diamond lattice showed enhanced energy absorption as a result of its extended plateau stress response, whereas the octet truss structure exhibited higher resistance to compressive loads at specific density ranges. By integrating computer-aided engineering analysis to map stress distribution within an injection mold under operational clamping forces and material pressure, the researchers strategically placed these lattice structures inside the mold. This design achieved a significant weight reduction of approximately 79% compared to a solid Ti-6Al-4V mold. Furthermore, the feasibility of the lightweight mold was confirmed through an injection molding test with polyvinyl chloride, successfully completing 400 shots without any damage to the mold structure. This work highlights the potential of lattice structures in creating lightweight, functional tools for manufacturing processes.

The pursuit of lightweight components also extends to the materials used in the MIM process itself, particularly aluminum alloys. Zou et al. [4] address recent developments in the MIM of aluminum alloys, focusing on aspects related to feedstock formulation, debinding behavior, and sintering performance. Aluminum alloys are attractive for lightweight applications because of their low density, high specific strength, and favorable thermal conductivity; however, their application in MIM is hindered by challenges associated with densification during sintering and the presence of stable surface oxide layers that restrict diffusion. Particular emphasis is placed on feedstock design, including the use of spherical gas-atomized powders with suitable particle size distributions and the development of tailored binder systems, such as wax-based and polyoxymethylene-based formulations. Debinding methods, including solvent, catalytic, and thermal debinding, were analyzed for their effectiveness in removing the binder without causing defects. Sintering in controlled atmospheres, particularly nitrogen was identified as crucial for disrupting the oxide layer and promoting densification, often aided by alloying elements like magnesium and tin. Optimizing these factors has been shown to enable the production of aluminum alloy components with high relative density and adequate mechanical performance, supporting their application in lightweight systems for electronics, medical, and automotive sectors.

Another significant development in lightweight component manufacturing is the creation of hollow metal parts through advanced MIM techniques. Kosugi and Terada [53] described the development of a mass production technology for hollow parts using a novel process that incorporates a soluble core. This technique allows for the fabrication of integrated three-dimensional hollow shapes that were previously unattainable with conventional molding methods, which often required assembling multiple parts. The process involves molding a core from a material with a higher melting point than the MIM feedstock, followed by insert molding the metal feedstock around the core. After molding, the core is extracted, and the green part undergoes standard debinding and sintering. Key challenges addressed included optimizing the core material and MIM feedstock to ensure a sufficient melting point difference, minimizing burrs by precisely controlling the clearance between the core and the mold, and establishing quality assurance methods for critical hollow section dimensions. The successful application of this technology was demonstrated with a ball screw part made from 17-4PH stainless steel, resulting in a lightweight component with complex internal geometry. This hollow MIM process offers substantial benefits for weight reduction, design flexibility, and reduced environmental impact, making it suitable for demanding industries such as aerospace and automotive.

The application of computer-aided engineering and intelligent modeling plays an important role in optimizing manufacturing processes for lightweight components. In this context, Huang et al. [54] addressed the optimization of injection molding parameters for automobile lock components fabricated from glass fiber-reinforced polyamide 66, with the aim of reducing warpage and improving temperature distribution. Using a Taguchi robust design with an L18 orthogonal array, the researchers analyzed eight control factors, including holding pressure, mold temperature, and cooling liquid type. The signal-to-noise ratios for warpage and average temperature were calculated, and gray relational analysis was employed to normalize these values. The normalized data were processed using a fuzzy inference system to obtain a multiple performance characteristic index, allowing the identification of process parameters that improved multiple quality metrics simultaneously. Different cooling channel configurations, including conventional, square, and conformal designs, were also evaluated. The conformal cooling approach, which more closely follows the part geometry, delivered the most favorable performance, reducing warpage by 12.2% and lowering the average temperature by 0.46% relative to the conventional cooling design. These results reported illustrate how process optimization combined with advanced cooling strategies can improve the dimensional stability and quality of lightweight polymer components commonly used in automotive applications.

The microstructure and resultant mechanical properties of MIM parts are critical for their performance in high temperature applications, which is relevant for lightweight components in demanding environments. Suzuki et al. [20] investigated the microstructure and high temperature mechanical properties of IN718 fabricated by MIM and compared them with specimens produced by selective laser melting. The metal injection molded IN718 exhibited a mixed grain structure and contained precipitates such as Al_2O_3 and Nb-rich carbides, which formed due to oxygen contamination and increased carbon content during the process. Tensile testing at 650 °C revealed that the yield strength of as-built metal injection molded specimens was similar to that of as-built selective laser melted specimens. However, after a standard heat treatment, the metal injection molded material showed significantly lower yield strength and ductility. This was attributed to the consumption of strengthening elements like Nb by the formation of carbides, which suppressed the precipitation of the strengthening gamma prime and gamma double prime phases.

4.5 Conclusions

MIM can be understood as a materials-focused manufacturing approach in which powder morphology, binder design, and the coupled debinding and sintering stages together control densification, microstructural evolution, and functional performance across a broad range of alloy and composite systems. It demonstrates that key attributes such as density, grain structure, and impurity levels are rooted in feedstock design and thermal history rather than in molding alone, with fine spherical powders enhancing sintering kinetics while imposing strict constraints on binder chemistry and contamination control. Multicomponent binder systems and their controlled removal establish the rheological stability and interconnected porosity required for defect-free shaping and densification, while powder loading, surface modification, and rheological metrics define robust processing windows. Sintering behavior, validated experimentally and through simulation, is shown to depend on particle size, loading, and thermal gradients, enabling targeted property tailoring via alloying, atmosphere control, and post-sintering treatments. These principles extend consistently to micro MIM and hybrid MIM additive routes, culminating in a view of MIM as a coordinated processing chain increasingly supported by digital twins, intelligent control, and surface engineering strategies to expand its application space.

References

1. Ni J, Yu M, Han K (2018) Debinding and sintering of an injection-moulded hypereutectic Al–Si alloy. Materials 11(5):807
2. Basir A et al (2023) Micro-injection molding and debinding behavior of hydroxyapatite/zirconia bi-materials fabricated by two-component micro-powder injection molding process. Materials (Basel) 16(19):6375
3. Attia UM et al (2014) Creating movable interfaces by micro-powder injection moulding. J Mater Process Technol 214(2):295–303
4. Zou H et al (2025) Research progress on metal injection molding of aluminum alloys. Powder Metall Technol 43(1):20–34
5. Wermuth DP et al (2021) Mechanical properties, in vitro and in vivo biocompatibility analysis of pure iron porous implant produced by metal injection molding: a new eco-friendly feedstock from natural rubber (Hevea brasiliensis). Mater Sci Eng C 131:112532
6. Sadaf M et al (2024) Comparative analysis of binder systems in copper feedstocks for metal extrusion additive manufacturing and metal injection moulding. J Mater Res Technol 29:4433–4444
7. Park SJ et al (2022) Lightweight injection mold using additively manufactured Ti-6Al-4V lattice structures. J Manuf Process 79:759–766
8. Ma H et al (2024) Effect of surface modification of metal powders (by aluminum stearate) on the properties of metal injection molding feedstock. Materials (Basel) 67(1):40–50
9. Liao Y et al (2023) Effect of sintering temperature on microstructure and mechanical properties of Fe–4Ni–0.8Mo–0.6C steel small-module gears fabricated by micrometal injection molding. Steel Research International 94(8):2200886. https://doi.org/10.1002/srin.202200886
10. Wen G et al (2013) Development and design of binder systems for titanium metal injection molding: an overview. Metall Mater Trans A 44(3):1530–1547

11. Lim K et al (2024) Formulating titanium powder feedstocks for metal injection moulding from a clean binder system. Powder Technol 433:119214
12. Liang C et al (2024) Microstructures evolution and properties of titanium vacuum sintering on metal injection molding 316 stainless steel via dip coating process. J Manuf Process 131:1073–1085
13. German RM (2013) Progress in titanium metal powder injection molding. Materials 6(8): 3641–3662
14. Ahmad MA et al (2021) Rheological evaluation of titanium alloy metal injection molding (MIM) feedstock. Saudi J Eng Technol 6(11):392–398
15. Ren W et al (2024) Investigating evolution of microstructures and mechanical properties in metal injection molding Ti6Al4V with sintering processing by quasi in-situ strategy. Powder Technol 439:119694
16. Hwang I-S et al (2023) Characterization of mechanical properties and grain size of stainless steel 316L via metal powder injection molding. Materials 16(6):2144
17. Kutsbakh A et al (2025) Modeling of feedstock formability to optimize mold design and prevent possible defects during metal injection molding. J Manuf Mater Process 9(6):203
18. Eickhoff R et al (2022) Feedstock development for material extrusion-based printing of Ti6Al4V parts. Materials 15(18):6442
19. Jiménez A et al (2021) Powder-based laser hybrid additive manufacturing of metals: a review. Int J Adv Manuf Technol 114(1):63–96
20. Suzuki H et al (2025) Microstructure and high temperature mechanical properties of IN718 fabricated by metal injection molding. J Jpn Soc Powder Powder Metall 72:S679
21. Mulge P, Kalashetty S (2025) Sustainable engineering of mechanical properties in ABS composites through glass fiber and granite powder reinforcements via injection molding. Solid State Phenom 369:3–15
22. Kurusu RS et al (2023) Surface properties of molds for powder injection molding and their effect on feedstock moldability and mold adhesion. Int J Adv Manuf Technol 126(1):381–390
23. Heaney DF (2012) 3—Powders for metal injection molding (MIM). In: Heaney DF (ed) Handbook of metal injection molding. Woodhead Publishing, Sawston, pp 50–63
24. Wolff M et al (2016) Metal injection molding (MIM) of magnesium and its alloys. Metals 6(5): 118
25. Chang G et al (2023) Printing, debinding and sintering of 15-5PH stainless steel components by fused deposition modeling. Addit Manuf 16(19):6372
26. Wu Y, Guo K, Ni J (2024) Characterization of particle size effects on sintering shrinkage and porosity in stainless steel metal injection molding using multi-physics simulation. Materials 17(23):5691
27. Safari M, Ghambari M, Akhtar M (2015) Experimental and theoretical investigation of powder–binder mixing mechanism for metal injection molding. Adv Manuf Technol 30(1):29–36
28. Momeni V et al (2024) Binder system composition on the rheological and magnetic properties of Nd-Fe-B feedstocks for metal injection molding. Appl Sci 14(13):5638
29. Gonzalez-Gutierrez J et al (2016) Models to predict the viscosity of metal injection molding feedstock materials as function of their formulation. Metals 6(6):129
30. Momeni V et al (2023) Research progress on low-pressure powder injection molding. Materials 16(1):379
31. Teferi FT, Tsegaw AA (2022) Metal injection molding (MIM) process and potential remedies for its defects: a review. In: Advances of science and technology. Springer International Publishing, Cham
32. Ledniczky G et al (2025) Impact of recycling on polymer binder integrity in metal injection molding. Sci Rep 15(1):21600
33. Ouyang M et al (2021) Effects of bonding treatment and ball milling on W-20 wt% Cu composite powder for injection molding. Materials 14(8):1897

34. Siedlecki K, Słoma M, Skalski A (2023) Comparison between micro-powder injection molding and material extrusion additive manufacturing of metal powders for the fabrication of sintered components. Materials 16(23):7268
35. Kieback B et al (2016) Micro parts manufacturing by powder metallurgy (micro PM)
36. Johnson JL, Heaney DF, Myers NS (2012) 21—Metal injection molding (MIM) of heavy alloys, refractory metals, and hardmetals. In: Heaney DF (ed) Handbook of metal injection molding. Woodhead Publishing, Sawston, pp 526–567
37. Novák M et al (2023) On the possibilities of merging additive manufacturing and powder injection molding in the production of metal parts. Rapid Prototyp J 30(11):50–58
38. Wang Z et al (2021) A study on intelligent manufacturing industrial internet for injection molding industry based on digital twin. Complexity 2021:1–16
39. Nasiri S et al (2024) Digital twin modeling for smart injection molding. J Manuf Mater Process 8(3):102
40. Post J, Groen M, Klaseboer G (2017) Physical model based digital twins in manufacturing processes. In: Forming technology forum 2017. University of Groningen, Enschede
41. Wang B et al (2023) Functionalization of aluminum alloy surface with reactive epoxide silane to induce ultra-high strength polyamide 6 / aluminum alloy composite joint. Appl Surf Sci 626: 157231
42. Wiratkapun K et al (2024) Surface modification and enhanced wear performance through severe shot peening treatment in 316L stainless steel manufactured by metal injection moulding. J Mater Res Technol 29:4072–4085
43. Clark R et al (2024) Towards sustainable injection moulding using 3D printed conformal cooling channels: a comparative simulation study. Prog Addit Manuf 9(5):1341–1352
44. Coronado LOB et al (2025) 3D-printed injection mold inserts for sustainability and circular economy. IOP Conf Ser Earth Environ Sci 1500(1):012050
45. Elkaseer A et al (2018) Replication of overmolded orthopedic implants with a functionalized thin layer of biodegradable polymer. Polymers (Basel) 10(7):707
46. Wiranegara H, Syahputra D, Arief T (2021) Injection process parameter analysis of metal injection molding for green part orthopedic implants. IOP Conf Ser Mater Sci Eng 1034:012177
47. Lin C-M, Wu J-J, Tan C-M (2020) Processing optimization for metal injection molding of orthodontic braces considering powder concentration distribution of feedstock. Polymers (Basel) 12(11):2635
48. Shu C et al (2022) Biocompatibility of vascular stents manufactured using metal injection molding in animal experiments. Trans Nonferrous Metals Soc China 32(2):569–580
49. Zlatkov BS et al (2008) Recent advances in PIM technology I. Sci Sinter 40:79–88
50. Supriadi S et al (2023) Review of metal injection moulding technology and its development in Indonesia. AIP Conf Proc 2837(1):060006
51. Moon AP et al (2021) Opportunity and challenges of iron powders for metal injection molding. ISIJ Int 61(7):2015–2033
52. Kong Y et al (2025) Research on intelligent analogy design method of cylindrical gear metal powder injection molding process based on knowledge-driven. Int J Adv Manuf Technol 136(3):1681–1702
53. Kosugi T, Terada H (2025) Development of mass production technology for hollow parts by MIM process. J Jpn Soc Powder Powder Metall 72(Suppl):S647–S650
54. Huang W-T et al (2021) Application of intelligent modeling method to optimize the multiple quality characteristics of the injection molding process of automobile lock parts. Polymers (Basel) 13(15):2515

Chapter 5
Products and Applications of Powder Metallurgy

Abstract A distinctive combination of microstructural control and functional integration underpins the widespread adoption of powder-based processing routes across tribological, magnetic, energy, biomedical, and transportation applications. In mechanical components such as gears and bearings, wear resistance and fatigue behavior are governed by controlled porosity, phase distribution, and interfacial chemistry, with pore morphology and surface treatments defining crack initiation thresholds. Self-lubricating systems exploit synergistic combinations of solid lubricants and reinforcements to form stable tribofilms that suppress friction under severe contact conditions. Magnetic performance is regulated by density, residual stress, and grain-scale domain control, with advanced sintering enabling low coercivity and reduced losses through refined microstructures. Energy and biomedical functionalities arise from diffusion-controlled architectures that enhance transport kinetics, surface bioactivity, and corrosion resistance, revealing a consistent link between densification pathways, defect mitigation, and application-specific performance.

Keywords Microstructural control · Functional integration · Controlled porosity · Application-specific performance

5.1 Introduction

The production of structural, functional, and energy-related components is supported by a technologically mature manufacturing route in which controlled porosity, phase constitution, and microstructural refinement govern performance under mechanical, thermal, magnetic, and electrochemical loading [1–5]. The chapter describes powder metallurgy (PM) not as a mere forming technology but as a microstructure engineering platform that links particle scale processing to macroscopic service behavior in bearings, gears, magnetic systems, energy devices, and multifunctional materials [6–9]. Across these applications, the intrinsic presence of porosity and the ability to tailor it through composition, sintering, and post-processing establish both the main opportunity and the primary constraint governing fatigue, wear, conductivity, corrosion, and functional stability [6, 10–17].

R. Pérez-Bustamante, *Fundamental, Practical and Disruptive Applications of Powder Metals*, Topics in Mining, Metallurgy and Materials Engineering,
https://doi.org/10.1007/978-3-032-27130-3_5

Current practice indicates that powder metallurgy components reach performance levels comparable to wrought or cast materials when porosity morphology, phase distribution, and residual stress states are controlled. In Fe-based gear systems, crack initiation and propagation are governed by pore shape, size, and spatial distribution, while high density PM and surface densification strategies markedly increase fatigue strength by transforming irregular pore networks into isolated, rounded defects that reduce stress concentration [9, 18, 19]. For tribological systems such as self-lubricating bearings and brake pads, PM enables the incorporation of lubricants, ceramics, and intermetallic phases into a designed microstructure that balances hardness, friction, and wear across different sliding regimes. Similarly, in soft and hard magnetic materials, insulating coatings, nanocrystalline structures, and advanced sintering routes such as spark plasma sintering regulate domain wall motion, eddy current losses, and coercivity by controlling interparticle boundaries, residual stresses, and crystallite size [2, 18, 20].

Interfaces and pores exhibit a dual function, enabling lubricant retention, gas transport, and magnetic insulation while simultaneously acting as stress concentrators, degradation pathways, and pinning sites for cracks or magnetic domains. This tension is particularly evident in energy systems, where PM-derived porosity supports hydrogen diffusion and ionic transport in SOFC supports and hydride materials, while excessive pore connectivity compromises mechanical and electrical stability [2, 20, 21–23]. The same conflict arises in corrosion and oxidation, where densification and phase homogeneity improve passive film continuity but reduce functional permeability. The chapter examines how compositional design, nanostructuring, thermomechanical processing, and advanced sintering can be combined to regulate diffusion pathways, defect populations, and phase equilibria across a broad spectrum of powder metallurgy products.

5.2 Applications and Material Performance Analysis: Bearings, Cutting Tools, Gear Systems, and Magnetic Materials

5.2.1 Microstructural Tailoring for Wear and Fatigue Resistance

Powder metallurgy (PM) components are inherently characterized by the presence of porosity, which significantly influences their performance in demanding applications such as gears and braking systems. Microstructural tailoring through composition control and post-sintering treatments is therefore critical for enhancing wear and fatigue resistance. Lu et al. [18] carried out research related to the contact fatigue performance of Fe-based small-module gears fabricated using PM. Subsurface cracks, predominantly originating from pores, were identified as the primary failure mechanism leading to spalling during rolling contact fatigue tests. The experimental methodology involved fabricating gears with a specific chemical composition, primarily iron with additions of nickel, copper, and molybdenum, and subjecting them to fatigue testing under grease lubrication. The researchers employed advanced characterization techniques, including scanning electron microscopy (SEM) and electron backscatter diffraction (EBSD), to analyze the failure mechanisms. They found that microcracks tended to initiate at the sharp corners of irregularly shaped pores and propagate between adjacent pores, ultimately forming a connected network that led to material spalling. The EBSD analysis further indicated an increase in local misorientation and dislocation density around these pores after fatigue loading, highlighting the role of localized plastic deformation and stress concentration in crack initiation [18].

To mitigate the detrimental effects of porosity and enhance surface-dominated properties like wear resistance, surface engineering techniques such as boriding are effectively employed. Kayali and Yönetken [16] explored the wear behavior of borided materials produced by PM with varying chromium content. Their research demonstrated that the pack boriding process, conducted at 950 °C for 4 h, resulted in the formation of extremely hard boride layers, such as FeB, Fe2B, and CrB, on the surface of the PM components. The microstructural characterization showed that the morphology of the boride layer transformed from a sawtooth structure in the chromium-free sample to a flatter interface as the chromium content increased, which was attributed to the hindering effect of alloying elements on boron diffusion. The boriding process led to a dramatic increase in surface hardness, from a base material range of 47.5–240.6 HV0.05 up to a boride layer hardness ranging from 1641 to 1950 HV0.05. This substantial hardening directly translated into improved wear resistance, as evidenced by pin on disc wear tests. After boriding, a reduction in the friction coefficient was observed, accompanied by an increase in wear resistance of up to 25 times in the best-performing sample, which exhibited a predominantly single-phase Fe_2B layer. The adhesion of the boride layer, evaluated using the Daimler Benz Rockwell C test, was found to be acceptable, though the presence of multiple hard and brittle phases like FeB could lead to microcracking and delamination under mechanical stress, thereby reducing wear performance in some compositions (Fig. 5.1).

The performance of PM components is also critically dependent on their bulk composition and the optimization of constituent phases for specific tribological conditions. This is particularly evident in the context of PM brake pads for high speed trains [15]. The research details the research status of both copper-based and iron-based brake pad materials. For copper-based systems, which are more commonly used, the microstructure is tailored by incorporating various friction components like SiO_2, Al_2O_3, and ZrO_2, and lubricating components such as graphite and lead. The incorporation of ceramic fibers or particles has been shown to enhance properties such as thermal resistance and matrix strength, while potentially introducing challenges including increased porosity or damage to the brake disc if hard particles become detached. The wear mechanism was found to evolve with braking conditions; at low speeds, abrasive wear was dominant, while at high speeds, oxidative wear became prevalent due to the increased surface temperature that

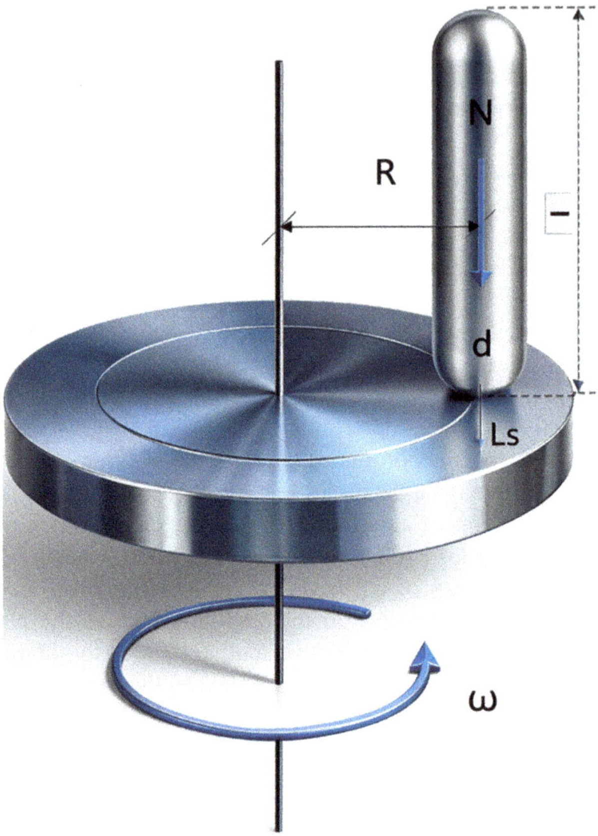

Fig. 5.1 Pin on disc tribometer motion diagram [24]

softens the copper matrix and promotes oxide film formation. The influence of braking pressure was also significant, with lower pressures resulting in adhesive wear and higher pressures leading to the formation and oxidative regeneration of a third-body layer on the friction surface, which affected both the friction coefficient and wear rate. Microstructural design plays a key role in balancing the need for a high and stable friction coefficient with the requirements of low wear and minimal damage to the counterpart material.

5.2.2 Tribological Performance of Powder Metallurgy Bearings Under Extreme Conditions

PM is a versatile manufacturing process particularly suited for producing self-lubricating bearings, which are designed to operate reliably under extreme conditions where continuous external lubrication is not feasible (see Cu bearings illustrated in Fig. 5.2). The capability of PM to generate controlled porosity within

Fig. 5.2 Copper bearings produced by PM

the material microstructure offers a clear advantage for bearing applications, as the interconnected pore network can be impregnated with lubricating oil to provide a self-sustaining lubrication mechanism during operation [25]. Research into copper-based and iron-based composites has demonstrated how specific material compositions and reinforcements can be engineered to optimize tribological performance under demanding service environments.

The influence of graphite content on the mechanical and tribological behavior of Cu-Sn-Zn/graphite metal matrix composites produced by PM has been examined in the context of bearing applications. The researchers fabricated composites with a bronze matrix of 88 wt% Cu, 9 wt% Sn, and 3 wt% Zn, with graphite additions of 0.5, 1.0, 1.5, and 2.0 wt%. The results revealed a nonlinear relationship between graphite content and performance. At 1.0 wt% graphite, the composite exhibited optimal properties, including a hardness of 55 BHN, a wear rate of 1.88×10^{-9} mm^3/mm, and a compressive strength of 538 MPa. This enhancement was attributed to liquid phase sintering, where Sn and Zn acted as coupling agents, improving the wetting between the copper matrix and graphite particles, filling pores, and forming strong interfacial bonds. However, when the graphite content exceeded 1.0 wt%, the mechanical properties deteriorated significantly. The hardness dropped to 30 BHN at

2.0 wt% graphite, and the wear rate increased, which was linked to insufficient wetting by the liquid phase, leading to weak interfaces and increased porosity.

Expanding the material scope, research on iron-based self-lubricating composites reinforced with nano zirconia (ZrO_2) and molybdenum disulfide (MoS_2) demonstrates an alternative approach for enhancing performance under extreme tribological contact [8]. The composites were based on an Fe 5Cu 3Sn system, with additions of 1 and 2 wt% nano ZrO_2 and a constant 0.5 wt% MoS_2. The incorporation of these reinforcements led to the formation of hard intermetallic phases such as $ZrFe_2$ and $FeMo_2S_4$ during sintering, which significantly improved the microhardness, reaching a maximum of 2.12 GPa for the composite with 2 wt% ZrO_2. Tribological tests using a ball on disc tribometer revealed that the composite with 2 wt% ZrO_2 and 0.5 wt% MoS_2 achieved an exceptionally low coefficient of friction of 0.128 and a low wear rate. The MoS_2 solid lubricant was smeared during sliding to form a protective tribolayer, reducing friction and wear. Simultaneously, the nano ZrO_2 particles contributed to a mending effect, where they filled surface grooves and generated a polishing effect, thereby reducing surface roughness and enhancing wear resistance. This synergy between a hard ceramic reinforcement and a solid lubricant presents a viable strategy for developing iron-based bearing materials capable of withstanding high loads and reducing friction in extreme environments.

While the aforementioned studies focus on bearing-specific alloys, research into PM Ti alloys for biomedical applications provides additional insights into the microstructural control achievable through PM, which is directly relevant to bearing performance [7]. Although the primary focus of this work was on mechanical properties and antibacterial characteristics for implants, the processing structure property relationships are analogous. For Ti-Al-Cu alloys, sintering conditions and post-processing steps, such as hot forging, were shown to markedly reduce porosity and promote refinement of microstructural features, including prior β-Ti grain size and $\alpha + \beta$ lamellar structures. In particular, forging increased the modulus of Young of the sintered alloys from an average of 104 to 118 GPa due to porosity reduction. This principle of enhancing density and refining microstructure through thermomechanical processing is directly transferable to bearing materials. A denser microstructure with fewer and smaller pores improves load-bearing capacity and fatigue resistance, which are essential for bearings operating under high stress and cyclic loading conditions. The ability of PM to be combined with subsequent deformation processes offers a pathway to further tailor the properties of self-lubricating bearings for extreme applications.

5.2.3 Thermomechanical Behavior of Powder Metallurgy Materials

Thermomechanical fatigue (TMF) behavior in PM superalloys is of particular relevance given the demanding service conditions experienced by components such as turbine disks in aeroengines. In this context, Shen et al. examined thermomechanical fatigue behavior and lifetime modeling in a PM superalloy, with emphasis on the effect of phase angle between mechanical strain and temperature cycling [13]. Thermomechanical fatigue tests were conducted under a range of mechanical strain amplitudes and phase angles, including in-phase and out-of-phase conditions, as well as intermediate phase angles of 45°, 90°, and 135°. The results demonstrated a significant dependency of TMF life on the phase angle, with the tensile mean temperature identified as a primary factor influencing life for these alternative phase angles. A modified energy-based life prediction model was proposed, which incorporated the effects of phase angle and tensile mean temperature, showing improved accuracy over traditional models. In addition, machine learning approaches, including support vector machine and random forest regression models, were applied for fatigue life prediction. The random forest model, combined with the modified energy model, was subsequently used to reliably predict TMF lifetime for non-tested phase angles, demonstrating the potential of data-driven approaches for complex fatigue load conditions.

Zhang et al. [14] examined thermomechanical fatigue crack growth behavior in the third-generation nickel-based PM superalloy FGH4099, considering both coarse-grained and fine-grained microstructures under in-phase and out-of-phase TMF loading. Crack propagation mechanisms were found to be strongly dependent on the applied phase angle. Under out-of-phase conditions, crack growth was predominantly transgranular, whereas in-phase loading led to intergranular crack propagation at low stress intensity factor ranges, transitioning to transgranular growth after passing a specific transition region. The fine-grained microstructure exhibited a lower crack growth resistance compared to the coarse-grained material at elevated temperatures, which was attributed to factors such as increased oxidation along the more abundant grain boundaries. Advanced characterization techniques, including SEM and EBSD, were employed to analyze crack paths and local misorientations. It was found that crystallographic slip controls the transgranular propagation, while the presence of twin boundaries and other special grain boundaries significantly influences crack deflection. A competition mechanism between crystallographic slip and grain boundary effects was identified, which helps explain the lower crack growth rates observed under out-of-phase TMF compared to in-phase or isothermal fatigue conditions.

The TMF behavior under hot deformation conditions is another vital aspect for PM materials, particularly for near-alpha titanium alloys used in high temperature applications. Piao et al. [26] characterized the hot deformation behavior of a near-alpha titanium alloy, Ti 1100, prepared via a TiH_2-based PM process. Through isothermal compression tests conducted across a temperature range of 1123–1323 K and strain rates from 0.01 to 1 s^{-1}, the study analyzed flow stress behavior, strain hardening exponents, strain rate sensitivity, and developed processing maps. The flow stress curves exhibited different behaviors, with softening responses dominant at lower strain rates and hardening effects observed at higher strain rates and temperatures. The strain hardening exponent was found to be negative at lower temperatures, indicating flow softening, and turned positive at higher temperatures, suggesting work hardening. The processing maps identified optimal processing windows at high temperatures of 1273–1323 K and low strain rates of 0.01–0.165 s^{-1}, where power dissipation efficiency was high. Instability domains, characterized by localized plastic flow and adiabatic shear bands, were identified at lower temperatures and higher strain rates. Microstructural observations confirmed that dynamic recrystallization was the main softening mechanism at lower temperatures, while dynamic globularization of the alpha phase and phase transformation governed softening at higher temperatures.

5.2.4 Load Transmission and Microcrack Initiation in Powder Metallurgy Gear Systems

The transmission of mechanical loads through PM gear systems is fundamentally influenced by the presence of porosity, which acts as a primary site for stress concentration and microcrack initiation. Under cyclic loading conditions, such as those experienced in gear operation, cracks consistently initiate at the edges of pores due to the elevated local stress states that exceed the nominal applied stress [27]. This behavior was relevant to gear applications, in which tooth root bending and flank contact pressures produced complex stress fields. A crossed cylinder cyclic sliding test configuration was employed to examine crack initiation and early propagation in case-hardened sintered steels. A characteristic load threshold for crack initiation was identified, corresponding to a maximum Hertzian contact pressure of approximately 4300 MPa, below which cracks with lengths of 10–20 μm developed after about 4000 cycles. EBSD analysis revealed that these microcracks initially propagated along the plane of the local maximum shear stress. A key finding was the strong tendency for cracks to follow prior austenite grain boundaries when present within a certain angular range from the initiation site, indicating that the intergranular path offers lower resistance to crack growth compared to transgranular propagation through the martensitic matrix.

The detrimental effect of porosity on load-bearing capacity is further corroborated by research on high density, high performance powder metallurgy (HPM) gears (Fig. 5.3). Lin et al. [11] conducted fatigue tests on spur gears, comparing traditional PM, HPM, and wrought forged gears. A direct correlation between material density and fatigue performance was observed. Traditional PM gears with a density of 7.15 g/cm^3 exhibited a fatigue strength of 304.8 MPa, whereas HPM gears, achieving a density of 7.62 g/cm^3, showed a 129.4% increase in fatigue strength to 699.1 MPa [11]. Finite element analysis (FEA) confirmed that the maximum stress under load occurred at the tooth root, aligning with the observed fracture locations. Microstructural analysis provided insight into the failure mechanisms; traditional PM samples contained numerous irregular, sharp-edged pores that served as potent initiation sites for microcracks. In contrast, the HPM process resulted in smaller,

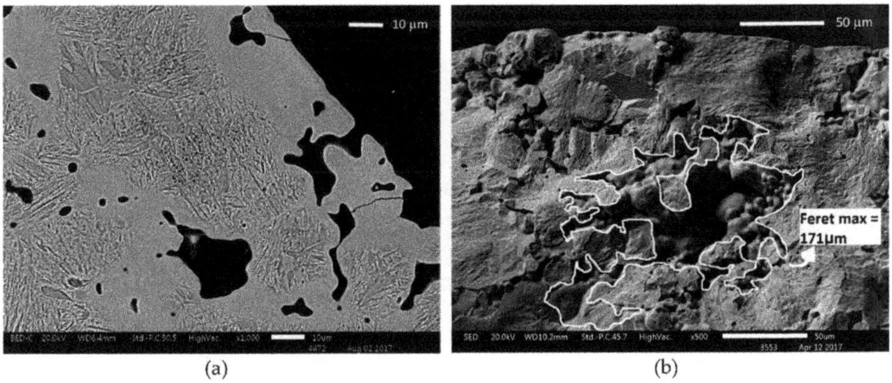

Fig. 5.3 SEM images showing (**a**) a layered surface with small cracks and pores and (**b**) a rough surface containing a large irregular pore [10]

more spherical, and isolated pores, which significantly reduced stress concentration and slowed the rate at which microcracks could link to form a critical crack.

The interaction between surface treatment and microcrack behavior represents a studied aspect of load transmission in PM gears. Scholzen et al. [9] investigated the influence of case hardening depth (CHD) and surface densification on the load capacity of sintered gears manufactured from Astaloy Mo85. Both experimental observations and simulation results indicated that increasing the case hardening depth had a predominantly positive effect on the load-bearing capacity of both the tooth root and flank. A higher CHD led to the introduction of deeper compressive residual stress profiles beneath the surface. These compressive stresses are beneficial as they superimpose with the applied tensile stresses from operational loads, effectively reducing the net stress intensity at critical locations and thereby retarding crack initiation and propagation. Computed tomography (CT) examinations of tested gears showed that for variants with a medium CHD, cracks initiated from surface pitting and propagated as a main crack behind the fully densified zone toward the tooth tip. However, in variants with a high CHD, no such main crack was identified, which was attributed to the more favorable stress state provided by the deeper compressive residual stress field.

5.2.5 *Magnetic Domain Engineering in Powder Metallurgy Soft and Hard Magnets*

PM serves as an important manufacturing route for advanced magnetic materials as illustrated in Fig. 5.4, as it enables precise control over microstructure and magnetic properties that are difficult to achieve using conventional casting methods. This control is essential for magnetic domain engineering, which directly influences the performance of both soft and hard magnets. In the context of soft magnetic

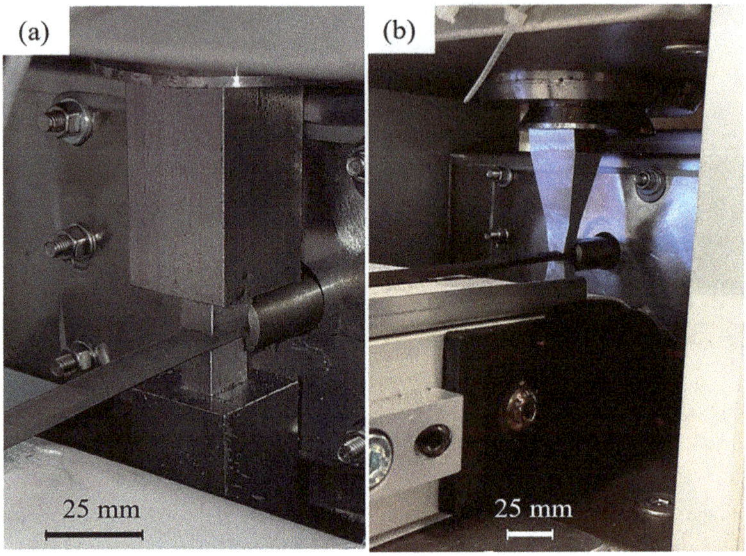

Fig. 5.4 Production of anisotropic NdFeB permanent magnets with in situ magnetic particle alignment using powder extrusion: (**a**) extrusion process with an alignment tool and (**b**) extrusion process with a different alignment tool [28]

composites (SMCs), the primary objective is to minimize energy losses, particularly eddy current losses at high frequencies. Fe-based amorphous soft magnetic composites (ASMCs) are prominent in this area due to their high electrical resistivity, which inherently reduces intraparticle eddy current losses. One effective approach to reducing interparticle eddy current losses involves applying a uniform insulating coating to the surfaces of amorphous powder particles. Research by Lu et al. [19] extensively reviews various coating systems, including organic materials like epoxy resin (EP) and poly para xylylene (PPX), inorganic materials such as phosphates, SiO_2, and TiO_2, and magnetic ferrites like Mn-Zn and Ni-Zn types. While nonmagnetic coatings increase electrical resistivity, they also lead to magnetic dilution, resulting in a marked reduction in the permeability of the composite core. This demagnetizing effect has been interpreted using simulations based on the Eshelby equivalent inclusion method and Mori-Tanaka theory, which show that the presence of a nonmagnetic matrix leads to a substantial reduction in effective permeability. Consequently, the use of high resistivity magnetic ferrites as an insulating layer is identified as a superior approach, as it effectively reduces core loss without severely compromising magnetic permeability, thereby favorably influencing the magnetic domain wall motion [19].

In PM, the consolidation process strongly influences the final density and residual stress state of the magnet, both of which affect the control and stability of magnetic domains. Conventional cold pressing, while efficient for mass production, often results in relatively low densities and introduces residual stresses that act as pinning sites for magnetic domain walls, increasing coercivity (Hc) and hysteresis loss. To address this, advanced sintering techniques are employed. Spark plasma sintering (SPS) is particularly notable for its ability to produce high density bulk amorphous/nanocrystalline composites with minimal porosity in a short time, thus inhibiting excessive grain growth and crystallization. SPS was used to fabricate FeSiBP/NiZn ferrite bulk amorphous soft magnetic composites with a density of 6.97 g cm^{-3}, with local electrical discharge and plasma effects, enhanced by nanoparticle addition, promoting densification and formation of a three-dimensional microcellular metal-ceramic interface. This dense microstructure with low porosity is crucial for facilitating unrestricted magnetic domain wall movement, leading to lower coercivity and improved soft magnetic properties. Furthermore, annealing is a critical post-processing step for stress relief. For instance, annealing FeSiBCr ASMCs at an appropriate temperature reduced Hc from 520 to 39.6 A/m and increased effective permeability, directly linked to the reduction of internal stresses that pin magnetic domains [18].

In the realm of high entropy alloys (HEAs) produced via PM, magnetic domain engineering is closely tied to phase formation and elemental distribution. In an equiatomic FeCoNiSi-based HEA with additions of Cu, Mn, and Ti, the relationship between phase structure and magnetic behavior has been examined, showing a strong dependence of magnetic response on phase constitution [2]. The FeCoNiSi-Cu HEA, consisting primarily of a face-centered cubic (FCC) phase, exhibited the highest saturation magnetization (Ms) of 59.039 emu/g among the studied alloys. This was attributed to the presence of a phase rich in ferromagnetic elements (Fe, Co, Ni) and the isolation of nonferromagnetic Si at the grain boundaries due to the low miscibility between Cu and Si. This microstructure minimizes the pinning effect on magnetic domains, allowing for easier domain wall motion. In contrast, the FeCoNiSi-Mn and FeCoNiSi-Ti HEAs showed nearly negligible magnetization saturation. For the Mn-containing alloy, this was explained by the failure of Mn to transform from an antiferromagnetic to a ferromagnetic state within the specific chemical environment of the HEA, which included Si. The presence of Si, with its zero magnetic moment, was found to act as a pinning site, decreasing the correlation between the FeCoNi matrix and the antiferromagnetic Mn atoms, thereby disrupting long-range magnetic order and domain alignment.

The microstructural characteristics at the nanoscale, such as crystallite size and lattice strain, are fundamental to magnetic domain engineering in nanostructured magnets. Research by Ben Ammar et al. [20] on the mechanically alloyed $Fe_{30}Co_{20}Ni_{20}Mn_{20}Ti_{10}$ HEA revealed that the alloy evolved into a two-phase system consisting of a body-centered cubic (BCC) supersaturated solid solution and an HCP-Fe_2Ti intermetallic after extended milling. The final average crystallite size was refined to approximately 10 nm. According to the random anisotropy model, when the crystallite size (D) is smaller than the ferromagnetic exchange length (Lex), the effective magnetocrystalline anisotropy is averaged out, leading to superior soft magnetic properties. This nanocrystalline structure resulted in a low coercivity of 25 A/m, characteristic of soft magnetic behavior. The low saturation magnetization of 28 emu/g was attributed to the presence of paramagnetic Ti atoms in solid solution and a high density of crystal defects and interfaces, which disrupt magnetic exchange coupling and act as pinning sites for magnetic domains, limiting the overall magnetic moment.

5.3 Applications in Energy

5.3.1 *Diffusion Pathways and Ionic Conductivity in Solid Oxide Fuel Cell Powder Metallurgy Components*

Research on diffusion pathways and ionic conductivity in solid oxide fuel cell (SOFC) components produced by PM has emphasized the influence of microstructural features, oxide scale development, and compositional adjustments on the efficiency of charge and mass transport. Metallic supports and interconnects prepared via PM methods exhibit complex diffusion mechanisms, largely governed by porosity, grain boundary characteristics, and oxide layer evolution during operation. In the research conducted by Al-Kattan et al. [22], the synthesis of a porous metallic alloy for SOFC applications through the reduction of an oxide precursor demonstrated that ionic and gas diffusion was enhanced by the formation of an interconnected porosity network of approximately 48% after reduction at 800 °C. This architecture enabled gas diffusion through the support while maintaining sufficient electrical conductivity. The porous Fe-Co-Ni alloy obtained through low temperature reduction provided a suitable pathway for hydrogen diffusion and oxygen ion transport, which are essential for electrochemical performance in third-generation SOFCs. The use of oxide precursors enabled improved control over reduction kinetics, promoting uniform metal formation and the development of continuous conductive pathways while limiting excessive grain coarsening. The presence of mixed spinel and corundum phases in the precursor also contributed to heterogeneous reduction stages, influencing local diffusion rates and eventual pore connectivity critical for ionic conduction.

Complementary investigations by Öztürk et al. [29] on Crofer 22 interconnects produced via PM techniques confirmed that diffusion and conductivity behavior are tightly linked to oxidation phenomena and the resulting oxide scale microstructure. During prolonged oxidation at 800 °C, the development of a dual-layer oxide structure, an inner chromia (Cr_2O_3) layer and an outer spinel $(Fe, Cr, Mn)_3O_4$, acted as a diffusion barrier that limited chromium outward migration and oxygen inward diffusion. This structure stabilized the electrical properties by reducing interfacial resistance growth over time. Area-specific resistance (ASR) values decreased initially and stabilized at approximately 10 $m\Omega \cdot cm^2$ after 525 h, demonstrating efficient electron transport and limited ionic resistance through the oxide scale. The PM samples showed a slightly higher initial ASR due to microstructural inhomogeneities and pore distribution, but these differences diminished after extended exposure, as oxide growth filled surface pores and refined diffusion pathways.

Further insight into diffusion and ionic transport in PM-based ferritic stainless steels was provided by Liu et al. [23], who investigated the influence of chromium content on mechanical, oxidation, and electrical properties of PM ferritic stainless steels used as SOFC interconnects. Increasing chromium content from 17 to 26 wt.% led to a systematic reduction in oxidation mass gain and ASR, indicating enhanced diffusion resistance of the oxide scales. The high chromium content promoted the formation of continuous, adherent Cr_2O_3 films that effectively inhibited cation and anion diffusion, stabilizing electrical performance over 300 h at 800 °C. Despite a minor reduction in ductility, the densified oxide barrier improved ionic selectivity by minimizing mixed conduction. These Cr-enriched oxides exhibited fine-grained morphologies, contributing to short circuit diffusion suppression while maintaining sufficient electronic pathways for charge transport. Compared to the reference ZMG232 alloy, the PM Fe-Cr22 and Fe-Cr26 compositions showed comparable ASR values and diffusion behavior, confirming that optimized PM alloys can replicate the diffusion control mechanisms found in conventionally processed materials.

5.3.2 Hydrogen Storage in Powder Metallurgy Alloys: Thermodynamics and Kinetics

The thermodynamic and kinetic behavior of hydrogen storage in PM alloys reveals significant advancements across multiple alloy systems, particularly those involving titanium hydrides, magnesium-based composites, and metal organic frameworks (MOFs). In titanium hydride systems, the influence of hydrogen content on sintering behavior and densification is essential for understanding their hydrogen storage capability. Mei et al. [30] demonstrated that both saturated (TiH2) and unsaturated (TiH1.5) hydrides undergo dehydrogenation between 400 and 850 °C, accompanied by a self-purification effect arising from the reaction between released hydrogen and surface oxides, leading to the formation of H_2O and the removal of surface impurities. This self-cleaning mechanism enhances sintering density and mechanical integrity, allowing TiH2 compacts to achieve the highest densification and compressive strength after sintering at 1300 °C compared to pure titanium compacts. The densification progresses through three distinct stages, slow, rapid, and full, corresponding to temperature increases and holding times, with the hydrogen-rich samples consistently showing superior performance due to the role of hydrogen in activating diffusion and eliminating oxide barriers that typically inhibit particle bonding [30].

Related results for magnesium-based alloys indicate that the incorporation of selected intermetallic phases and carbonaceous additives leads to a marked improvement in hydrogen sorption thermodynamics and kinetics. Ding et al. [31] synthesized an Mg-$MgCNi_3$ composite via PM, observing that the inclusion of $MgCNi_3$ reduced the onset dehydrogenation temperature from 651 K for milled MgH_2 to

520 K, thus enhancing hydrogen release kinetics. The reaction pathway demonstrated that $MgCNi_3$ actively participated in the formation of Mg_2NiH_4 and carbon species during hydriding–dehydriding cycles, where Mg_2NiH_4 facilitated dehydrogenation of MgH_2 and the carbon phase served as a dispersive matrix preventing particle agglomeration. The composite exhibited a reversible hydrogen capacity of 4.81 wt% within 20 min at 593 K and retained 98% of its capacity after 10 cycles, showing superior thermodynamic stability and structural resilience. The synergistic catalytic effects of Mg_2NiH_4 and carbon significantly lowered activation energy barriers for hydrogen desorption, as confirmed by differential scanning calorimetry (DSC) and Kissinger kinetic analyses.

Meanwhile, the integration of metal organic frameworks (MOFs) in hydrogen storage demonstrates a different yet complementary strategy that emphasizes structural design and morphological control rather than alloy modification. Ren et al. [32] advanced the engineering of zirconium-based MOFs (Zr-MOFs) through a centrifugal granulation technique that efficiently shaped MOF powders into spherical pellets (0.5–15 mm diameter) using sucrose as a binder. This method improved mechanical stability without compromising crystallinity, as confirmed by X-ray diffraction (XRD) and thermogravimetric analysis (TGA). The shaped pellets maintained integrity under repeated drop and abrasion tests, exhibiting less than 5% breakage after 60 min of tumbling at 25 rpm, indicating excellent mechanical robustness. Although the inclusion of sucrose reduced specific surface area and slightly diminished hydrogen storage capacity, the resulting macroporous interparticle network enhanced hydrogen diffusion and adsorption kinetics [32].

5.3.3 Frontiers in PM: High Entropy Alloys and Nanostructured Powders

Within the context of PM applied to HEA (Fig. 5.5), both refractory and transition metal systems have demonstrated significant potential for energy-related

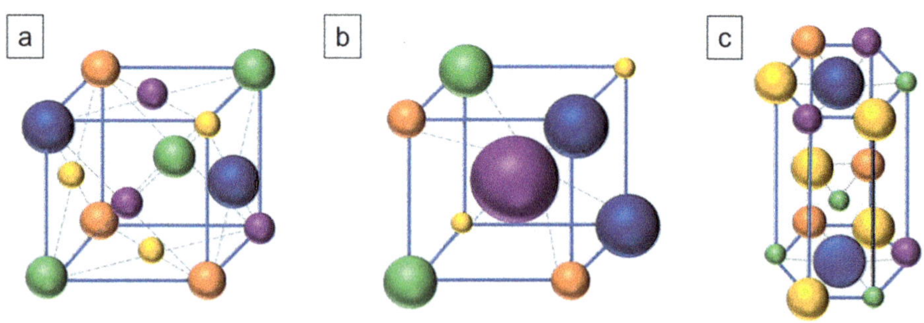

Fig. 5.5 Illustrations showing the lattice structure of (**a**) FCC, (**b**) BCC, and (**c**) HCP HEAs. Each color represents a different element [33]

applications, particularly when processed from nanostructured or mechanically alloyed powders. The development of these materials constitutes a key area in PM, driven by their ability to combine mechanical strength, thermal stability, and microstructural tunability under controlled solid-state synthesis conditions. The synthesis and optimization of refractory high entropy alloys (RHEAs) through mechanical alloying (MA) and advanced sintering have enabled the control of phases, grain refinement, and the reduction of defects, aspects that are crucial for applications such as solid oxide fuel cells, hydrogen storage, and high temperature structural systems.

PM methods provide unique advantages for the fabrication of RHEAs and nanostructured alloys. Zhang et al. [14] emphasized that the MA process allows for the formation of highly homogeneous nanocrystalline powders with atomic scale mixing among refractory elements such as W, Nb, Ta, Mo, and Ti, which is not achievable by melting techniques. The repeated cold welding and fracturing of particles in high energy ball milling enable the formation of solid solutions even among immiscible elements, leading to powders with grain sizes often below 20 nm. These nanosized structures enhance sintering activity, lower the densification temperature, and result in superior mechanical integrity in the consolidated materials. However, contamination by oxygen, carbon, and nitrogen during milling has been identified as a recurring issue, leading to the formation of TiC, TiO, or TiN precipitates in Ti-containing alloys after sintering, which in turn can act as strengthening phases but may also influence ductility and fracture behavior depending on their distribution and morphology.

Moravcikova Gouvea et al. [34] reported an extensive experimental investigation of the $Al_{0.3}NbTa_{0.8}Ti_{1.5}V_{0.2}Zr$ refractory high-entropy alloy, produced by MA and subsequently consolidated by SPS. Milling duration played a key role in balancing chemical homogeneity and contamination. A milling time of 10 h resulted in a favorable combination of microstructural uniformity, grain refinement, and limited carbide and oxide formation. The use of SPS at 1200 °C under 50 MPa pressure allowed the consolidation of dense materials with predominantly BCC_B2 phase structure, consistent with CALPHAD predictions. Prolonged milling times exceeding 30 h led to increased oxygen incorporation associated with wear of the steel milling media, indicating a strong dependence of the structural and chemical stability of the final alloy on processing parameters. The refinement achieved through controlled alloying not only improved hardness but also increased the elastic modulus, which is particularly advantageous for load-bearing and thermal stress-resistant components in energy devices.

The characterization of FeCoNiCr-based high entropy systems further extends the understanding of nanostructured PM alloys. Basem et al. [35] reported FeCoNiCrCux alloys processed via planetary ball milling and vacuum sintering, focusing on the effects of copper incorporation on phase constitution and functional properties. The introduction of copper led to a transition from a single FCC solid solution to a dual-phase FCC and Cu-rich FCC structure, modifying both mechanical and magnetic characteristics. The progressive substitution of copper refined the crystallite size while increasing lattice strain, which contributed to dislocation generation and enhanced microstructural uniformity. The copper addition also stabilized the FCC phase thermally and increased the Curie temperature, evidencing the role of alloying in tuning the functional response of PM-based HEAs for energy storage and conversion systems.

The integration of advanced densification techniques, such as SPS and hot pressing, enables the production of high density materials with controlled defect populations. As reported by Zhang et al. [14], SPS provides rapid heating and uniform current distribution, which limit grain coarsening and reduce the tendency for undesired phase segregation. Compared to conventional sintering, this approach ensures higher densification efficiency and allows the preservation of nanostructured features. The consolidated bulk alloys typically reach relative densities above 98%, with minimal residual porosity and largely isotropic mechanical properties. Controlled heating rates, as described by Moravcikova Gouvea et al. [34], help to limit excessive carbide precipitation and preserve the equilibrium BCC_B2 microstructure predicted by thermodynamic modeling. This level of thermal control allows elastic constants and hardness to be adjusted without compromising toughness, which is relevant for components operating in hydrogen environments or fuel cell electrodes where mechanical and thermal loads coexist.

Moreover, the evolution from elemental powders to dense RHEA components illustrates how PM bridges the gap between design and functional performance. In MA systems such as WMoTaNbV or $Al_{0.3}NbTa_{0.8}Ti_{1.5}V_{0.2}Zr$, phase formation kinetics is dictated by diffusion and strain accumulation within the powder particles, resulting in defect-mediated solid solution formation. Subsequent SPS processing eliminates internal porosity while preserving the chemical uniformity achieved during milling. These processing steps lead to materials with improved oxidation resistance and retained mechanical performance at elevated temperatures, supporting their use in thermal energy conversion and catalytic systems. The integration of CALPHAD simulations with experimental phase mapping enables accurate prediction of microstructural evolution and establishes a clear connection between computational alloy design and PM optimization.

5.3.4 Refractory High Entropy Alloys

RHEAs (Fig. 5.6) synthesized through PM represent one of the most promising material systems for high temperature energy applications. The combination of MA and SPS has been shown to generate dense ultrafine-grained microstructures with high thermal stability, superior specific strength, and improved oxidation resistance when compared with alloys produced through conventional casting techniques. The synthesis pathway enables fine control over microstructural evolution and elemental homogeneity, which are crucial for ensuring reliable mechanical performance in extreme environments such as those present in solid oxide fuel cells or hydrogen storage systems [21].

The process begins with the MA of elemental refractory powders such as Mo, Nb, Ta, Ti, and V, producing solid-state solutions with uniform element distribution.

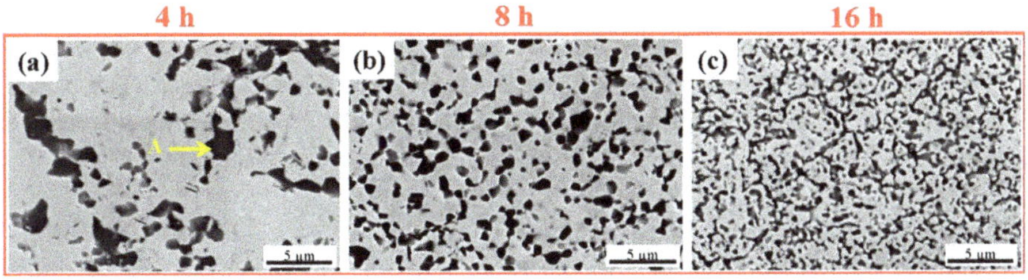

Fig. 5.6 SEM images of the sintered WVTaTi RHEA for different milling times: (**a**) 4 h, (**b**) 8 h, and (**c**) 16 h [36]

During milling, repetitive impacts between balls and particles induce severe plastic deformation, cold welding, and fracturing, which promote interdiffusion and eventually form nanocrystalline or amorphous precursor powders. The subsequent SPS step consolidates these powders at elevated temperatures under vacuum and pressure, creating fully dense bodies with ultrafine grains typically below 1 μm. This combination of techniques minimizes contamination, reduces compositional segregation, and avoids the dendritic structures commonly observed in cast refractory alloys. The resulting materials exhibit a body-centered cubic solid solution matrix often accompanied by stable oxide or carbide precipitates that pin the grain boundaries and enhance mechanical strength through dispersion hardening [37].

A representative example of these advances is the MoNbTaTiV system synthesized by MA and sintered at 1873 K, which yields an ultrafine-grained microstructure with an average grain size of approximately 0.6 μm and fine Ti-rich precipitates along the grain boundaries. Subsequent hot deformation experiments reveal that the deformation behavior is highly sensitive to temperature and strain rate. At low temperatures and high strain rates, deformation is dominated by dislocation motion accompanied by the segregation of Ta and Mo atoms at grain boundaries. As the deformation temperature increases or the strain rate decreases, grain boundary sliding becomes predominant, reducing dislocation density and eliminating elemental segregation. This transition between deformation mechanisms is essential for applications requiring plastic stability at elevated temperatures, such as thermal barrier or electrode materials in high-efficiency energy devices [37].

The influence of atomic segregation in these alloys is directly related to their energy storage and conversion performance. Elemental partitioning at grain boundaries can modify diffusion paths and interfacial cohesion, which affects creep resistance and ionic transport. In MoNbTaTiV alloys, Ta and Mo atoms with higher atomic weights accumulate near grain boundaries driven by dislocation movement, while lighter Ti and V atoms diffuse inward due to concentration gradients. This phenomenon, though initially generated during deformation, may be exploited to tailor grain boundary chemistry for specific functional behaviors such as enhanced grain boundary diffusion or controlled sliding for superplastic deformation. The dynamic recovery of segregation with increased temperature further suggests that thermomechanical treatments can be designed to control boundary mobility and therefore optimize microstructure stability for long-term operation [38].

The mechanical and thermal responses of these ultrafine-grained RHEAs are characterized by their exceptional stability under high temperature exposure. In particular, an $AlCr_{0.3}FeMoNbTiV_2$ alloy fabricated through the same PM route exhibited a body-centered cubic matrix with an average grain size of 0.32 μm and secondary phases of titanium carbide and Laves phases dispersed uniformly throughout the structure. Even after 16 h at 1350 °C, the alloy retained its phase constitution, with only moderate coarsening to about 1.5 μm. Compression tests at temperatures between 950 and 1100 °C indicated that the main deformation mechanism was grain boundary sliding, associated with an apparent activation energy of 527.7 kJ mol^{-1} and a stress exponent of approximately 2. This behavior is typical of ultrafine-grained systems where boundary-mediated processes dominate, and it confirms the capacity of such materials to maintain integrity under long-term high temperature loading. The capacity for stable grain boundary sliding suggests that these materials could accommodate the cyclic stresses expected in fuel cell interconnects or hydrogen diffusion membranes without premature failure [38].

Beyond their mechanical robustness, PM refractory HEA display advantageous thermophysical and chemical properties. The high melting points of constituent elements such as Mo and Ta confer thermal resilience, while the inclusion of lighter elements like Ti and V reduces overall density and improves the specific strength-to-weight ratio. The combination results in specific yield strengths exceeding those of conventional superalloys under similar high temperature conditions, reaching values above 150 MPa per gram per cubic centimeter at 1000 °C. Moreover, the homogeneous dispersion of fine oxides and carbides produced in situ during sintering not only stabilizes the microstructure but also improves oxidation resistance by forming protective surface layers. These features contribute to stable performance in oxidative and hydrogen-rich environments typical of advanced energy conversion systems [21].

Despite these advantages, several challenges remain in scaling the PM route for large component fabrication. The MA process is energy-intensive and sensitive to contamination from the milling media or process control agents. Even small levels of oxygen or carbon impurities can influence phase stability and mechanical response. Additionally, SPS limits part size due to tooling constraints and nonuniform current distribution. However, continuous efforts in optimizing milling parameters, employing inert atmospheres, and designing hybrid sintering techniques are addressing these limitations. Appropriate selection of milling time, rotational speed, and ball-to-powder ratio allows alloying efficiency to be balanced with contamination control, resulting in highly reproducible powder batches suitable for near-net-shape fabrication [21].

5.4 The Role of Powder Metallurgy in Consumer Electronics, Healthcare, and Transportation

5.4.1 Electrical and Thermal Conductivity of Powder Metallurgy Components in Miniaturized Devices

The electrical and thermal conductivity of PM components in miniaturized devices is determined by the complex interplay between composition, microstructure, and processing parameters, which together control the transport of charge and heat at the microscale, as illustrated in Fig. 5.7. Moustafa et al. [17] investigated aluminum-based composites reinforced with graphene and silver, fabricated via PM, emphasizing how the incorporation of metallic and carbonaceous phases affects the formation of conductive networks and phonon pathways. The introduction of silver into the aluminum matrix facilitated the formation of the Ag_2Al phase, enhancing densification and interfacial bonding, while graphene improved both mechanical and thermal properties through its exceptional intrinsic conductivity. The hybrid Al-5 vol.% Ag-2.5 vol.% G composition exhibited a thermal conductivity of 278.86 W/mK, corresponding to a 20.6% increase compared to pure aluminum, confirming the synergistic contribution of metallic and graphene fillers to electron and phonon transport. However, the electrical conductivity slightly decreased because of increased scattering at the Al-Ag_2Al interfaces and lattice distortion introduced by graphene sheets. The fine dispersion of nanoscale graphene and silver within the aluminum matrix created multiple conduction channels, facilitating rapid heat dissipation essential for compact heat sinks and microelectronic packaging applications where high thermal flux management is required.

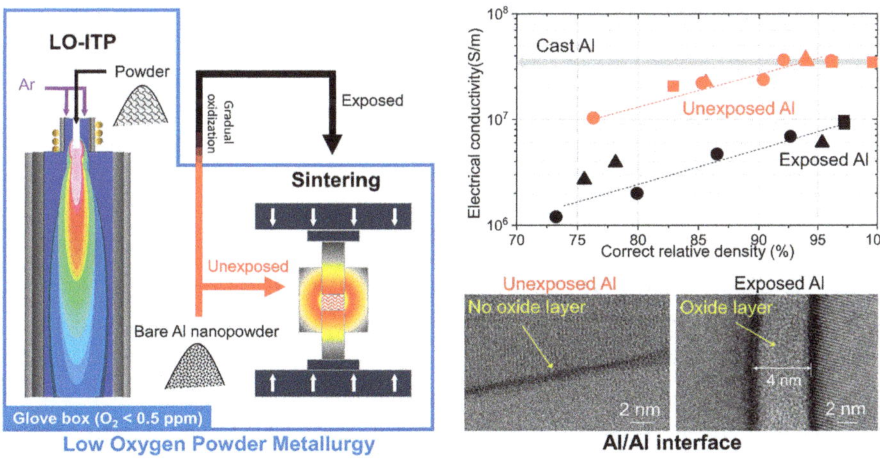

Fig. 5.7 Highly conductive Al/Al interfaces in ultrafine-grained Al compact prepared by low oxygen PM technique [39]

The same principles of interfacial control and network formation are critical in polymer-based composites designed for flexible electronic devices. Katheria et al. [6] fabricated ethyl methacrylate (EMA) polymer composites reinforced with hybrid fillers comprising MXene and Fe_3O_4 decorated g-C_3N_4 using a solvent mixing route. MXene, a two-dimensional carbide with metallic conductivity, provided highly efficient electron pathways, while Fe_3O_4-g-C_3N_4 contributed to magnetic and dielectric losses and improved interfacial thermal coupling. The resulting EMA/MXene/ Fe_3O_4-g-C_3N_4 composite displayed an electrical conductivity of 1.5×10^{-2} S/cm and a thermal conductivity of 0.98 W/mK, showing that a hierarchical filler design enhances charge transport and phonon propagation in polymer matrices. The presence of Fe_3O_4 nanoparticles improved interfacial polarization and stability, and g-C_3N_4 nanosheets promoted uniform filler dispersion, reducing contact resistance. These synergistic effects created percolative networks that lowered interfacial impedance and facilitated electron hopping between conductive domains. Moreover, the system achieved an electromagnetic interference (EMI) shielding effectiveness of −36 dB, confirming that electrical pathways were sufficiently continuous to support strong reflection and absorption of electromagnetic waves, a property directly linked to high electrical conductivity.

At a more structural scale, Nie et al. [40] employed SPS to optimize Cu/Invar bimetal matrix composites, emphasizing the balance between thermal conductivity, coefficient of thermal expansion (CTE), and mechanical integrity for electronic packaging. The composites combined the high conductivity of copper with the low CTE of the Fe-Ni-Co Invar alloy. The SPS technique, by applying pulsed current and high pressure, minimized diffusion at the Cu-Invar interface and achieved dense microstructures at reduced temperatures. Increasing copper content from 30 to 50 wt.% resulted in a continuous copper network, elevating the thermal conductivity to 90.7 W/mK but slightly reducing tensile strength due to the decreased Invar fraction. Conversely, raising the sintering temperature to 800 °C improved interfacial diffusion and mechanical bonding but reduced thermal conductivity because of enhanced Fe and Ni diffusion into Cu, which increased electron scattering. The optimized process (50 wt.% Cu, 700 °C, 50 MPa) yielded the best combination of thermal conductivity and mechanical stability, demonstrating that interface diffusion control is essential for maintaining high thermal transport in PM-based metallic composites.

5.4.2 *Biofunctional Surface Modification of Powder Metallurgy Implants*

Biofunctional surface modification of powder metallurgy implants has gained importance in enhancing osseointegration and in the development of antibacterial coatings for biomedical devices (Fig. 5.8). Titanium and stainless steel systems fabricated through PM have demonstrated unique advantages in tailoring surface

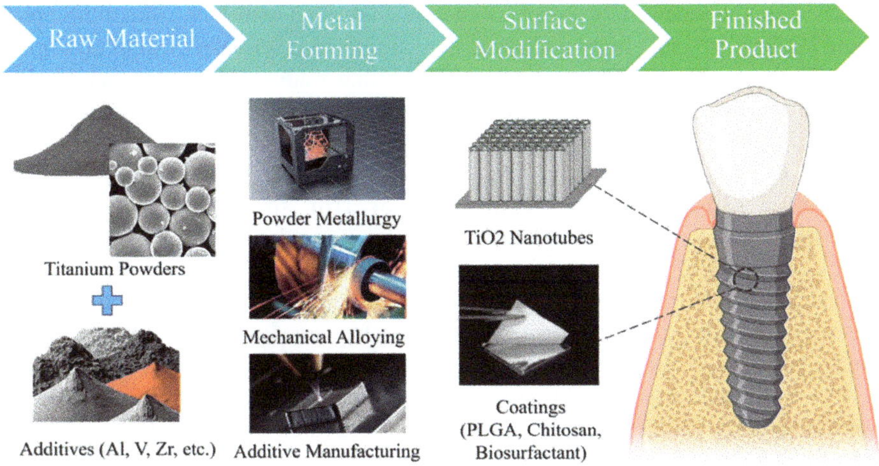

Fig. 5.8 Scheme demonstrating the processes of producing dental implants [41]

topography, porosity, and chemical functionality to promote bone integration and control bacterial adhesion. The combination of structural porosity and surface bioactivity defines the long-term clinical performance of orthopedic and dental implants, and the studies analyzed here demonstrate that PM routes allow precise control over these attributes [42].

Rodriguez Contreras et al. [42] emphasized that the success of porous titanium implants relies strongly on the morphological characteristics obtained through PM with space holder techniques, which enable the production of interconnected porosity ranging from 150 to 500 μm and micropores below 10 μm that enhance ion exchange and the nucleation of apatite phases. These features directly influence cell migration, nutrient transport, and osteogenic differentiation. The presence of open cell structures allows vascular infiltration and the anchoring of osteoblasts within the inner volume of the implant. Moreover, the surface roughness inherent to the PM space holder (SH) process provides a favorable environment for protein adsorption, which in turn promotes osteoblastic attachment and matrix mineralization. Optimization of compaction pressure and sintering parameters governs both mechanical strength and pore morphology, which together influence the stability of bone implant interfaces. By adjusting parameters such as binder type, sintering temperature, and space holder particle size, the final surface chemistry and microstructure can be tuned to emulate the stiffness and mechanical response of natural bone, thus reducing stress shielding while supporting biointegration.

Li et al. [43] expanded this approach by introducing magnesium into Ti-Nb alloys via mechanical alloying and spark plasma sintering, resulting in a Ti-Nb-Mg system with intrinsically bioactive surfaces. Magnesium acted not only as a β-phase stabilizer but also as an in situ agent for generating nanoscale pores and increased roughness upon partial degradation. This process resulted in hydrophilic surfaces and enhanced cell spreading, as verified through contact angle and fluorescence assays. The degradation of Mg enriched the surface with biocompatible ions, producing localized alkalinity and a transient antibacterial environment while simultaneously favoring osteoblast proliferation. Unlike traditional coating approaches that often suffer from poor adhesion or delamination, the homogeneous distribution of Mg within the PM matrix produced a self-regulated surface modification as Mg diffused or dissolved, eliminating the need for an external coating layer.

Ali et al. [3] examined the modification of 316L stainless steel through the addition of boron, titanium, and niobium during powder metallurgy processing, leading to improved mechanical integrity and corrosion resistance in physiological environments. The addition of niobium and titanium produced protective oxide layers on the alloy surface that acted as passive barriers against ion release, significantly enhancing biocompatibility. These oxides not only improved corrosion resistance but also provided active sites for cellular adhesion, a feature critical for osseointegration. Furthermore, boron promoted activated sintering and contributed to a refined microstructure, which improved hardness and compressive strength. The resulting surface morphology was characterized by a uniform passive film that limited bacterial colonization. Compared with conventional coatings, such integrated alloy modifications ensured higher durability and uniformity under cyclic physiological loading, showing the ability of PM routes to yield biomaterials with both mechanical reliability and functionalized surfaces.

5.4.3 *Crashworthiness and Fatigue Life of Powder Metallurgy Alloys in Transportation*

The improvement of crashworthiness and fatigue life in PM alloys has become a significant focus in the transportation sector. In this context, layered aluminum foams with graded cell sizes have been examined, showing characteristic macrostructural features across different orientations, including cell size variations and localized powder aggregation (Fig. 5.9). Studies on titanium, aluminum, and steel alloys produced through PM routes demonstrate that optimizing processing parameters, microstructural evolution, and heat treatment procedures can substantially enhance the mechanical performance of lightweight components subjected to cyclic loading and impact stresses. In titanium PM systems, Guo et al. [45] established a two-step hot isostatic pressing (HIP) approach for Ti55 near-alpha alloys that effectively promoted in situ globularization of lamellar α structures, achieving a remarkable fatigue strength of approximately 600 MPa, about 25% higher than conventionally HIPed counterparts. The improvement originated from the refinement and randomization of equiaxed α grains, which reduced the basal texture intensity and hindered slip localization during cyclic deformation. The introduction of a deformation stage in the low $\alpha + \beta$ region followed by recrystallization in the high $\alpha + \beta$ regime generated strain-induced dislocation networks that accelerated boundary splitting and dynamic recrystallization. Consequently, fatigue cracks in the

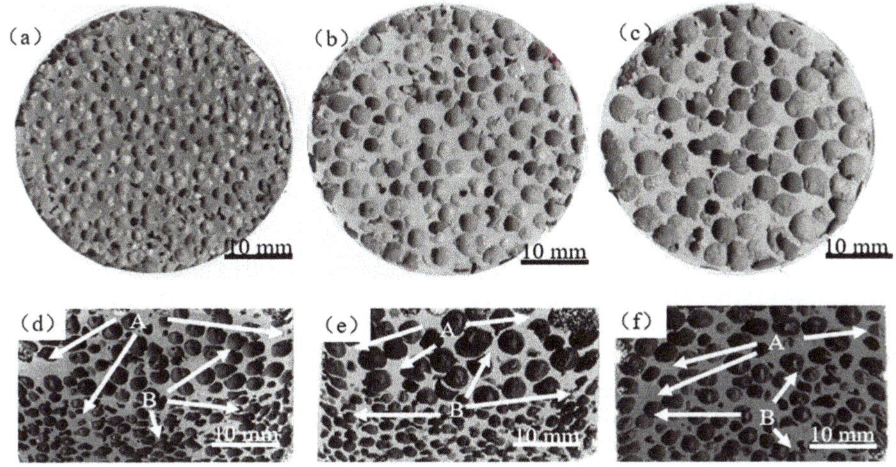

Fig. 5.9 Macrostructure of two-layered cell size gradient Al foams in both orthogonal directions; transverse direction: Al foams with a cell size of (**a**) 2 mm, (**b**) 3 mm, (**c**) 4 mm, longitudinal direction: Al foams with a cell size of (**d**) 3–2 mm, (**e**) 4–2 mm, (**f**) 4–3 mm, A, aggregation of aluminum powders, B, a weak connection of cells [44]

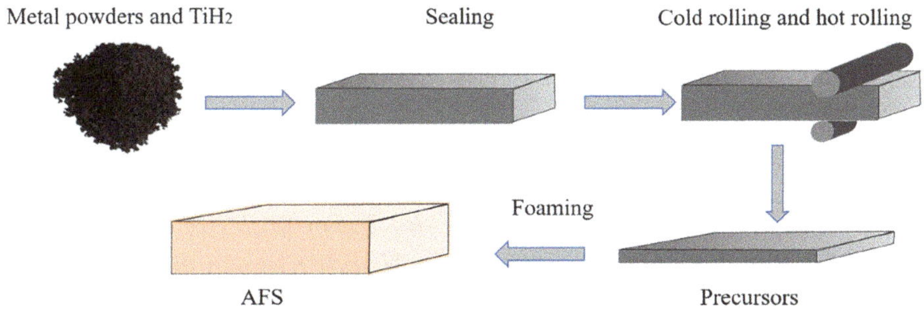

Fig. 5.10 Process flow chart of an AFS made by the PM method [12]

HIPed Ti55 propagated more slowly due to the higher resistance to transgranular fracture along the α lamellae, providing an advantageous balance between strength and endurance for aerospace transportation components.

Aluminum-based PM sandwich structures further exemplify the potential of tailored microstructural control for enhanced fatigue performance (Fig. 5.10). Aluminum foam sandwich panels produced by rolling composite powder metallurgy were compared with adhesive-bonded structures, with metallurgical bonding generating a uniform, defect-free interface. Diffusion of Si, Cu, and Mg between foam core and face sheets produced intermetallic phases such as Al_2Cu, Mg_2Si, and $Al_4Cu_2Mg_8Si_7$, contributing to interfacial cohesion [11, 45]. These metallic bonds increased interfacial load transfer capacity and prevented early delamination, leading to a distinct fatigue behavior dominated by shear failure rather than interface debonding as observed in adhesive joints. Under three-point bending fatigue at 15 Hz and stress ratio $R = 0.1$, the PM-processed AFS exhibited a markedly longer fatigue life and greater deformation stability. The cyclic deflection curves showed gradual stiffness degradation, while adhesive-based samples displayed rapid deflection and premature cracking. This superior performance was attributed to the metallurgical interface continuity, which distributed cyclic stresses evenly and reduced the likelihood of localized plasticity or stress concentrations. The powder metallurgy route increased interfacial bonding and improved crashworthiness, with integrated metallic cores and face sheets providing higher energy absorption under impact and cyclic loading conditions typical of transport applications.

Developments in powder metallurgy tool steels indicate that microstructural adjustment through heat treatment influences fatigue behavior in transport components. Morri et al. [46] analyzed Böhler K890 Microclean steel, a powder metallurgy cold work tool steel, and evaluated the effects of conventional and cryogenically assisted tempering treatments on fatigue resistance under unnotched and notched conditions. The addition of a cryogenic step at −80 °C between tempering stages promoted homogeneous precipitation of nanoscale secondary carbides within the martensitic matrix, increasing tensile strength by approximately 10% and elongation by 7%. These microstructural refinements minimized retained austenite and improved carbide dispersion uniformity, which led to an enhancement of about 15% in unnotched and 25% in notched fatigue strength compared to conventionally treated specimens. The cryogenically treated material also exhibited a lower fatigue notch factor by approximately 15%, indicating reduced sensitivity to geometric discontinuities.

5.4.4 *Corrosion, Oxidation, and Long-Term Stability of Powder Metallurgy Parts in Harsh Environments*

Corrosion, oxidation, and long-term stability of PM parts in harsh environments have been extensively evaluated through comparative studies on metallic systems such as Ti6Al4V, CoCrCuFeNi, and Mg Zn Ca/Fe composites (corrosion tests illustrated in Fig. 5.11). The electrochemical and microstructural behavior of these alloys reveals that their stability is determined by processing parameters, phase distribution, and oxide film formation dynamics. Zhou et al. [1] analyzed PM Ti6Al4V alloys of varying densities and found that corrosion resistance increased with densification due to reduced porosity and enhanced passive TiO_2 layer continuity. The alloys exhibited low corrosion current densities in 3.5 wt.% NaCl solution, indicating high passivity. Despite residual porosity in samples with lower sintering temperatures, those with nearly full density achieved corrosion performance comparable to wrought Ti6Al4V alloys. The protective oxide layer composed primarily of TiO_2, TiO, and Ti_2O_3 evolved spontaneously upon immersion, and the transformation from Ti_2O_3 to TiO_2 provided improved resistance against localized

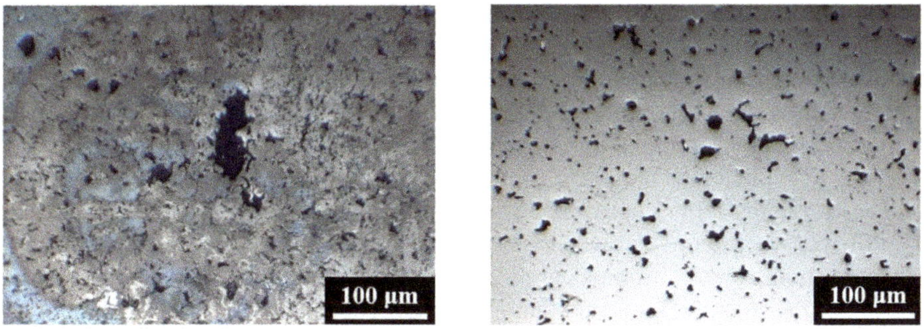

(a) (b)

Fig. 5.11 The surfaces of (**a**) 304L and (**b**) 304L + 0.6B after corrosion tests [4]

attack. Electrochemical impedance spectroscopy further confirmed that densification reduced charge transfer resistance, correlating mechanical integrity with corrosion performance in marine environments.

Mukanov et al. [47] explored the corrosion mechanisms in CoCrCuFeNi HEAs synthesized by PM. The introduction of copper altered the microstructural and electrochemical stability of the system, demonstrating the complex influence of elemental partitioning. When copper content was below 9 at.%, the alloy maintained a single-phase FCC structure with uniform elemental distribution. Beyond this solubility limit, the segregation of Cu-rich FCC_2 phases induced galvanic coupling with the CoCrFeNi matrix. Corrosion experiments in 3.5% NaCl revealed that alloys containing 5–10 at.% Cu exhibited lower corrosion resistance due to the simultaneous formation of Cu_2O and Cr_2O_3. However, at higher Cu concentrations (15–20 at.%), coarsening of Cu-rich grains decreased the interfacial area between the anodic and cathodic regions, thus mitigating galvanic effects. The observed trend implies that microstructural refinement achieved via PM enhances uniform passivation compared to cast alloys, which commonly exhibit dendritic segregation. The authors identified an abrasion-oxidative wear mechanism consistent with the presence of mixed oxides on the surface, suggesting that controlled Cu addition and phase balance are essential for maintaining stability under chloride-rich conditions.

The performance of Mg-based PM materials in corrosive and biological environments presents another dimension to the analysis of long-term stability. Li et al. [5] investigated $Mg_{66}Zn_{30}Ca_4$/Fe bulk metallic glass composites fabricated via SPS, focusing on the influence of Fe addition on corrosion kinetics and mechanical durability. The incorporation of Fe formed a secondary phase uniformly distributed within the Mg-Zn-Ca matrix, reducing the effective surface area exposed to aggressive media and enhancing corrosion resistance. Potentiodynamic polarization tests in Hank's solution showed that increasing Fe content shifted the corrosion potential toward more noble values, while the current density decreased, indicating an improvement in passivation behavior. The amorphous Mg-Zn-Ca phase promoted the formation of a compact, adherent corrosion film that limited Mg dissolution. The long-term immersion tests confirmed a reduction in corrosion rate as Fe content rose from 5% to 20%, demonstrating the role of Fe in stabilizing interfacial oxides and hindering localized pitting. The composites maintained structural coherence due to the absence of oxide layers at Fe-matrix interfaces, thus preventing galvanic degradation commonly associated with multiphase metallic systems. The presence of an amorphous matrix promoted uniform degradation, a feature of particular interest for biomedical applications requiring controlled dissolution.

These systems show improved corrosion and oxidation stability when processed by PM relative to conventional melting routes, associated with enhanced microstructural homogeneity, refined grain size, and reduced segregation. In Ti6Al4V alloys, the dense oxide films and minimal porosity control anodic dissolution under saline exposure. For CoCrCuFeNi HEAs, as illustrated in Fig. 5.12, compositional optimization balances mechanical strength and corrosion resistance by manipulating FCC phase coexistence. Meanwhile, in Mg-Zn-Ca/Fe composites, amorphization and Fe distribution delay the onset of electrochemical degradation. The corrosion mechanisms across these alloys share a dependence on passive film formation, TiO_2 in titanium alloys, Cr_2O_3 and Cu_2O in HEAs, and $Mg(OH)_2$ combined with Fe oxides in Mg-based composites. These films govern oxidation kinetics, barrier integrity, and the capacity for self-healing under cyclic exposure to electrolytes [5].

5.4.5 Integrating Powder Metallurgy with Smart Materials for Next-Generation Multifunctional Systems

Integrating powder metallurgy with smart materials has enabled significant advances in the design and functionality of next-generation multifunctional systems, as exemplified by the processing route for porous NiTi shape memory alloys illustrated in Fig. 5.13. The synthesis and processing of shape memory alloys (SMAs) through PM have enabled precise microstructural control, improved mechanical properties, and enhanced performance in demanding applications across consumer electronics, healthcare, and transportation. By leveraging PM, researchers have been able to tailor the phase transformations and functional behavior of alloys such as Fe-Mn-Si-Cr-Ni, Cu-Al-Ni, and NiTi, achieving a balance between mechanical strength, functional recovery, and manufacturability suitable for intelligent systems integration [49].

In the case of Fe-Mn-Si-Cr-Ni alloys, Pricop et al. [49] demonstrated that combining MA with PM overcame the inherent limitations of ingot metallurgy, such as compositional segregation and cracking during solidification. MA-enhanced chemical homogeneity and controlled grain size, which promoted the formation of ε hexagonal close-packed martensite over the less desirable α' body-centered cubic phase. Adjusting parameters such as milling time, ball-to-powder ratio, and surfactant concentration allowed researchers to optimize density and porosity, achieving materials with better sinterability and mechanical strength. The resulting Fe-Mn-Si-Cr-Ni compacts, after hot rolling and thermal treatment, exhibited significant shape

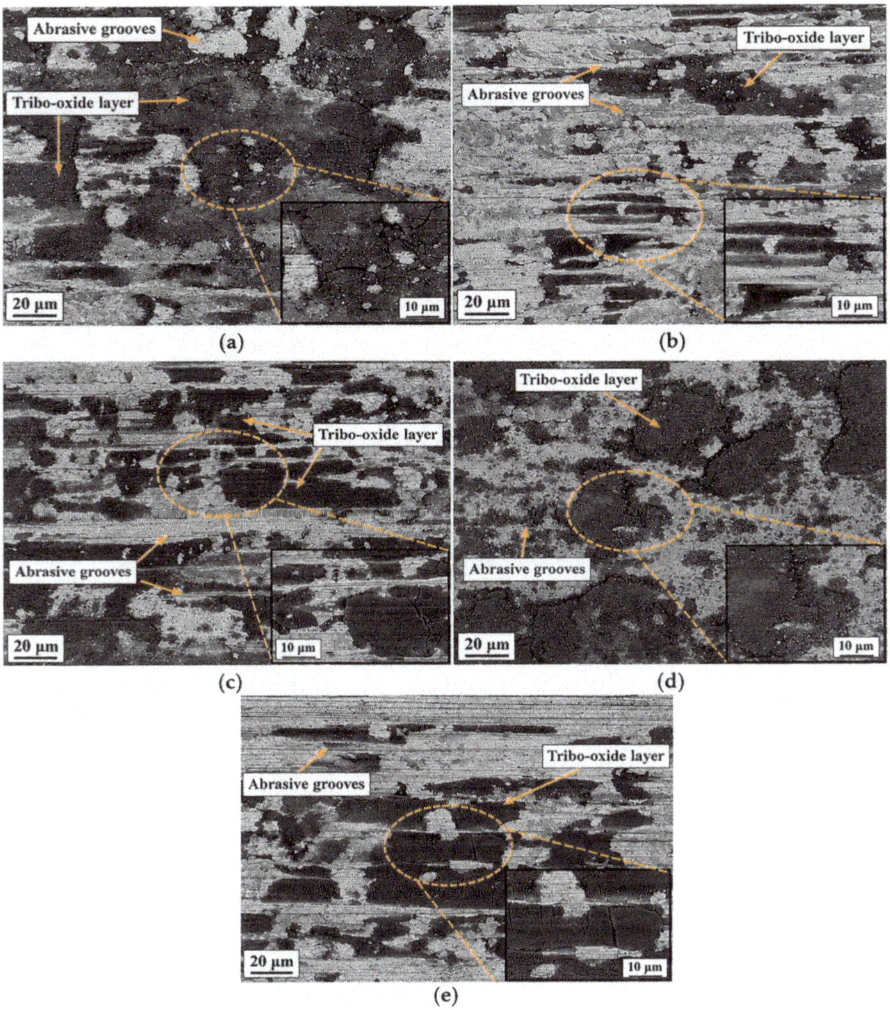

Fig. 5.12 SEM images of the abrasive grooves in the samples: (**a**) CoCrFeNi; (**b**) CoCrCu0.25FeNi; (**c**) CoCrCu0.5FeNi; (**d**) CoCrCu0.75FeNi; (**e**) CoCrCuFeNi [47]

memory effects associated with the reversible transformation between martensite and austenite. Such alloys displayed internal friction peaks linked to antiferromagnetic–paramagnetic transitions, suggesting their suitability for damping and vibration control applications in transportation and energy systems.

Similarly, San Juan et al. [50] advanced the understanding of Cu-Al-Ni shape memory alloys fabricated by PM. Using gas atomization followed by HIP and hot rolling, they achieved nearly full density materials with refined grain structures. The refinement was key to reducing intergranular fracture and improving ductility, addressing the primary drawbacks of these high temperature alloys. The

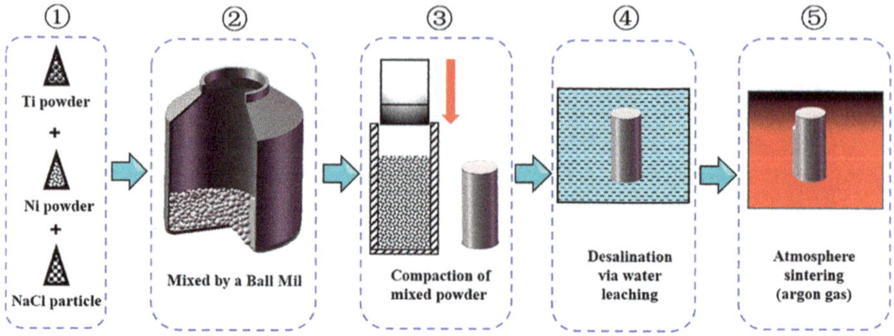

Fig. 5.13 Schematic illustration of the processing route for preparing porous NiTi SMA [48]

thermomechanical processing yielded a final grain size near 100 μm and a low thermal hysteresis (~10 °C), comparable to single crystals. Mechanical testing revealed superelastic behavior with full strain recovery below 4% and partial recovery beyond that limit. The shape memory effect, with recoverable strains up to 7%, placed these powder-processed materials at performance levels close to monocrystals but with significantly improved ductility. The combined effects of atomization, HIP compaction, and rolling resulted in strong interparticle bonding and texturing, thereby enhancing both the mechanical and functional properties essential for smart actuators and sensors in aerospace and robotic systems.

Ni-Ti alloys, a widely used shape memory alloy system, have also benefited from advances in powder metallurgy processing. Jabur et al. [51] examined the effects of compaction pressure and chromium and aluminum additions on the structural and functional performance of powder metallurgy-processed Ni–Ti alloys. The processed alloys closely replicated the phase structures and transformation behavior of commercial thermally activated Ni-Ti archwires, combining martensitic and austenitic phases at room temperature. Compacting pressures up to 800 MPa and sintering at 950 °C under argon produced dense samples with reduced porosity and excellent phase homogeneity. Additions of Cr and Al further enhanced corrosion resistance and hardness without significantly altering transformation temperatures. These improvements underscore the potential of PM Ni-Ti alloys for biomedical applications such as orthodontic wires, stents, and implantable actuators, where biocompatibility, fatigue resistance, and controlled pseudoelasticity are vital.

5.5 Conclusions

The chapter presents powder metallurgy as a unified, microstructure-driven design framework spanning structural, magnetic, energy, and multifunctional applications, where porosity, phase constitution, and interfacial chemistry dictate functional performance. In load-bearing components, fatigue, wear, and tribological response are governed by pore morphology and phase stability, with high density processing, surface densification, and engineered secondary phases mitigating stress concentrations and controlling friction. For magnetic, electronic, and conductive materials, PM enables property optimization through nanostructuring, particle coatings, and advanced sintering routes that regulate domain behavior, transport pathways, and interfacial diffusion. The same structure–property principles extend to energy and advanced alloy systems, where controlled porosity, phase chemistry, and nanostructure determine ionic transport, hydrogen storage, and high temperature stability. Collectively, these examples position powder metallurgy as a materials-centered manufacturing approach in which microstructural architecture, rather than geometry, governs performance across diverse application domains.

References

1. Zhou Y et al (2021) Electrochemical corrosion behavior of powder metallurgy Ti6Al4V alloy. J Mater Eng Perform 30(1):556–564
2. Abolkassem S et al (2024) Microstructure, mechanical, and magnetic properties of powder metallurgy FeCoNiSi–Cu, FeCoNiSi–Mn, and FeCoNiSi-Ti equiatomic HEAs manufactured by spark plasma sintering. J Mater Res Technol 33:9426–9438
3. Ali S et al (2022) Microstructure and mechanical properties of modified 316L stainless steel alloy for biomedical applications using. Powder Metall 15(8):2822
4. Ku M-H et al (2022) Improved mechanical and corrosion properties of powder metallurgy austenitic, ferritic, and martensitic stainless steels by liquid phase sintering. Materials 15(16): 5483
5. Li K et al (2022) Mechanical properties and corrosion resistance of powder metallurgical Mg-Zn-Ca/Fe bulk metal glass composites for biomedical application. J Mater Sci Technol 103:73–83
6. Katheria A et al (2025) MXene and Fe3O4 decorated g-C3N4 incorporated high flexible hybrid polymer composite for enhanced electrical conductivity, EMI shielding and thermal conductivity. Next Mater 6:100292
7. Bolzoni L, Yang F (2019) Development of Cu-bearing powder metallurgy Ti alloys for biomedical applications. J Mech Behav Biomed Mater 97:41–48
8. Parveez B, Wani MF (2021) Tribological behaviour of nano-zirconia reinforced iron-based self-lubricating composites for bearing applications. Tribol Int 159:106969
9. Scholzen P et al (2023) Influence of heat treatment and densification on the load capacity of sintered gears. Powder Metall 66(2):86–93
10. Fontanari V et al (2019) Tooth root bending fatigue strength of high-density sintered small-module spur gears: the effect of porosity and microstructure. Metals 9(5):599
11. Lin T-Y et al (2025) Fatigue analysis of high-density high-performance powder metallurgy (HPM) spur gears. J Jpn Soc Powder Powder Metall 72(Suppl):S249–S254
12. Liu S et al (2023) Fatigue of an aluminum foam sandwich formed by powder metallurgy. Materials 16(3):1226
13. Shen J et al (2022) Thermomechanical fatigue behavior and lifetime modeling of powder metallurgy superalloy considering phase angle effect. Int J Fatigue 164:107164
14. Zhang L et al (2024) Thermo-mechanical fatigue crack growth in a nickel-based powder metallurgy superalloy. Int J Fatigue 186:108413
15. Yu C et al (2022) Research progress on friction and wear properties of powder metallurgy brake pad. Adv Mech Eng 14(12):16878132221126854

16. Yusuf K, Ahmet Y (2021) Investigation of wear behavior of borided materials produced by the powder metallurgy method in different compositions. Prot Met Phys Chem Surf 57(4):771–778
17. Moustafa EB et al (2023) Influence of graphene and silver addition on aluminum's thermal conductivity and mechanical properties produced by the powder metallurgy technique. Metals 13(5):836
18. Lu S et al (2023) Recent advances and future developments in Fe-based amorphous soft magnetic composites. J Non-Cryst Solids 616:122440–1427
19. Shi Q et al (2025) Preparation of spherical WMoNbTa refractory high entropy alloy powder by radio frequency plasma and its laser powder bed fusion densification. J Mater Res Technol 38: 1539–1551
20. Ben Ammar C et al (2024) Microstructural and magnetic characteristics of high-entropy FeCoNiMnTi alloy produced via mechanical alloying. Metals 14(11):1302
21. Martin P, Aguilar C, Cabrera JM (2024) A review on mechanical alloying and spark plasma sintering of refractory high-entropy alloys: challenges, microstructures, and mechanical behavior. J Mater Res Technol 30:1900–1928
22. Al-Kattan D et al (2018) From powdered oxide to shaped metal: an easy way to prepare a porous metallic alloy for SOFC. Fuel Cells 18(1):18–26
23. Liu Z et al (2025) Influence of Cr content on the mechanical, oxidation, and electrical properties of powder metallurgy ferritic stainless steel for solid oxide fuel cell interconnect. J Mater Eng Perform 34(13):13259–13270
24. Salguero J et al (2018) Application of pin-on-disc techniques for the study of tribological interferences in the dry machining of A92024-T3 (Al–Cu) alloys. Materials 11(7):1236
25. Susilowati SE, Fudholi A, Sumardiyanto D (2022) Mechanical and microstructural characteristics of Cu–Sn–ZnGr metal matrix composites processed by powder metallurgy for bearing materials. Results Eng 14:100377
26. Piao R et al (2022) Characterization of hot deformation of near alpha titanium alloy prepared by TiH2-based. Powder Metall 15(17):5932
27. Holmberg A et al (2023) Crack initiation and early propagation in case hardened sintered PM steels under cyclic load. Powder Metall 66(2):164–175
28. Rathfelder S et al (2025) Production of anisotropic NdFeB permanent magnets with in situ magnetic particle alignment using powder extrusion. Materials 18(15):3668
29. Öztürk B et al (2018) Oxidation, electrical and mechanical properties of Crofer®22 solid oxide fuel cell metallic interconnects manufactured through powder metallurgy. Int J Hydrog Energy 43(23):10822–10833
30. Mei L et al (2018) Effects of hydrogen content on powder metallurgy characteristic of titanium hydrides. Int J Hydrog Energy 43(14):7102–7107
31. Ding Z et al (2019) MgCNi3 prepared by powder metallurgy for improved hydrogen storage properties of MgH2. Int J Hydrog Energy 44(16):8347–8356
32. Ren J et al (2015) A more efficient way to shape metal-organic framework (MOF) powder materials for hydrogen storage applications. Int J Hydrog Energy 40(13):4617–4622
33. Bridges D et al (2023) Novel frontiers in high-entropy alloys. Metals 13(7):1193
34. Moravcikova-Gouvea L et al (2021) Tailoring a refractory high entropy alloy by powder metallurgy process optimization. Materials 14(19):5796
35. Basem A et al (2024) Characterization of FeCoNiCr high-entropy alloys manufactured by powder metallurgy technique. J Mater Res Technol 30:88–100
36. Chen C-L, Lin J-H (2023) Effects of alloying elements and mechanical alloying on characteristics of WVTaTiCr refractory high-entropy alloys. Materials 16(18):6194
37. Liu Q et al (2024) Deformation-induced grain boundary segregation in a powder-metallurgy ultrafine-grained MoNbTaTiV refractory high-entropy alloy. Heliyon 10(17):e37392
38. Martin P et al (2024) Microstructure and constitutive modeling of an ultrafine-grained refractory high-entropy alloy fabricated by powder metallurgy. J Mater Res Technol 30:7910–7926
39. Kim D et al (2021) Highly conductive Al/Al interfaces in ultrafine grained Al compact prepared by low oxygen powder metallurgy technique. Nanomaterials 11(5):1182

40. Nie Q-q et al (2021) Process optimization, microstructures and mechanical/thermal properties of Cu/Invar bi-metal matrix composites fabricated by spark plasma sintering. Trans Nonferrous Metals Soc China 31(10):3050–3062
41. Wu H et al (2023) Mechanical and biological properties of titanium and its alloys for oral implant with preparation techniques: a review. Materials 16(21):6860
42. Rodriguez-Contreras A et al (2021) Powder metallurgy with space holder for porous titanium implants: a review. J Mater Sci Technol 76:129–149
43. Li Z et al (2021) Characteristics of Ti-Nb-Mg alloy by powder metallurgy for biomedical applications. Mater Charact 173:110953
44. Huang W et al (2021) Compressive properties and failure mechanisms of gradient aluminum foams prepared by a powder metallurgy method. Metals 11(9):1337
45. Guo RP et al (2023) Achieving superior fatigue strength in a powder-metallurgy titanium alloy via in-situ globularization during hot isostatic pressing. Scr Mater 228:115345
46. Morri A, Ceschini L, Messieri S (2022) Effect of different heat treatments on tensile properties and unnotched and notched fatigue strength of cold work tool steel produced by powder metallurgy. Metals 12(6):900
47. Mukanov S et al (2023) The effect of copper on the microstructure, wear and corrosion resistance of CoCrCuFeNi high-entropy alloys manufactured by powder metallurgy. Materials 16(3):1178
48. Hu L et al (2023) Microstructure characteristics of porous NiTi shape memory alloy synthesized by powder metallurgy during compressive deformation at room temperature. Metals 13(11): 1806
49. Pricop B et al (2020) Powder metallurgy: an alternative for FeMnSiCrNi shape memory alloys processing 7
50. Péréz-Sáez R et al (2000) Advanced shape memory alloys processed by powder metallurgy. Adv Eng Mater 2:49–53
51. Jabur AS, Al-Haidary JT, Al-Hasani ES (2013) Characterization of Ni–Ti shape memory alloys prepared by powder metallurgy. J Alloys Compd 578:136–142

Chapter 6
Cutting-Edge Technologies in Powder Metallurgy

Abstract Solid-state and hybrid powder-based technologies expand the manufacturing envelope of powder metallurgy by decoupling bonding from melting and enabling kinetically driven consolidation, programmable microstructures, and time-dependent functionality. Cold spray deposition exploits supersonic particle impact and adiabatic shear instability to achieve dense, oxide-free metallic and composite deposits, with critical velocity thresholds governed by particle size, thermal diffusivity, and impact temperature, yielding deposition efficiencies approaching 80% and electrical conductivities exceeding 90% IACS after annealing. Surface activation, nozzle gas dynamics, and particle–substrate interactions dictate adhesion and coating integrity, supporting applications in corrosion protection, structural repair, and additive buildup without phase degradation. Filament-based metal extrusion extends powder metallurgy into low cost additive routes, where powder dispersion, binder chemistry, and rheological control determine print stability, shrinkage, and final density, typically reaching 90 to 95% after debinding and sintering. Hybridization with laser-based processes enables defect-free joining of dissimilar alloys and functionally graded architectures while suppressing brittle intermetallic formation. Emerging integration of smart materials, particularly shape memory alloys, introduces 4-dimensional behavior in which microstructure, phase stability, and geometry are co-programmed during fabrication, allowing thermally, electrically, or magnetically triggered actuation. Coupled with data-driven optimization and machine learning, these approaches redefine powder metallurgy as a framework for adaptive, multifunctional metallic matter rather than static structural fabrication.

6.1 Introduction

Recent advances in manufacturing science have enabled the convergence of solid-state deposition methods, filament-based additive manufacturing, and programmable material concepts, opening new pathways for the design and fabrication of metallic and hybrid structures. Within this context, this chapter examines cold spray (CS),

R. Pérez-Bustamante, *Fundamental, Practical and Disruptive Applications of Powder Metals*, Topics in Mining, Metallurgy and Materials Engineering,
https://doi.org/10.1007/978-3-032-27130-3_6

fused filament fabrication (FFF), and 4-dimensional printing as complementary strategies in which microstructural evolution, geometric definition, and functional response are determined by powder attributes and process control [1–14]. Cold spray represents a kinetic consolidation route in which metallic or composite particles, accelerated to 300–1200 m s^{-1} through Laval nozzles, bond to substrates through adiabatic shear instability and intense plastic deformation rather than melting, preserving feedstock phases and minimizing oxidation [14, 15]. The existence of velocity thresholds and incubation regimes establishes a physical linkage between gas dynamics, particle size, and surface condition, governing deposition efficiency and coating integrity [11, 12, 16].

Building on this process structure relationship, filament-based additive manufacturing extends powder metallurgy into extrusion-driven architectures, in which metallic powders incorporated within thermoplastic binders are shaped through material deposition and subsequently debound and sintered to yield dense components. Rheology, dispersion, and polymer compatibility define printability and final density, with solution mixing and optimized binder systems enabling metal loadings exceeding 60 vol% and shrinkage control during sintering. These feedstocks translate microstructural design into layerwise fabrication, allowing multimaterial and graded structures that cannot be produced by conventional press-and-sinter routes [4, 5, 17–19].

This chapter presents how 4-dimensional printing is implemented through time-dependent material responses in shape memory alloys (SMAs), magnetically responsive composites, and microstructure-programmed metals. These responses are enabled by additive manufacturing routes such as laser powder bed fusion (L-PBF), cold spray additive manufacturing, and electrochemical deposition, which provide the geometric and compositional control needed to encode phase transformations, thermal-expansion mismatch, and dislocation architectures within printed structures [3, 20–22]. These powder and filament-based processes couple solid-state consolidation with programmable microstructures, resulting in metallic components whose mechanical, electrical, and functional behavior is governed by particle scale interactions and time-dependent phase evolution, with thermomechanical actuation described through computational modeling, finite element simulation, and data-driven alloy and process design.

6.2 Cold Spray Technology: Deposition of Coatings and Bulk Materials

6.2.1 Fundamentals of Cold Spray Deposition

Under the Cold Spray (CS) process, the deposition of solid materials is achieved through the kinetic energy of particles that impact a substrate at supersonic velocities. The fundamentals of this technology rest on a complex interplay between gas

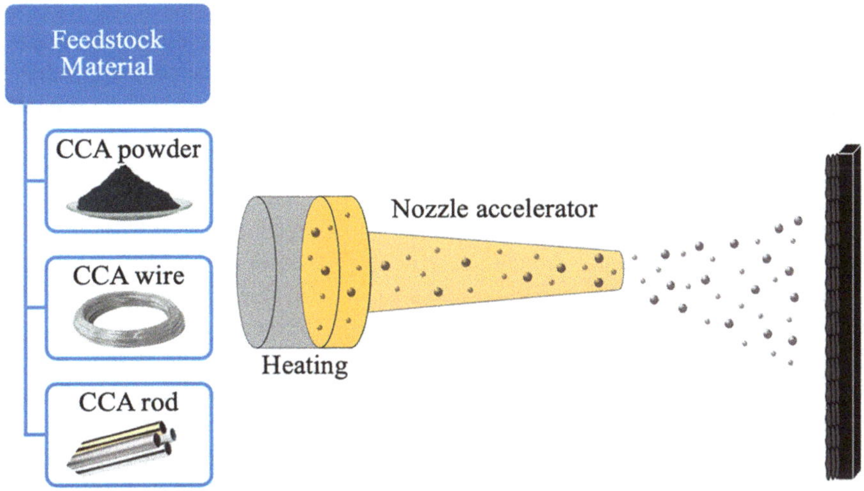

Fig. 6.1 Schematic diagram showing the thermal spray process [23]

dynamics, particle acceleration, impact phenomena, and interfacial bonding mechanisms. Cold spraying differs from conventional thermal spraying in that it operates below the melting temperature of the materials involved, allowing for coatings that retain the original phase and microstructure of the feedstock powder while minimizing oxidation and residual stress (Fig. 6.1). The process was first developed in Novosibirsk in the 1980s and has since evolved into a mature technique for depositing dense metallic and composite coatings with broad industrial relevance [14].

The central concept underlying CS deposition involves accelerating fine metallic or composite particles within a supersonic gas stream, typically generated through a converging-diverging Laval nozzle. The particles, which range in size from 5 to 100 μm, reach impact velocities between 300 and 1200 m/s depending on the gas pressure, temperature, and nozzle design. Upon collision with a substrate, if the particle velocity exceeds a material-specific critical value, intense localized plastic deformation occurs at the interface, resulting in bonding through mechanical interlocking and metallurgical contact. The theoretical framework for this critical velocity, developed at Helmut Schmidt University, relates it to the tensile strength, melting temperature, and specific heat of the particle material although it does not directly account for the surface state of either the substrate or the particles [14]. Experimental results, however, reveal that oxide films and surface roughness play decisive roles in determining whether adhesion occurs.

Klinkov et al. [24] established that CS occupies a unique domain in the general classification of impact phenomena. Within a velocity range of approximately 400–1200 m/s, ductile metallic particles can adhere strongly to metallic, ceramic, or even glass substrates without melting. At lower velocities, particles merely rebound, while at higher velocities material destruction dominates due to the generation of shock waves. This specific window of adhesion results from a balance between adiabatic shear instability and localized thermal softening at the particle substrate interface, enabling bonding without bulk melting. Klinkov and his coauthors concluded that no other known impact regime reproduces the adhesion behavior characteristic of CS deposition, emphasizing the uniqueness of its kinetic bonding mechanism.

A defining aspect of CS deposition is the presence of an incubation period during which the substrate surface undergoes mechanical and chemical activation. During this stage, the initial particles do not adhere but instead clean and deform the substrate surface. As impacts continue, the substrate becomes increasingly favorable for bonding. The duration of this induction period is inversely proportional to the particle flux and strongly dependent on impact velocity. Experimental data show that as particle velocity increases, the incubation time decreases until it vanishes once a second critical velocity is reached, beyond which particles can adhere immediately even on unactivated surfaces [12].

The deposition efficiency, defined as the ratio between the mass of material adhered to the substrate and the mass of powder fed into the gas stream, serves as a key metric for assessing process performance. Klinkov and Kosarev [12] developed an analytical model for coating formation kinetics, distinguishing between three sequential stages: the induction period, the formation of the first continuous layer, and the subsequent buildup of coating thickness. In the first stage, only erosion and surface conditioning occur. The second stage begins once the first continuous layer of adhered particles is formed, establishing a metallurgical base that significantly enhances the probability of subsequent adhesion. The third stage, referred to as the buildup phase, is characterized by linear growth in coating mass since the impacting particles now interact primarily with a surface of similar material. These kinetic insights highlight the dynamic and evolving nature of the deposition interface throughout the spraying process.

The relationship between deposition efficiency and impact conditions reflects the interplay between gas dynamic parameters and material properties. Grigoriev et al. [14] reported that the use of elongated prechambers in nozzle design improves heat transfer between particles and the carrier gas, allowing higher particle temperatures and, consequently, improved deformability upon impact. Additionally, optimized bell-shaped nozzles can generate more homogeneous velocity distributions, enhancing coating uniformity. These improvements in nozzle design have extended the range of sprayable materials, including alloys and composites with higher critical velocities. By increasing the stagnation pressure up to 5 MPa and the gas temperature to 1273 K, researchers have achieved deposition efficiencies approaching 80% for certain aluminum-based systems while maintaining the low oxidation characteristic of CS.

The concept of surface activation has been extensively investigated as a means of enhancing deposition efficiency and coating quality. Experiments conducted by Grigoriev et al. (2015) using a dual powder injection system demonstrated that adding ceramic particles such as alumina to metallic feedstock reduced the critical velocity required for deposition. The high velocity abrasive particles modified the substrate surface through microscale mechanical treatment, removing oxide layers and increasing roughness, thereby facilitating improved interfacial bonding. This synergistic behavior of metal ceramic mixtures, known as the activating effect, was particularly effective for aluminum-based powders but less pronounced for harder materials such as titanium or stainless steel [12, 14, 24].

Furthermore, laser-assisted surface activation has been proposed as a method to optimize substrate conditions before or during spraying. By locally removing oxides and heating the substrate, laser treatment creates a clean, chemically active, and roughened surface that enhances mechanical interlocking and metallurgical bonding. Comparative studies showed that coatings deposited on laser-treated substrates exhibited higher adhesion strength and improved microstructural integrity compared with those produced on grit-blasted surfaces, demonstrating the critical importance of pre-impact surface conditioning [12, 14, 24].

6.2.2 Process Parameters and Their Influence on Material Bonding

The efficiency of material bonding in CS deposition (schematic process CS and bonding illustrated in Figs. 6.2 and 6.3, respectively) is strongly governed by the interaction between process parameters and the fundamental physics of particle impact. The control of variables such as gas pressure, temperature, particle size, and velocity directly dictates the ability of particles to adhere to the substrate, influencing coating density, microstructure, and overall quality. The establishment of a general parameter window has been essential for predicting the bonding conditions across various materials, allowing optimization of CS processes based

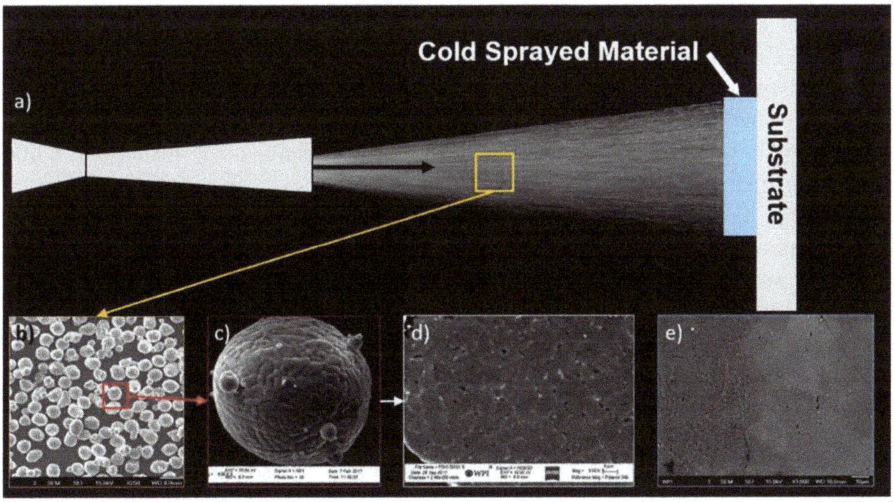

Fig. 6.2 Schematic of the CS process, from powder to coated surface [25]

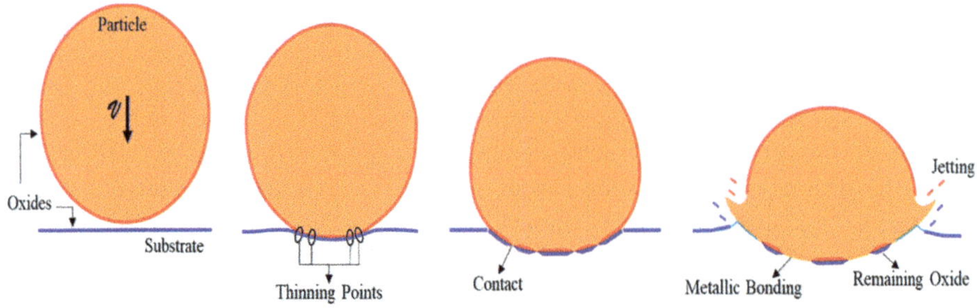

Fig. 6.3 Schematic showing the bonding mechanism of a typical CS process [23]

on material-specific behaviors. As reported by Schmidt et al. [9], particle bonding occurs when the impact velocity exceeds a threshold defined by material and process characteristics. This threshold velocity depends on factors such as density, melting temperature, yield strength, and impact temperature. Adiabatic shear instabilities, characterized by rapid and localized plastic deformation, contribute to the formation of metallurgical bonds between the particle and the substrate. The Johnson Cook model implemented in their finite element simulations demonstrated how these instabilities form at high strain rates approaching $10^9\ s^{-1}$, leading to jetting phenomena and oxide layer disruption, prerequisites for bonding in CS coatings.

The effect of particle size on bonding is particularly critical. Smaller particles, though easier to accelerate, cool too rapidly after impact due to their higher surface-to-volume ratio, which hinders localized adiabatic shear and results in incomplete bonding. Conversely, larger particles experience lower cooling rates and sustain higher interface temperatures, allowing sufficient thermal softening and diffusion across the contact zone. Schmidt et al. quantified this through simulations showing that a 5 μm copper particle impacting at 600 m/s failed to bond, while 25 and 50 μm particles at similar velocities achieved localized melting conditions necessary for adhesion. Experimental validation using cold spray and explosive compaction associated particle size with variations in threshold velocity and deposition efficiency. Furthermore, tensile and microstructural tests correlated higher coating strength with process conditions that promote sufficient shear instability while preventing fracture or rebound.

Assadi et al. [26] further developed this framework by proposing a model that relates material bonding to measurable process parameters through the concept of a parameter selection map. They proposed that both the particle impact velocity and the critical bonding velocity can be expressed in terms of key variables such as gas temperature (T_0), pressure (p_0), and particle diameter (d_p). Their analysis established that the most critical factor for bonding is the ratio between the actual particle velocity upon impact (v_{pi}) and the critical velocity (v_{cr}). When this ratio, denoted as $\gamma = v_{pi}/v_{cr}$, exceeds unity, bonding is achieved; when it falls below unity, only erosion occurs. Numerical modeling based on drag force equations and 1-dimensional isentropic gas flow demonstrated that both gas temperature and pressure have nonlinear but predictable effects on particle acceleration. A 10%

increase in T_0 or p_0 resulted in roughly 2.9% and 2.1% increases in impact velocity, respectively [11].

Assadi et al. [26] formulated a generalized relationship for the threshold velocity as a function of particle size, density, heat capacity, and the temperature difference between melting and impact conditions. Their model integrated the concept of adiabaticity, represented by a dimensionless parameter $\beta = d_p v_{pi}/\alpha$, where α is the thermal diffusivity. This parameter quantifies the degree to which particle deformation remains adiabatic, bridging microstructural bonding mechanisms with macroscopic process variables. For large particles or low diffusivity materials, β tends toward infinity, representing nearly adiabatic conditions and a lower critical velocity. In contrast, small particles with higher diffusivity experience nearly isothermal deformation, requiring higher impact velocities for bonding. Through experimental validation with copper powders ranging from 5 to 100 μm, their model successfully predicted critical velocities between 550 and 850 m/s, consistent with observed deposition behavior in laboratory trials.

In this regard, Klinkov and Kosarev [12] examined deposition efficiency as a quantitative indicator of bonding behavior. Their analytical and experimental analysis distinguished three stages in the cold spray process: an induction phase associated with surface activation, an initial coating phase marked by the formation of the first particle layer, and a steady-state buildup phase. The induction period, characterized by surface erosion rather than deposition, reflects the time required to achieve sufficient substrate activation and local heating for effective adhesion. They defined two distinct critical velocities: the lower limit (v_{cr1}), below which only erosion occurs, and the upper limit (v_{cr2}), beyond which particles adhere even to unactivated surfaces. For aluminum particles impacting copper substrates, vcr_1 and vcr_2 were determined as 550 m/s and 850 m/s, respectively, aligning closely with the predictions of the generalized models of Schmidt and Assadi. Their kinetic model also highlighted that experimental conditions such as substrate motion speed, powder feed rate, and exposure time significantly influence apparent DE values, demonstrating the need to correct measurements to isolate intrinsic bonding efficiency from process artifacts.

6.2.3 Applications in Protective Coatings and Structural Repair

In recent years, CS technology has emerged as a versatile method for fabricating protective coatings and repairing structural components, offering significant advantages over traditional high temperature deposition processes. The fundamental principle of this method lies in the acceleration of powder particles at supersonic velocities toward a substrate, where bonding occurs through severe plastic deformation and adiabatic shear instability rather than melting. This solid-state nature eliminates phase transformation, oxidation, and thermal residual stresses, thereby maintaining the intrinsic properties of the feedstock materials while producing dense, adherent coatings suitable for structural reinforcement and surface protection.

The development of sustainable CS coatings has led to their increasing application in corrosion, oxidation, and wear protection, as well as in remanufacturing damaged components. Pathak and Saha [10] demonstrated that CS coatings can serve both protective and restorative purposes, producing either thin or thick deposits while minimizing residual stresses. Unlike thermal spraying processes such as high velocity oxy fuel or plasma spray, the CS process allows coating of thermally sensitive materials, including aluminum, copper, and titanium without degradation. High pressure CS systems can reach particle velocities of up to 1400 m/s, enabling the deposition of dense and homogeneous coatings as well as the fabrication of 3-dimensional structures through additive manufacturing (AM). These capabilities make CS suitable not only for surface protection but also for structural restoration of aerospace and defense components, where precision and mechanical integrity are crucial.

The protective efficiency of CS coatings depends strongly on their microstructural characteristics and bonding mechanisms. Ranjan and Das [27] reported that coatings produced by this method exhibit minimal porosity, enhanced hardness, and improved adhesion strength compared with traditional thermal spray coatings. Their comprehensive review revealed that CS coatings significantly enhance corrosion and oxidation resistance in metallic substrates such as aluminum, steel, and titanium alloys. The dense microstructure obtained by supersonic impact results in high coating compactness and reduced pathways for corrosive media, thereby extending service life in harsh environments. In marine and power generation sectors, Cu-Al bronze or Ni-based composite coatings fabricated via CS have been shown to withstand corrosion and erosion under aggressive operational conditions, making them particularly valuable for ship propellers, turbine blades, and boiler components.

Wu et al. [16] extended this line of work by producing Zn-graphene-Al composite coatings on low carbon steel using low pressure cold spray (LPCS). Graphene was incorporated as a reinforcing phase to enhance cathodic protection and self-healing capability. Graphene uniformly distributed within the Zn-Al matrix established conductive channels that maintained electron transfer between the zinc anode and steel substrate even after localized damage. The coating with 0.2 wt.% graphene demonstrated superior electrochemical performance, with lower corrosion potential and increased polarization resistance after prolonged immersion in saline solution. The formation of compact corrosion products within scratches confirmed an active self-healing mechanism, which is critical for long-term structural protection in marine and industrial environments. The successful dispersion of graphene using an in situ reduction method and its incorporation through CS deposition indicated a new direction in designing multifunctional corrosion-resistant coatings.

Beyond corrosion resistance, CS has been employed to produce high performance protective coatings against directed energy and laser-based threats, as illustrated by the cross sectional microstructures of cold sprayed metallic coatings processed under different nitrogen propellant gas temperatures and pressures

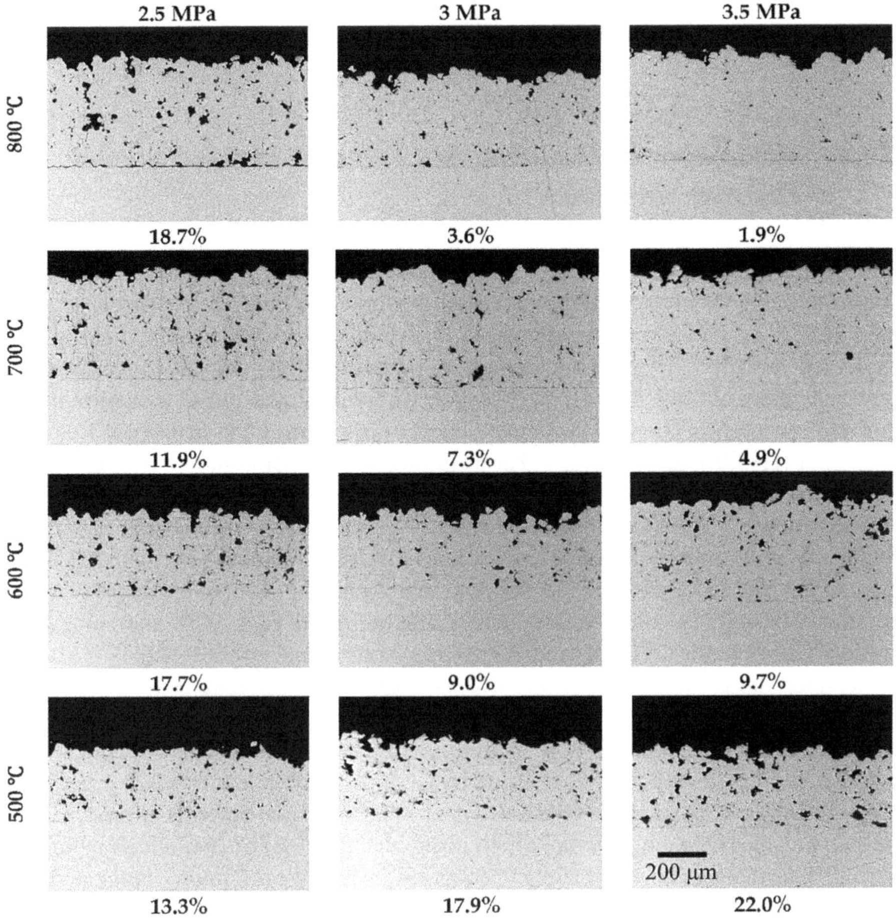

Fig. 6.4 Cold sprayed AISI 316L coating cross sectional microstructures using various nitrogen propellant gas temperatures and pressures; the used AISI 316L powder had particle diameters of 10–45 μm [28]

(Fig. 6.4). Khan et al. (2025) demonstrated the fabrication of ZrC Cu metal matrix composite coatings on Al-6061 substrates via high pressure CS for laser protection applications. The incorporation of zirconium carbide, an ultra-high temperature ceramic, imparted exceptional resistance to laser ablation, while copper provided a reflective and thermally conductive matrix. Microstructural analyses revealed dense, oxidation-free coatings with strong interfacial bonding between the metallic and ceramic phases. Among various compositions, Cu 30% ZrC achieved a reflectivity of 75% at 1080 nm, equivalent to 85% of bulk copper, and exhibited no laser-induced damage under high intensity irradiation. This performance was attributed to the combination of copper's high thermal conductivity and the stability of ZrC at temperatures exceeding 3000 °C. Such coatings offer potential for aerospace and defense applications requiring resistance to intense thermal fluxes from laser exposure.

6.2.4 Advantages and Limitations Compared to Conventional Thermal Spraying

Cold spray technology presents distinct advantages and limitations relative to conventional thermal spraying processes as illustrated in Fig. 6.5. A primary advantage of CS is its solid-state deposition mechanism, which avoids melting the feedstock powder. This characteristic is particularly beneficial for oxidation-sensitive materials like copper, as it prevents oxidation and phase transformations that are common in thermal processes like Arc Spraying (AS) and High Velocity Oxy Fuel (HVOF) spraying [29]. The resulting CS copper coatings exhibit significantly lower oxygen content, typically between 0.04 and 0.09 weight%, compared to 1.46 weight% in AS coatings and 0.89 weight% in HVOF coatings. This low oxygen content directly translates to superior electrical conductivity, with CS coatings achieving about 63% of the International Annealed Copper Standard (IACS) conductivity in the as-sprayed state, which can be improved to 92% after annealing at 600 °C. In contrast, HVOF and AS coatings only reached 39% and 19% IACS, respectively, in the as-sprayed state, with post-annealing improvements being more modest. The bonding mechanism in CS, attributed to high strain rate deformation and localized shear instabilities at particle interfaces, produces dense coatings with high hardness, comparable to cold rolled bulk copper, and excellent bond strength on softer substrates like aluminum and copper. The use of helium as a process gas further enhances coating quality, increasing the fraction of metallically bonded particle interfaces to approximately 75% and significantly improving bond strength on harder substrates like steel, demonstrating the process capability to achieve near-bulk-like properties.

Fig. 6.5 Advantages of CS Additive Manufacturing [13]

However, CS also presents certain limitations. Successful bonding in the cold spray process depends on reaching a particle velocity threshold that is specific to the material, requiring careful selection of process parameters and, in some cases, the use of high cost gases such as helium to achieve optimal performance. Furthermore, the mechanical properties of CS coatings can be anisotropic and inferior to bulk material when processed with less effective propellant gases like nitrogen. Specifically, CS copper coatings sprayed with nitrogen exhibited brittle behavior in tensile tests with very low elongation to failure in the as-sprayed state, whereas helium sprayed coatings demonstrated ductile behavior and strength similar to heavily deformed bulk copper [30]. While annealing can improve the properties of nitrogen sprayed coatings by promoting diffusion and recrystallization, this adds an extra processing step that is not required for thermal spray coatings to achieve similar mechanical performance from the outset in some cases. The deposition of ceramic materials like hydroxyapatite (HA) presents another challenge for CS. While thermal spray techniques like Atmospheric Plasma Spray (APS) are well established for HA, their high operating temperatures lead to decomposition of HA into amorphous phases and other calcium phosphate compounds, which increases dissolution rates in body fluids and can compromise long-term stability [8]. HVOF spraying, with its lower temperatures, produces HA coatings with higher crystallinity than APS, but some amorphous phase formation still occurs. CS offers a significant advantage here by preserving the crystallinity of the original HA feedstock due to its low operating temperatures, resulting in coatings with up to 89% crystallinity. This high crystallinity is correlated with enhanced in vitro cell proliferation and differentiation over longer culture periods. A limitation observed with CS-HA coatings, however, was a slower initial cell attachment compared to APS and HVOF coatings, which was attributed to differences in surface microfeatures and extreme superhydrophilicity of the CS coating surface, whereas APS coatings with moderate wettability and more surface microfeatures facilitated faster cell anchoring. This indicates that while CS excels at maintaining chemical and phase integrity, the final surface topography crucial for biological response may differ from and require optimization compared to conventional thermal spray coatings.

6.2.5 Emerging Trends: Hybrid Systems and Additive Manufacturing Integration

Emerging hybrid AM platforms integrate multiple additive manufacturing processes (Fig. 6.6). Among these, CS and selective laser melting (SLM) are at the forefront of efforts to combine highprecision, fusion-based fabrication with high-deposition-rate, solid-state processing. Hybridizing these complementary methods enables optimized component design and performance by leveraging the geometric accuracy of SLM together with the rapid buildup capability and robust metallurgical bonding provided by CS. Yin et al. [31] demonstrated the potential of this integration in the fabrication

Fig. 6.6 Machining and Additive Manufacturing Platform. (Courtesy of Meltio)

of 316L stainless steel hybrid components, showing that a CS layer deposited over an SLM substrate could be effectively consolidated through post-deposition heat treatment, achieving significant improvements in interface adhesion and mechanical properties. The CS process produced dense deposits due to particle kinetic consolidation, while the SLM portion exhibited refined subgrain structures resulting from rapid solidification. Post-annealing led to enhanced diffusion across the CS–SLM interface, transforming discrete boundaries into metallurgically bonded regions and increasing tensile strength of CS layers by over 200% when helium was used as the carrier gas.

The same principle was successfully extended to dissimilar material systems such as Al-Ti6Al4V functionally graded materials (FGMs) through the integration of SLM and CS, where Ti6Al4V was first processed by SLM and Al or Al Al_2O_3 composites were subsequently deposited via CS [20]. This hybrid configuration addressed challenges in joining non-weldable alloys, as the solid-state nature of CS prevented the formation of brittle intermetallic phases like Al_3Ti and Al_2Ti typically produced during high temperature fusion processes. The SLM Ti6Al4V region exhibited an acicular martensitic alpha prime microstructure, whereas the cold spray aluminum layer preserved the grain morphology of the feedstock while maintaining high density and machinability. The resulting FGMs achieved nearly full density (above 99%) and displayed a smooth, defect-free interface without pores or cracks. The integration of alumina reinforcements further enhanced hardness, demonstrating that hybrid CS–SLM systems can produce lightweight, wear-resistant FGMs with promising structural applications in aerospace and automotive components.

Hybrid approaches have been extended to multimaterial configurations by combining cold spray AM with post-machining steps [7]. The bonding behavior of bimetallic deposits produced by hybrid CSAM was examined across a range of metal pairings, including Al–Al, Cu–Cu, and Cu–Al, with particular attention to interfacial adhesion. It was reported that the mechanical interlocking effect at rough interfaces, achieved through grit blasting, significantly enhanced adhesion compared to smooth or machined surfaces. The Al-Al interfaces exhibited the highest tensile and shear strengths (>30 MPa), whereas the Cu-Al combinations had inferior bonding (<16 MPa) due to the disparity in hardness and plasticity between materials. Morphological analyses confirmed that high surface roughness increased effective contact area and promoted localized embedding between layers, while excessive machining reduced adhesion due to loss of asperities. Microhardness measurements showed that work hardening during CS deposition increased the hardness of Cu and Al deposits relative to their cast states, contributing to improved load-bearing capacity at interfaces.

6.3 Filament-Based Technologies: 3D Printing of Powder-Infused Thermoplastics

6.3.1 Principles of Filament-Based Additive Manufacturing in Powder Metallurgy

The fundamental principles of filament-based AM in PM are grounded in the adaptation of thermoplastic extrusion mechanisms to the processing of metal and metal polymer composites (Fig. 6.7). Fused Filament Fabrication (FFF) or Fused Deposition Modeling (FDM) provides the framework for this adaptation, where a filament composed of metallic powders and polymeric binders is melted and deposited layer by layer to form a solid structure. The technique depends on controlled thermal and rheological behavior to ensure structural integrity and dimensional stability while preserving appropriate mechanical and thermal properties in the printed components. Hwang et al. [18] formulated composite filaments based on acrylonitrile butadiene styrene combined with copper and iron powders to examine their thermomechanical response during fused deposition modeling. The metal loading, ranging from 0 to 50 wt.%, directly affected tensile strength and thermal conductivity. As the metallic content increased, tensile strength decreased due to particle-induced stress concentrations, whereas thermal conductivity improved significantly. It was examined the role of processing parameters such as extrusion temperature and infill density. The viscosity of ABS, calculated using the Hagen-Poiseuille equation, decreased with increasing temperature, leading to improved layer adhesion and higher tensile strain in the final printed specimens. Optimal results were observed at temperatures between 190 °C and 220 °C, balancing the filament flow rate and mechanical strength. Moreover, varying infill density revealed a proportional relationship between material compaction and tensile strength, highlighting the influence of internal architecture on performance.

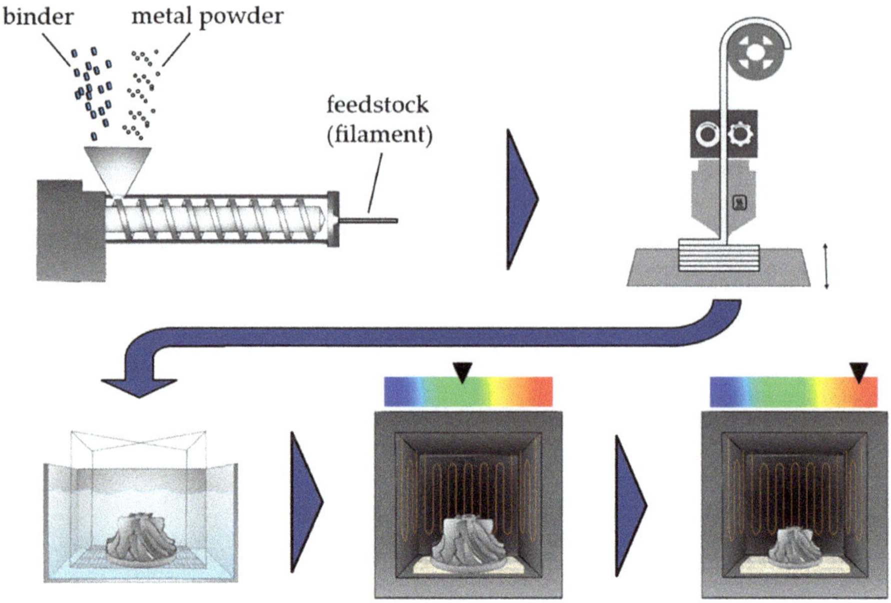

Fig. 6.7 Typical processing scheme for indirect AM of metal parts via filament-based MEX MSt/M [32]

Gibson et al. [33] expanded the concept of FFF beyond polymer composites by demonstrating the direct 3D printing of bulk metallic glasses (BMGs), exploiting their thermoplastic-like softening behavior in the supercooled liquid region. BMGs, such as Zr44Ti11Cu10Ni10Be25, exhibit continuous softening upon heating, allowing extrusion at viscosities between 10^2 and 10^5 Pa s, comparable to that of conventional thermoplastics like ABS. This rheological similarity enabled BMGs to be processed via FFF under ambient conditions, producing fully dense and amorphous metallic structures. The researchers optimized extrusion at 460 °C, balancing the need for sufficient viscosity to enable flow with the requirement to prevent crystallization, which occurs above 476 °C. Their design incorporated a pulsed Joule heating system that locally reheated prior layers, achieving metallurgical bonding through interfacial strain without the need for inert atmospheres. Tensile testing of printed BMG parts revealed fracture strengths up to 1220 MPa parallel to the print layers and 790 MPa perpendicular to them, underscoring the mechanical robustness of this approach. The work demonstrated that BMG-based FFF effectively bridges the processability gap between metals and thermoplastics, providing a pathway toward metallic components with high strength and amorphous microstructures using extrusion-based AM.

Roshchupkin et al. [34] outlined the principal methodologies and material systems involved in metal FFF, describing the process as a sequence of material compounding, filament extrusion, printing, binder removal, and sintering. Metal powders such as stainless steels (316L and 17 4 PH), titanium alloys, and tungsten carbides are typically used, with optimal loadings between 50% and 65% by volume to balance strength and printability. The binder system, usually composed of waxes, thermoplastics, and additives like stearic acid, serves to ensure homogeneity and flexibility during filament formation and extrusion. Proper control of powder morphology, particularly high sphericity and narrow particle size distribution, was emphasized to maintain flow stability and prevent clogging during deposition. Post-processing involves solvent and thermal debinding to remove the polymer matrix, followed by sintering to densify the metallic phase. The authors reported typical shrinkage rates of 16–20% and final densities of up to 95% relative to bulk materials, depending on alloy composition and processing conditions.

6.3.2 Powder Selection, Dispersion, and Polymer Compatibility

Powder selection, dispersion, and polymer compatibility are decisive factors in filament-based additive manufacturing of metal polymer composites (Fig. 6.8). The combination of powder morphology, particle size distribution, and polymer chemistry governs feedstock homogeneity, rheological behavior, and final part density. Hasib et al. [17] examined PLA reinforced with gas-atomized Ni-Cu powders, showing that spherical particles offered superior flowability, packing efficiency, and dispersion uniformity compared with irregular powders. Gas-atomized powders reduced mechanical interlocking, enabling shear thinning behavior up to 100 s^{-1} and allowing extrusion at high solid loadings without jamming. Solution mixing, based on dissolving PLA in dichloromethane prior to blending with the metal powder, yielded a homogeneous feedstock sustaining metal contents of up to 63.4 vol%, whereas physical mixing was limited to approximately 54 vol%. The improved interfacial adhesion between metal and polymer reduced

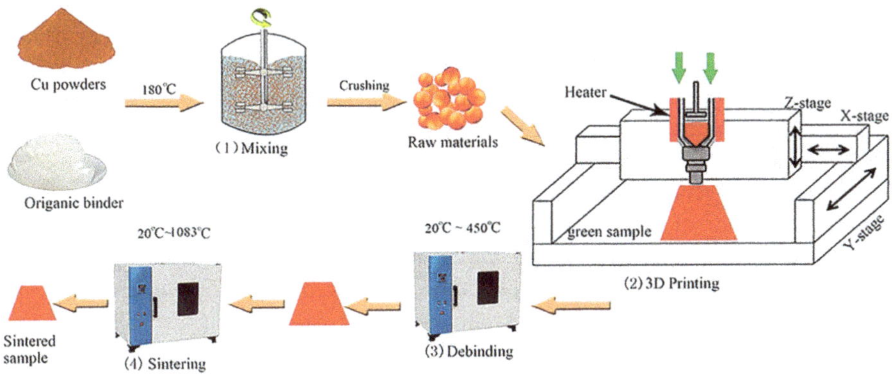

Fig. 6.8 Schematic of the technological process [36]

melt fracture, enhanced extrudability, and decreased sintering shrinkage by 76% relative to physically mixed samples.

Feedstock rheology directly reflects the degree of powder dispersion and binder compatibility. Hasib et al. modeled zero shear viscosity with the Krieger Dougherty relation but noted deviations at high shear rates due to complex particle polymer interactions. The transition from viscoelastic to elastic behavior with increasing metal content limited processability. Proper dispersion delayed this transition, sustaining shear thinning under extrusion conditions typical of 10–1000 s^{-1}. This rheological control ensured uniform filament diameter and continuous winding. Costa and coworkers [35], emphasized that achieving this stability requires careful control of powder characteristics, morphology, density, and surface area, as well as binder composition. They indicated that filament feedstocks generally contain 60 vol % metal powders dispersed in polymeric matrices to ensure flow stability, though higher concentrations demand advanced mixing and surface treatments.

Polymer selection determines the wetting and adhesion behavior at the powder interface. Common thermoplastics such as PLA, PA, and PEEK are used for their thermal stability and controlled degradation. Jacob et al. [4] reported binder systems comprising backbone polymers, plasticizers, and surfactants, noting that these multicomponent formulations enhance particle polymer compatibility and flexibility of the green filament. Poorly matched systems cause phase separation and brittleness during extrusion, leading to voids and poor sintering response. They also emphasized that binder decomposition must occur uniformly during debinding to prevent cracking and maintain dimensional accuracy.

6.3.3 *Process Optimization: Extrusion, Printing Parameters, and Post-Processing*

The optimization of FFF processes involves precise control over extrusion dynamics, printing parameters, and post-processing techniques to ensure enhanced mechanical and surface properties in powder-infused thermoplastic composites (different sintering parameters are illustrated in Fig. 6.9). The extrusion stage plays a critical role in material quality, as the rheological behavior and thermal stability of the molten filament determine the consistency and accuracy of deposition. Shaqour et al. [37] showed that maintaining an appropriate thermal gradient between the cold and hot zones of the extrusion head reduces filament buckling and flow instability, promoting homogeneous melting and stable viscosity during deposition. The shear thinning behavior of thermoplastic melts was linked to nozzle geometry, heating block design, and feed rate, all of which influence extrusion pressure and melt uniformity. Improper temperature regulation in the heating block can cause excessive viscosity or premature solidification, directly affecting layer adhesion and dimensional accuracy of the printed structures.

Fig. 6.9 Sintered samples with different sintering parameters [36]

Bouzaglou et al. [38] expanded upon the interrelationship between printing parameters and material performance by analyzing the effects of raster angle, infill density, build orientation, and extrusion temperature on dimensional precision and mechanical anisotropy. Thin layer deposition combined with lower extrusion temperatures was associated with improved dimensional accuracy, whereas increased layer thickness and positive air gaps reduced printing time. Tensile strength and elastic modulus of fused filament fabricated parts varied by up to 60% depending on the combination of these parameters, reflecting their coupled influence on filament bonding and cooling rate. Mechanical anisotropy was further influenced not only by build direction but also by the relationship between extrusion velocity, cooling rate, and nozzle diameter, which governs interlayer diffusion and fusion. Optimization studies using Taguchi design and signal-to-noise ratios revealed that layer thickness, raster orientation, and build direction are dominant factors in surface quality and dimensional fidelity, suggesting that parameter interactions must be modeled collectively rather than independently.

Post-processing treatments contribute to improved structural integrity and reduced anisotropy in printed components. Malagutti et al. [39] introduced a two-stage approach combining remelting and compaction within a granular salt mold to mitigate porosity and weak interlayer bonding. This treatment reduced void content and strengthened interlayer fusion while preserving geometric integrity. Tensile testing revealed that transverse strength and stiffness increased to values comparable with those measured along the deposition direction, resulting in nearly isotropic behavior in poly (lactic acid)/poly(hydroxyalkanoate) composites filled with wood fibers. The method involved annealing at 185 °C under a pressure of 10 bar for 5 min, followed by controlled cooling to 60 °C. The applied compaction pressure effectively counteracted dimensional distortions typically observed during unconstrained remelting and allowed recrystallization of polymer chains, improving diffusion across printed interfaces and homogenizing microstructural density.

The optimization of post-processing was also investigated by Karamimoghadam et al. [3], who integrated CO_2 laser cutting with FFF printing to refine surface morphology and reduce stairstep defects common in layered deposition. Application of a statistical response surface methodology identified infill % age and extrusion temperature as the main factors affecting surface roughness prior to laser treatment, whereas laser power and scanning speed governed the precision of surface ablation. An infill density of 30% combined with an extrusion temperature of 190 °C resulted in a minimum surface roughness of approximately 2.15 μm, while optimized laser conditions, corresponding to 300 W power and a scanning speed of 5 mm s^{-1}, further reduced top edge irregularities to about 275 μm. The hybrid method improved surface finish and geometrical accuracy by minimizing the heat-affected zone, which typically contributes to warping and microcracking. This approach illustrates the synergy between additive and subtractive processes in enhancing the final morphology and functional performance of FFF-fabricated components.

6.3.4 *Applications in Prototyping and Functional Components*

The application of FFF for metallic parts has demonstrated significant utility in the domains of prototyping and the creation of functional components, leveraging its design freedom and cost-effectiveness. This technology, often referred to as metal FFF or fused deposition modeling and sintering (FDMS, Fig. 6.10), utilizes filaments loaded with high metal powder content to produce green parts that are subsequently debound and sintered into dense metallic components [6]. The process enables the fabrication of complex geometries that are challenging or impossible to achieve with traditional manufacturing methods like machining, making it particularly suitable for prototyping and low volume production. Industrial applications include the fabrication of an end-user brake pedal from AISI 316L stainless steel, illustrating the capability of the technology to produce functional, load-bearing components. In the medical sector, this approach has been employed to develop personalized devices, such as antimicrobial finger prostheses incorporating copper

(a) (b)

Fig. 6.10 Cross section SEM analysis of the Ultrafuse 316L filament: (**a**) magnification 50×; (**b**) magnification 2500×. The internal structure is composed of metal particles with dimensions varying from 10 μm to 1 μm fully embedded in polymeric strands [1]

and maxillofacial implant prototypes from titanium alloy, highlighting its versatility for customized solutions.

The functional potential of metal FFF is illustrated by the work of Mousapour et al. [40], who demonstrated the feasibility of producing multimaterial components. Three distinct sample configurations, mixed, coupled, and graded, were fabricated using filaments based on high carbon iron and 316L stainless steel. The mixed configuration, characterized by a homogeneous distribution of both metal powders, reached relative densities of up to 92% after sintering at 1360 °C for 6 h, supporting the feasibility of forming alloy-like structures directly from printed composites. This configuration also exhibited enhanced mechanical performance, with a microhardness of 373.39 HV0.2, exceeding the values measured in the individual regions of the coupled samples. These results indicate that FFF enables the development of novel material combinations with improved performance for functional components. However, the process also introduced challenges such as differential shrinkage and distortion, with the coupled sample exhibiting a stair-shaped distortion due to varying shrinkage behavior between the two materials, a critical consideration for the dimensional accuracy of prototypes and final parts.

The shift from prototyping toward large-scale production of functional components requires attention to occupational safety, particularly with respect to emissions generated during the printing phase. Karayannis et al. [2] examined the release of ultrafine particles and volatile organic compounds during fused filament fabrication using various polylactic acid-based materials intended for microfluidic device substrates. The methodology involved an assessment of emissions from different filament compositions, including a PLA blend, a PLA polybutylene adipate terephthalate compound containing inorganic filler, and a PLA composite incorporating reduced graphene oxide. The emission profile was shown to depend strongly on both material composition and printing parameters. Notably, the rGO-reinforced filament exhibited the highest UFP emission potential at nozzle temperatures near the optimal printing range (215–225 °C), with peak concentrations exceeding 100,000 particles per cubic centimeter, whereas the PLA/PBAT blend demonstrated the lowest emissions under the same conditions.

Furthermore, it was highlighted that the geometry of the printed object influences emission levels. Printing a complex microfluidic device (Type B object) led to significantly higher emission peaks compared to simpler test objects, which was attributed to the intricate pathways and small cavities characteristic of such components. This observation has implications for the industrial use of fused filament fabrication in the production of advanced functional components, as it indicates that emission control strategies should be adapted to the specific design being manufactured. A workflow integrating emission assessment with standard printability tests was proposed, guiding the selection of materials and process parameters that balance part functionality, print quality, and reduced occupational exposure. Results demonstrated that while a fully enclosed printer with high efficiency particulate air (HEPA) filtration effectively reduced operator exposure, brief disruptions of the enclosure during high emitting process phases, such as print initiation or failure intervention, could lead to significant short-term exposure peaks, necessitating the use of personal protective equipment like FFP3 respirators in such scenarios.

6.3.5 Challenges and Future Directions

Scalability in filament-based AM remains a primary obstacle for the widespread industrial adoption of powder infused thermoplastics. According to Alarifi [41], achieving consistent mechanical performance while increasing deposition speed between 40 and 150 mm per second remains difficult because higher rates reduce layer adhesion and interfacial bonding quality. The complex rheology of reinforced filaments introduces additional variability in flow behavior and melt homogeneity, leading to anisotropic properties. To overcome these limitations, precision in nozzle temperature control, optimized layer thickness, and adaptive extrusion strategies are required. Moreover, Alarifi emphasized that scaling up production demands balancing efficiency with product reliability through artificial intelligence-driven calibration and monitoring systems that continuously adjust parameters to prevent structural defects and void formation.

Parallel to the problem of scalability, sustainability challenges define the long-term viability of FFF. Fico et al. [5] identified that although AM reduces waste compared to subtractive processes, it still relies heavily on petroleum-based thermoplastics, which hinder circular economy objectives. The incorporation of recycled and bio-derived polymers, such as polylactic acid and post-consumer blends, partially mitigates this issue, yet repeated reprocessing cycles degrade molecular weight and rheological uniformity. This degradation reduces interlayer diffusion and weakens mechanical integrity. Additionally, hybrid formulations containing metal or ceramic powders enhance stiffness but compromise biodegradability. Fico and collaborators proposed closed-loop recycling strategies and in situ filament re-extrusion systems as a means to decrease transportation and energy costs though contamination and inconsistent feedstock purity remain unresolved. For a sustainable transition, they recommended standardizing feedstock certification and promoting bio-based fillers like cellulose and lignin, despite their moisture sensitivity and limited processing temperature windows.

The integration of multiple materials within a single build process represents the next frontier in AM. Raj et al. [42] demonstrated that multimaterial extrusion enables fabrication of functional devices, particularly sensors that combine conductive and flexible polymers. However, it was observed that interfacial bonding between dissimilar materials often fails due to differential shrinkage and nonuniform thermal distribution during printing. The inclusion of conductive additives such as carbon black or graphene in thermoplastic matrices improved sensing capability but introduced porosity and irregular flow patterns at transition zones. Maintaining consistent electrical conductivity across interfaces thus requires synchronized temperature zoning and dynamic control of extrusion rates. Raj and colleagues also noted that alignment errors accumulate during hybrid deposition, limiting scalability and precision for complex architectures.

6.4 4D Printing: Smart Materials and Shape Memory Alloys in Powder Metallurgy

6.4.1 Conceptual Framework of 4D Printing in Powder Metallurgy

The concept of 4D printing in PM combines the core principles of AM with the time-dependent functional response of smart materials, particularly SMAs and related metallic systems (Fig. 6.11). This approach enables the fabrication of components that can transform their shape, properties, or functionality in response to external stimuli such as heat, magnetic fields, or electrical current. Zafar and Zhao [43] describe 4D printing as an evolution of AM where "time" is introduced as a fourth dimension, allowing printed structures to undergo programmed transformations after fabrication. This is achieved through the use of stimuli-responsive materials, which are central to the 4D printing process. In the context of PM, this often involves the use of metal powders processed via techniques such as powder bed fusion (PBF), including selective laser melting (SLM) and electron beam selective melting (EBSM), to create intricate, high performance metallic components.

One element of this framework involves the use of shape memory materials, particularly SMAs such as nickel titanium, which exhibit the shape memory effect. These materials can recover their original shape after deformation when exposed to a specific stimulus, such as heat. It was noted that SMAs operate through a reversible martensitic phase transformation, which is driven by changes in temperature or magnetic fields. The AM of these alloys, however, presents specific challenges.

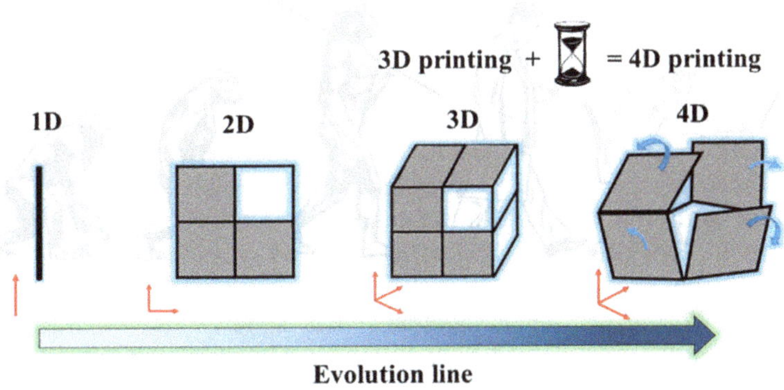

Fig. 6.11 Illustration of the evolution of 1D to 4D printing with time. 4D printing is the addition of time dimension to 3D printing [44]

Processing NiTi via SLM can lead to compositional changes due to the evaporation of nickel, which has a lower evaporation temperature than titanium, thereby altering the transformation temperatures of the alloy. Furthermore, the presence of impurities can be mitigated by conducting the process in an inert atmosphere, highlighting the importance of controlled processing environments in powder-based 4D printing [43].

Expanding the material palette, Chen et al. [45] demonstrate a multi-metal electrochemical AM (ECAM) approach for creating 4D metallic structures. Their system utilizes a meniscus-confined electrodeposition process to print bimetallic structures, such as copper-nickel strips, which exhibit thermally induced deformations. The differing coefficients of thermal expansion between copper and nickel generate internal stresses upon heating, causing the printed strips to bend. This phenomenon is programmable based on the thickness and arrangement of the deposited metal layers. Chen et al. (2019) report that increasing the thickness of the nickel layer on a copper base resulted in greater bending angles at elevated temperatures, up to 300 °C.

The integration of 4D printing within PM also necessitates advancements in computational modeling and design. Gobena and Woldeyohannes [19] emphasize that smart materials used in 4D printing possess capabilities such as self-sensing, self-actuation, and shape memory. These properties must be precisely modeled to predict the time-dependent shape-shifting behavior of printed structures and to prevent collisions during self-assembly processes. Finite element analysis (FEA) software, such as ABAQUS and COMSOL, has been employed to simulate the thermomechanical response of 4D-printed components, enabling the optimization of designs before physical fabrication. This is particularly important for complex structures like the self-folding origami and kirigami shapes that are common in 4D printing applications.

6.4.2 *Types of Smart Materials and Shape Memory Alloys Relevant to Powder Metallurgy*

Smart materials are fundamental in the advancement of 4-dimensional printing applied to PM because they enable the temporal transformation of printed structures under specific external stimuli. Among these materials, SMAs and shape memory polymers have emerged as the most representative examples due to their reversible deformation and ability to respond to thermal, electrical, magnetic, or chemical fields. Li et al. [46] defined intelligent materials as systems capable of perceiving external stimuli and producing an adaptive response such as shape alteration, actuation, or self-repair. When integrated into additive layered manufacturing, these materials allow the printed parts to evolve over time, expanding the field of programmable matter for applications in aerospace, biomedical devices, and robotic systems. The fourth dimension introduced in 4D printing corresponds to this

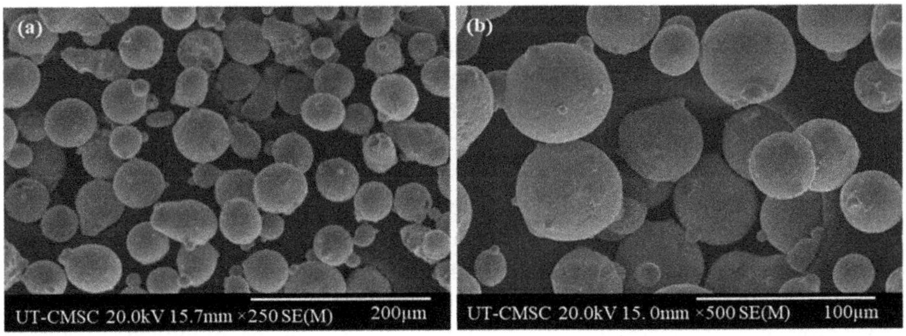

Fig. 6.12 SEM images (**a**) at 250× magnification and (**b**) a close-up view at 500× of Fe-Mn-Al-Ni SMA powder [22]

programmed transformation, which depends on the intrinsic responsiveness of the employed materials rather than on the mechanical motion of external systems.

Shape memory materials are characterized by the shape memory effect, which refers to the ability of a material that has been previously deformed to recover its original configuration when subjected to the appropriate stimulus. In SMAs such as NiTi or Fe-Mn-Al-Ni (Fig. 6.12), this process involves a reversible transformation between the martensitic and austenitic phases. Additive layered manufacturing based on organic polymer binders was applied to NiTi powders, producing components with densities approaching 95% of theoretical while retaining shape memory behavior, demonstrating the viability of powder-based fabrication routes for these alloys. On the other hand, shape memory polymers based on thermosetting polyurethane exhibit high recovery force, heat resistance, and low cost, which make them attractive for large morphing structures such as deployable aerospace elements and adaptive aircraft wings. The difference between one-way and two-way memory behaviors depends on whether the polymer network is trained to remember a single or two opposite shapes, controlled by thermal cycling and the reconfiguration of molecular chains [46].

This field has been extended through the development of magnetically controlled materials that enable remote, noncontact actuation. Wang et al. [47] described two primary mechanisms for magnetic control: deformation induced by magnetic torque and magnetothermal conversion, in which magnetic inclusions convert electromagnetic energy into localized heat to trigger shape memory transitions. Magnetic shape memory hydrogels and elastomers are typical examples that incorporate Fe_3O_4 nanoparticles or magnetic droplets. These composites combined mechanical compliance with magnetic responsiveness for biomedical uses such as implantable actuators and drug delivery systems. Chitosan polyolefin Fe_3O_4 networks exhibited magnetoacoustic actuation and self-healing behavior, while ferromagnetic droplets retained magnetization in the absence of external fields, enabling reconfigurable liquid magnets for adaptive printed systems [46, 47].

An additional technological direction integrates metallic and polymeric phases within a single structure to enhance the mechanical performance of 4D-printed architectures. Song et al. [48] proposed a thermoplastic shape memory polymer and metal heterointerface created through multimaterial digital light processing combined with electroless plating. In this process, Pd^{2+} ions embedded in an active polymer precursor catalyze the deposition of nickel, forming a multi-layered interface that includes a polymer base, a composite intermediate region, and a metallic layer. The resulting heterointerface provides both electroactivity and mechanical robustness, converting thermally actuated polymers into electrically controlled materials. Experimental characterization confirmed that the metallic coatings exhibited uniform resistivity and high ductility, ensuring precise electrothermal deformation under controlled voltage inputs. Moreover, the combination of polycaprolactone and acrylonitrile butadiene styrene resins produced a balance between flexibility and strength, while preserving the shape memory function required for programmable actuation.

6.4.3 Design Strategies for Responsive and Programmable Structures

Design strategies for responsive and programmable structures in PM have evolved through the convergence of AM, intelligent materials, and computational design, enabling structures that can dynamically alter their geometry or functionality under external stimuli. The development of these adaptive architectures integrates concepts from 4D printing, where temporal evolution is embedded during fabrication rather than as a post-processing step. This paradigm shift, discussed by Melly et al. [49], represents a transformation from static 3D-printed geometries to systems capable of self-actuation or functional transformation through material programmability (see Fig. 6.13). Their analysis emphasizes that such structures derive responsiveness primarily from smart materials, such as SMAs or polymers, which are activated by heat, light, or magnetic fields. Predictable and reversible deformation is governed by the regulation of intrinsic properties, including elastic modulus, thermal expansion, and phase transition characteristics. Under this methodology, design strategies integrate material choice, processing conditions, and stimulus control to achieve defined spatiotemporal responses in both motion and material properties.

The integration of material programmability into metallic systems extends the concept of 4D printing beyond polymeric applications. Gao et al. [50] demonstrated an advanced design strategy in stainless steel 316L, where the microstructure and, consequently, the mechanical properties were programmed directly during L-PBF. By controlling hatch spacing and laser remelting, they were able to manipulate dislocation density and chemical heterogeneity, effectively determining the alloy's thermal stability and its ability to recrystallize upon heat treatment. This control allows for site-specific microstructure programming, creating regions with distinct recrystallization states and mechanical behavior within a single part. The capacity to alternate between recrystallized and non-recrystallized domains enables mechanical

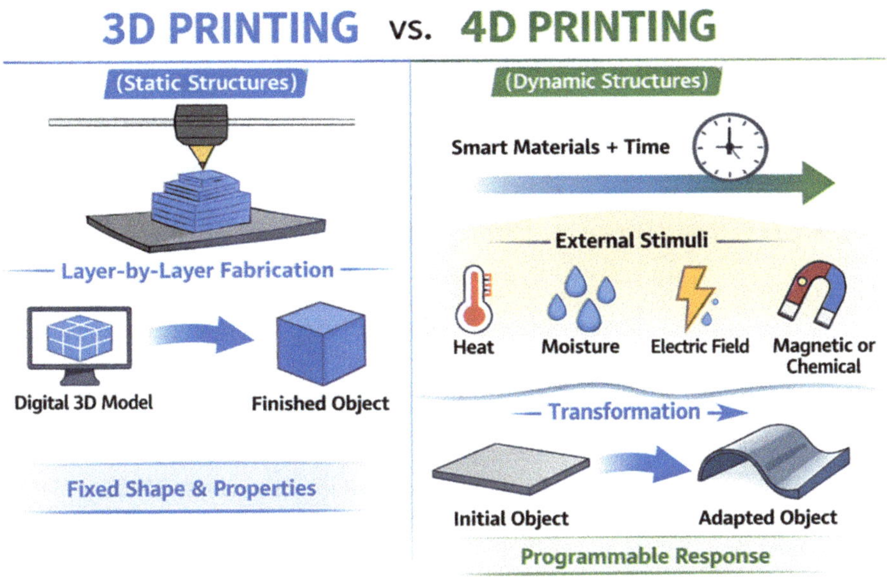

Fig. 6.13 Differences between 3D and 4D printing

gradients that enhance toughness and fatigue resistance. The methodology introduces programmable microstructure architectures that respond predictably to thermal stimuli, effectively transforming static components into programmable functional entities. The implication is a step toward metal-based responsive systems where control over dislocation structures replaces external mechanical deformation as the means of achieving adaptive response.

Similarly, Tan et al. [51] extended the concept of programmability to in situ 4D printing through the machine learning (ML)-assisted design of Fe Ni Ti Al maraging steel. By leveraging intrinsic heat treatment effects inherent to laser additive manufacturing (LAM), they developed a composition capable of forming Ni_3Ti precipitates dynamically during deposition, eliminating the need for post-heat treatment. The use of ML surrogate modeling and optimization algorithms enabled precise tuning of alloying elements to maximize strengthening precipitates while minimizing brittle phases. The resulting alloy achieved superior tensile strength through synchronous geometric formation and microstructural evolution, embodying an energy-efficient realization of in situ 4D printing. This approach represents a new design philosophy where alloy composition and process parameters are co-optimized to exploit cyclic thermal histories for time-dependent microstructural transformations. The resulting structures, therefore, possess built-in responsiveness, as the same process parameters that define geometry also dictate internal phase evolution and mechanical adaptation.

6.4.4 Applications in Biomedical Devices, Aerospace, Robotics, and Artificial Intelligence

Recent advancements in PM and AM have accelerated the integration of SMAs and 4D printing technologies into key sectors such as biomedical engineering, aerospace, robotics, and artificial intelligence. The integration of NiTi-based SMAs and other smart materials into 4D printing has transformed traditional manufacturing into a dynamic process capable of producing responsive, adaptive, and intelligent systems, as illustrated in Fig. 6.14. These developments relied on the superelastic and shape memory responses of SMAs together with the geometric precision and programmable material distribution enabled by AM [52].

In the biomedical domain, NiTi alloys have demonstrated remarkable biocompatibility, corrosion resistance, and mechanical resilience, making them ideal for implants, stents, orthodontic devices, and smart prosthetics. Laser Powder Bed Fusion (LPBF) has emerged as a key technique for fabricating NiTi structures with complex architectures and tailored transformation temperatures. Safaei et al. [53] detailed how LPBF enables microstructural refinement and control of phase distribution to enhance superelastic behavior and fatigue resistance in biomedical components. Additive processes also facilitate the production of porous scaffolds that mimic bone structures, enabling cell growth and vascularization. Bajpai et al. [54] emphasized that 4D printing extends these functionalities by integrating shape

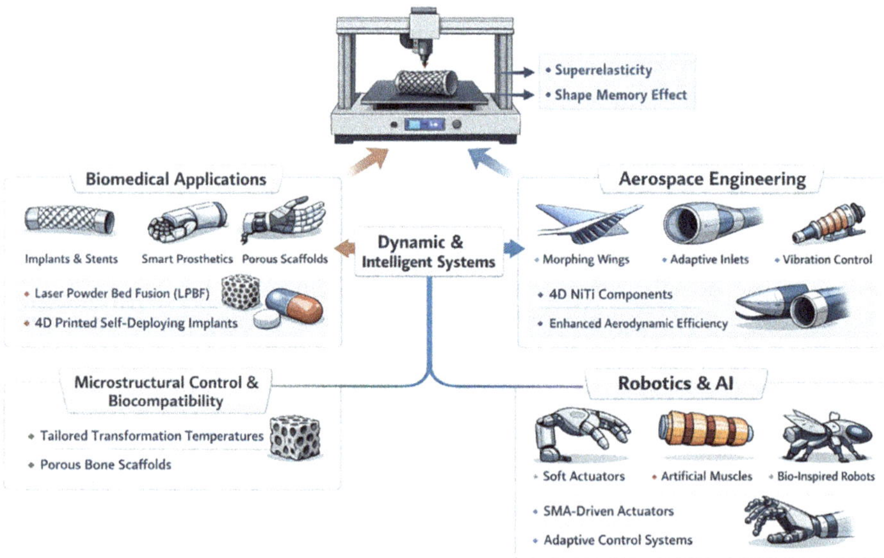

Fig. 6.14 Schematic overview of the integration of NiTi-based SMAs and 4D printing, highlighting their functional properties (superelasticity and shape memory effect) and applications in biomedical engineering, aerospace systems, and robotics through adaptive, stimulus-responsive components

memory polymers and alloys capable of responding to stimuli such as heat or pH, leading to self-deploying implants, adaptive drug delivery systems, and morphing tissue scaffolds.

In aerospace engineering, SMAs are being deployed for active control and efficiency optimization. Costanza and Tata [52] described their applications in morphing wings, deployable structures, and vibration isolation systems. These alloys enable actuators that replace traditional hydraulic systems, offering weight reduction and improved reliability. Kang et al. [55] presented a 4D-printed NiTi variable geometry inlet for aero engines, fabricated via laser-directed energy deposition. The component exhibited two-way shape memory behavior and adaptive deformation under thermal stimuli, improving aerodynamic efficiency by dynamically adjusting inlet geometry with temperature changes. Differential scanning calorimetry confirmed the stability of transformation temperatures, and fluid dynamics simulations demonstrated a total pressure recovery factor approaching unity, underscoring the design's aerodynamic potential.

In robotics, the integration of SMAs has redefined actuation and control strategies, enabling lightweight, flexible, and bioinspired designs. Ruth et al. [56] reviewed SMA-driven robots that leverage large actuation forces and self-sensing capabilities for soft robotic systems, artificial muscles, and compact actuators. SMA-based systems utilize thermal or electrical stimuli to induce motion, mimicking muscle contractions in human-like manipulators. Advances in control methodologies, ranging from proportional integral derivative (PID) and fuzzy PID schemes to pulse width modulation, have improved precision and responsiveness in robotic systems. Active biasing techniques using antagonistic SMA actuators further enhance linearity and reduce hysteresis, enabling real-time adaptive control in prosthetics and microrobots.

Artificial intelligence and machine learning are now being integrated into the design and optimization of 4D-printed and SMA-based systems. Mehrpouya et al. [21] demonstrated a predictive model using artificial neural networks (ANNs) to determine optimal laser parameters for NiTi AM, correlating process inputs such as laser power and scan speed with mechanical and functional outputs. The model achieved reliable prediction of transformation temperatures and strain recovery behavior, enabling real-time control of AM parameters. Similarly, Sivaraos et al. [57] developed an ANN-based model to predict strain recovery in NiTi SMAs under varying thermal and mechanical conditions with an error margin below 1%, significantly surpassing traditional physical models in predictive accuracy and computational speed. This approach allows the fine-tuning of material behavior for applications requiring adaptive control, such as robotic grippers or aerospace morphing systems.

The coupling of 4D printing and AI-based design represents a broader paradigm in smart materials development. Machine learning approaches enable the optimization of material architecture and the prediction of time-dependent responses in 4-dimensional printed smart materials, supporting the design of structures capable of autonomous adaptation in real time as described by Azher et al. [58]. By integrating predictive modeling and AM, intelligent systems can be produced that continuously adapt to environmental and operational conditions, spanning biomedical, aerospace, and robotic domains.

6.5 Conclusions

The chapter establishes that advanced powder-based technologies provide a unified framework for constructing dense, multifunctional, and programmable metallic systems. Cold spray defines a solid-state deposition regime in which critical velocity, surface activation, and adiabatic shear instability govern bonding, enabling oxidation-free, low stress coatings and additively built structures with near-bulk density and tailored interfaces. Filament-based additive manufacturing extends these principles into extrusion-driven architectures, where powder morphology, dispersion, and polymer compatibility control rheology, shrinkage, and final density after debinding and sintering. The integration of these processes with hybrid additive routes and 4-dimensional printing demonstrates that microstructure, phase constitution, and interfacial chemistry can be programmed during fabrication rather than imposed by post-processing. SMAs, magnetically responsive composites, and microstructure engineered steels exemplify how thermal, magnetic, and mechanical stimuli are converted into controlled deformation and functional adaptation. Computational modeling and machine-learning-guided parameter optimization link particle scale physics to macroscopic response.

References

1. Tosto C et al (2021) Hybrid metal/polymer filaments for fused filament fabrication (FFF) to print metal parts. Appl Sci 11(4):1444
2. Karayannis P et al (2022) Facilitating safe FFF 3D printing: a prototype material case study. Sustainability 14:3046. https://doi.org/10.3390/su14053046
3. Karamimoghadam M et al (2023) Influence of post-processing CO2 laser cutting and FFF 3D printing parameters on the surface morphology of PLAs: statistical modelling and RSM optimisation. Int J Lightweight Mater Manuf 6(2):285–295
4. Jacob J et al (2024) A review of fused filament fabrication of metal parts (metal FFF): current developments and future challenges. Technologies 12:267. https://doi.org/10.3390/technologies12120267
5. Fico D et al (2022) A review of polymer-based materials for fused filament fabrication (FFF): focus on sustainability and recycled materials. Polymers 14:465. https://doi.org/10.3390/polym14030465
6. Buj-Corral I, Fenollosa-Artés F, Minguella-Canela J (2024) Three-dimensional printing of metallic parts by means of fused filament fabrication (FFF). Metals 14:1291. https://doi.org/10.3390/met14111291
7. Wu H et al (2022) Bonding behavior of bi-metal-deposits produced by hybrid cold spray additive manufacturing. J Mater Process Technol 299:117375
8. Vilardell AM et al (2020) In-vitro comparison of hydroxyapatite coatings obtained by cold spray and conventional thermal spray technologies. Mater Sci Eng C 107:110306

9. Schmidt T et al (2006) Development of a generalized parameter window for cold spray deposition. Acta Mater 54(3):729–742
10. Pathak S, Saha GC (2017) Development of sustainable cold spray coatings and 3D additive manufacturing components for repair/manufacturing applications: a critical review. Coatings 7: 122. https://doi.org/10.3390/coatings7080122
11. Nourian A et al (2022) Effects of process parameters on cold spray deposition of Al-6061 alloy. J Therm Spray Technol 31(8):2517–2536
12. Klinkov SV, Kosarev VF (2006) Measurements of cold spray deposition efficiency. J Therm Spray Technol 15(3):364–371
13. Kafle A et al (2024) Advancements in cold spray additive manufacturing: process, materials, optimization, applications, and challenges. Materials 17(22):5431
14. Grigoriev S et al (2015) Cold spraying: from process fundamentals towards advanced applications. Surf Coat Technol 268:77–84
15. Khan SW et al (2025) Highly reflective ZrC-Cu-based metal matrix composite coatings deposited via cold-spray for laser protection applications. Opt Laser Technol 182:112171
16. Wu H et al (2020) Deposition of Zn-G/Al composite coating with excellent cathodic protection on low-carbon steel by low-pressure cold spraying. J Alloys Compd 821:153483
17. Hasib AG et al (2021) Rheology scaling of spherical metal powders dispersed in thermoplastics and its correlation to the extrudability of filaments for 3D printing. Addit Manuf 41:101967
18. Hwang S et al (2015) Thermo-mechanical characterization of metal/polymer composite filaments and printing parameter study for fused deposition modeling in the 3D printing process. J Electron Mater 44(3):771–777
19. Gobena ST, Woldeyohannes AD (2024) Comparative review on the application of smart material in additive manufacturing: 3D and 4D printing. Discov Appl Sci 6(7):353
20. Yin S et al (2018) Hybrid additive manufacturing of Al-Ti6Al4V functionally graded materials with selective laser melting and cold spraying. J Mater Process Technol 255:650–655
21. Mehrpouya M et al (2019) A prediction model for finding the optimal laser parameters in additive manufacturing of NiTi shape memory alloy. Int J Adv Manuf Technol 105(11): 4691–4699
22. Alhamdi I et al (2023) Fe-Mn-Al-Ni shape memory alloy additively manufactured via laser powder bed fusion. Crystals 13(10):1505
23. Klenam D et al (2023) Cold spray coatings of complex concentrated alloys: critical assessment of milestones, challenges, and opportunities. Coatings 13(3):538
24. Klinkov SV, Kosarev VF, Rein M (2005) Cold spray deposition: significance of particle impact phenomena. Aerosp Sci Technol 9(7):582–591
25. Sousa BC et al (2020) Understanding the antipathogenic performance of nanostructured and conventional copper cold spray material consolidations and coated surfaces. Crystals 10(6):504
26. Assadi H et al (2011) On parameter selection in cold spraying. J Therm Spray Technol 20(6): 1161–1176
27. Ranjan R, Das AK (2022) A review on surface protective coating using cold spray cladding technique. Mater Today Proc 56:768–773
28. Adachi S, Ueda N (2017) Effect of cold-spray conditions using a nitrogen propellant gas on AISI 316L stainless steel-coating microstructures. Coatings 7(7):87
29. Stoltenhoff T et al (2006) Microstructures and key properties of cold-sprayed and thermally sprayed copper coatings. Surf Coat Technol 200(16):4947–4960
30. Gärtner F et al (2006) Mechanical properties of cold-sprayed and thermally sprayed copper coatings. Surf Coat Technol 200(24):6770–6782
31. Yin S et al (2019) Hybrid additive manufacture of 316L stainless steel with cold spray and selective laser melting: microstructure and mechanical properties. J Mater Process Technol 273: 116248
32. Forstner T, Cholewa S, Drummer D (2024) Enhanced feedstock processability for the indirect additive manufacturing of metals by material extrusion through ethylene–propylene copolymer modification. Polymers 16(18):2658

33. Gibson MA et al (2018) 3D printing metals like thermoplastics: fused filament fabrication of metallic glasses. Mater Today 21(7):697–702
34. Roshchupkin S et al (2021) A brief review of main ideas of metal fused filament fabrication. Mater Today Proc 38:2063–2067
35. Costa JM, Sequeiros EW, Vieira MF (2023) Fused filament fabrication for metallic materials: a brief review. Materials 16:7505. https://doi.org/10.3390/ma16247505
36. Ren L et al (2017) Process parameter optimization of extrusion-based 3D metal printing utilizing PW–LDPE–SA binder system. Materials 10(3):305
37. Shaqour B et al (2021) Gaining a better understanding of the extrusion process in fused filament fabrication 3D printing: a review. Int J Adv Manuf Technol 114(5):1279–1291
38. Bouzaglou O, Golan O, Lachman N (2023) Process design and parameters interaction in material extrusion 3D printing: a review. Polymers 15:2280. https://doi.org/10.3390/polym15102280
39. Malagutti L et al (2022) A post-processing method for improving the mechanical properties of fused-filament-fabricated 3D-printed parts. PRO 10:2399. https://doi.org/10.3390/pr10112399
40. Mousapour M et al (2021) Feasibility study of producing multi-metal parts by fused filament fabrication (FFF) technique. J Manuf Process 67:438–446
41. Alarifi IM (2024) Revolutionising fabrication advances and applications of 3D printing with composite materials: a review. Virtual Phys Prototyp 19(1):e2390504
42. Raj A et al (2024) Advancements in material extrusion based three-dimensional printing of sensors: a review. Int J Interact Des Manuf (IJIDeM) 18(2):627–648
43. Zafar MQ, Zhao H (2020) 4D printing: future insight in additive manufacturing. Met Mater Int 26(5):564–585
44. Mahmood A et al (2022) On the evolution of additive manufacturing (3D/4D printing) technologies: materials, applications, and challenges. Polymers 14(21):4698
45. Chen X et al (2019) Multi-metal 4D printing with a desktop electrochemical 3D printer. Sci Rep 9(1):3973
46. Li X, Shang J, Wang Z (2017) Intelligent materials: a review of applications in 4D printing. Assem Autom 37(2):170–185
47. Wang H et al (2023) Recent research developments of 4D printing technology for magnetically controlled smart materials: a review. Magnetochemistry 9:204. https://doi.org/10.3390/magnetochemistry9080204
48. Song K et al (2025) Multi-material 4D printing and 3D patterned metallization enables smart architectures. Compos Part B 295:112218
49. Melly SK et al (2020) On 4D printing as a revolutionary fabrication technique for smart structures. Smart Mater Struct 29(8):083001
50. Gao S et al (2023) Additive manufacturing of alloys with programmable microstructure and properties. Nat Commun 14(1):6752
51. Tan C et al (2023) Machine learning customized novel material for energy-efficient 4D printing. Adv Sci 10(10):2206607
52. Costanza G, Tata ME (2020) Shape memory alloys for aerospace, recent developments, and new applications: a short review. Materials 13:1856. https://doi.org/10.3390/ma13081856
53. Safaei K et al (2021) Additive manufacturing of NiTi shape memory alloy for biomedical applications: review of the LPBF process ecosystem. JOM 73(12):3771–3786
54. Bajpai A et al (2020) 4D printing: materials, technologies, and future applications in the biomedical field. Sustainability 12:10628. https://doi.org/10.3390/su122410628
55. Kang J et al (2024) 4D printed NiTi variable-geometry inlet for aero engines. Virtual Phys Prototyp 19(1):e2382166

56. Ruth DJ et al (2022) Control aspects of shape memory alloys in robotics applications: a review over the last decade. Sensors 22:4860. https://doi.org/10.3390/s22134860
57. Sivaraos et al (2025) Artificial neural network for performance modelling of shape memory alloy. Int J Interact Des Manuf (IJIDeM) 19(9):6655–6671
58. Azher K et al (2025) Revolutionizing the future of smart materials: a review of 4D printing, design, optimization, and machine learning integration. Adv Mater Technol 10(12):2401369

Chapter 7
Sustainable Powder Metallurgy

Abstract Sustainable powder metallurgy is defined by the integration of solid-state recycling, energy-efficient consolidation, and circular economy principles that collectively decouple materials performance from primary resource extraction. Direct conversion of industrial scrap into powders through mechanical milling and atomization enables substantial reductions in energy demand while inducing microstructural refinement, lattice strain, and phase homogenization that are retained through subsequent sintering or additive manufacturing. Recycled powders derived from steels, aluminum alloys, and titanium demonstrate mechanical properties comparable to or exceeding those of virgin feedstocks when contamination, oxidation, and particle degradation are rigorously controlled. Energy consumption is dominated by sintering, where conventional furnace efficiencies remain below 30%, motivating the adoption of field-assisted routes such as spark plasma, microwave, and induction sintering that achieve near-full density at reduced temperatures and times while suppressing grain growth and deleterious reactions. Binder chemistry and decomposition pathways exert a nontrivial influence on microstructural evolution, enabling in situ reactions, oxide reduction, and tailored phase morphologies. Digitalization, traceability, and data-driven monitoring provide the infrastructure required to manage powder degradation, enable closed-loop reuse, and link life cycle assessment with process control. Together, these advances establish powder metallurgy as a materially circular manufacturing paradigm in which sustainability emerges from coupled control of energy flows, microstructure, and information across the full product life cycle.

7.1 Introduction

The integration of materials recycling, energy-efficient processing routes, and circular economy pathways across the metallic value chain defines sustainable powder metallurgy (PM). Industrial manufacturing produces metallic scrap in multiple forms, including machining chips, mill scale, grinding sludge, and additive manufacturing residues, which constitute secondary material streams alongside

R. Pérez-Bustamante, *Fundamental, Practical and Disruptive Applications of Powder Metals*, Topics in Mining, Metallurgy and Materials Engineering,
https://doi.org/10.1007/978-3-032-27130-3_7

conventional waste byproducts. Through mechanical milling and atomization, these streams can be converted into powders with controlled morphology and microstructure, enabling their direct reintegration into PM and additive manufacturing routes [1–13]. Planetary and attritor milling of carbon steels, tool steels, and aluminum alloys induce severe plastic deformation, crystallite refinement, and microstrain accumulation, producing submicron powders without the energy demand of melting. In parallel, gas atomization of recycled feedstock, including stainless steels and construction steel scrap, yields spherical powders with narrow size distributions and low oxygen content, supporting laser-based additive manufacturing with mechanical performance comparable to or exceeding that of virgin materials [14–16].

Quality control protocols are routinely implemented to address powder degradation during successive reuse cycles. Oxidation, spatter-related contamination, and particle coarsening lead to changes in flowability and packing density, with direct consequences for fusion behavior in laser powder bed systems. Monitoring of these effects relies on analytical techniques such as SEM, XRD, and particle size analysis, which also underpin corrective measures including sieving, blending, and plasma spheroidization to recover particle morphology and chemical purity [15, 17–20].

In this chapter, we discuss how energy efficiency is constrained by sintering and powder annealing, which dominate the energy footprint of PM owing to furnace efficiencies below 30% and losses associated with protective atmospheres and partial loading, whereas field-assisted routes such as spark plasma, microwave, and induction sintering enable rapid densification at reduced temperatures, limiting grain coarsening and oxidation while lowering specific energy consumption. Within a circular economy scheme, recycled powders, low temperature consolidation, and digitalized traceability operate as a coupled system in which waste streams are converted into high value products, with scrap conversion, powder quality, and energy input jointly governing closed-loop PM under preserved mechanical and functional performance [3, 21–23].

7.2 Recycling of Scrap Metal into Powders

7.2.1 Sources of Metallic Scrap Suitable for Powder Metallurgy

The generation of metallic scrap (Fig. 7.1) is an inherent byproduct of industrial manufacturing processes such as machining, trimming, and cutting, presenting a significant opportunity for recycling through PM techniques. A primary source of such scrap is ferrous chips, which are produced in substantial volumes and are highly suitable for conversion into powder. Verma et al. [20] demonstrated the direct recycling of industrial steel scrap, specifically extra low carbon and low carbon steel chips, into nanostructured powders using planetary ball milling. Steel chips that would otherwise be remelted, a process associated with high energy demand and

Fig. 7.1 Metal chips produced from lathe machining operations

additional waste generation, can instead be converted into submicron scale powders through mechanical pulverization The morphology of the as received chips was noted to influence the milling behavior; low carbon steel chips were more or less straight, while extra low carbon chips were more curved, yet both were successfully transformed into powder after milling for 5 h under a toluene environment to prevent oxidation.

In addition to conventional steel scrap, high value alloy residues are also used as feedstock in PM. Gonçalves et al. [24] investigated the recycling of Vanadis®8 tool steel scraps, a high performance steel used in tool and die manufacturing, through high energy ball milling (HEBM). The scraps, generated by machining an annealed billet without lubricants to avoid contamination, exhibited a helical morphology with an average thickness of 1.4 mm. Powder production after 30 h of milling at 400 rpm yielded a trimodal particle size distribution along with pronounced crystallite refinement. The process showed no evidence of contamination and introduced microstrain together with reduced crystallite size in both the ferritic matrix and vanadium carbide phases, supporting the use of complex high value alloy scrap as feedstock for PM recycling.

Aluminum-based scrap, as illustrated in Fig. 7.2, represents another material stream that can be repurposed through PM. Kumar and Bharti [25] described several approaches for recycling aluminum chips, highlighting the high material utilization and near-net-shape capability of the process. They outlined three primary approaches for aluminum scrap recycling: the direct recycling of unreinforced aluminum chips, the creation of a two-phase material by mixing scrap chips with primary aluminum powder, and the reinforcement of the scrap aluminum matrix with ceramic particles such as silicon carbide (SiC) or alumina (Al_2O_3). The physical and mechanical properties of the final product were shown to be highly dependent on the initial chip characteristics, including their shape, size, and weight fraction, as well as process parameters like compaction pressure.

Fig. 7.2 Scrap from aluminum extruded profiles

7.2.2 Technologies for Scrap Conversion into Powders

The transformation of metallic scrap into powders for sustainable PM involves the development of several recycling routes based on mechanical and atomization technologies. Among these, gas atomization and mechanical milling represent the most relevant and mature approaches for converting waste metals into feedstock suitable for additive manufacturing (AM) and sintering applications. Gas atomization techniques rely on the rapid solidification of molten recycled alloys into fine powders under controlled inert gas streams, whereas mechanical methods such as ball milling and attritor milling induce plastic deformation and fracture of solid metal chips into particles of suitable size and morphology for PM. Both processes contribute significantly to the transition toward a circular economy by reducing primary raw material dependence and promoting the reuse of industrial metallic waste.

In the atomization route, the production of powders from recycled materials has demonstrated high efficiency and strong alignment with the requirements of AM. Cacace et al. [15] examined the feasibility of using recycled AISI 316 L stainless steel to generate powders via vacuum-induced gas atomization, later applied in selective laser melting (SLM) and laser metal deposition (LMD) processes. Powders produced from recycled feedstock reached mechanical performance comparable to, and in some cases higher than, that of powders derived from virgin materials, with ultimate tensile strengths of 653 ± 1.79 MPa for selective laser melting and 675 ± 2 MPa for laser metal deposition, compared with 594 ± 0.89 MPa and 585 ± 3.1 MPa for their standard counterparts. These differences are associated with microstructural refinement arising from repeated melting and solidification cycles during recycling-based powder production. Control of atomization parameters such as gas pressure, melt superheat, and nozzle geometry influences the formation of spherical particles with narrow size distributions and low oxygen content, which supports flowability and uniform layer spreading in powder-based AM systems.

Bernsmann and Schleifenbaum [5] described the conversion of large-scale construction steel scrap into atomized powder feedstock for laser powder bed fusion.

The full recycling route was detailed, from demolition steel segregation and pre-cleaning through atomization using an ALD VIGA 2B system to the AM of dense components. Approximately 50 kg of dismantled structural steel were atomized, yielding 27.8 kg of powder below 90 μm, which proved suitable for PBF LB/M processing. The resulting parts exhibited relative densities near 99.9% and tensile strength values around 808 MPa, comparable with wrought steel, confirming the viability of using reclaimed construction scrap for structural AM applications. Key to this success was the optimization of the atomization process by adjusting gas flow rate, nozzle gap, and melt temperature between 1540 °C and 1580 °C, ensuring stable droplet formation and preventing oxidation. The integration of such high efficiency gas atomization systems with rigorous pretreatment steps, including sandblasting and chemical cleaning of scrap, enabled the production of powders with consistent morphology and minimal contamination.

Complementary to gas atomization, mechanical conversion methods such as ball milling provide an alternative route for powder synthesis from solid scrap without remelting, thus minimizing energy consumption and preserving alloying elements susceptible to volatilization or oxidation. According to Wei et al. [12], the ball milling process serves as a key mechanical alloying method capable of transforming machining chips into fine powders through repeated cold welding, fracturing, and rewelding under high energy impact conditions. Parameters such as ball-to-powder ratio, rotational speed, and milling time influence powder size reduction, morphology, and homogeneity. This technique enables the processing of various metals, including copper, steel, and aluminum, where the transition from ductile to brittle behavior under high strain rates facilitates effective fragmentation. Optimization of the process control agent and milling atmosphere limits cold welding and oxidation, yielding powders suitable for mold AM and rapid tooling. Mechanical milling also avoids melting steps, which reduces energy demand and associated emissions.

Batista et al. [2] described the parameters and mechanisms governing the transformation of machining chips into powders by milling. Attritor, planetary, and disc milling techniques were outlined as solid-state processes that impose severe plastic deformation, leading to the formation of nanostructured or submicron particles. Control variables such as milling time, rotational speed, ball size, and the use of process control agents directly affect the evolution of particle size, shape, and structural refinement, referred to as the 4S's: size, size distribution, shape, and structure. A ball-to-powder ratio increase enhances fragmentation efficiency, whereas cryogenic milling can assist ductile-to-brittle transition, promoting finer particles. The resulting powders have been found to exhibit microstructural refinement suitable for sintering and AM, representing an energy-efficient and sustainable alternative to conventional atomization. Jet milling and shaker milling were compared with planetary ball milling, showing faster fragmentation rates and offering pathways toward industrial scale processing of large volumes of metallic waste.

Recycling waste streams that contain fine metallic residues, such as grinding sludge, expands the scope of scrap to powder technologies. Jäger et al. [26] investigated the potential of recovering metallic fractions from grinding sludge, composed of tool steel particles, abrasives, and lubricants, through powder metallurgical processes. By applying Supersolidus Liquid Phase Sintering (SLPS) and Hot Isostatic Pressing (HIP), the metallic swarf derived from tool steel X155CrVMo12-1 was consolidated into dense components with mechanical properties comparable to conventionally atomized powders. Removal of contaminants and attainment of adequate powder purity were identified as factors affecting sintering densification and mechanical integrity. The combination of metallic sludge with gas atomized powders improved overall density and reduced processing costs, indicating that hybrid recycling strategies could further expand powder feedstock availability while maintaining high material performance standards.

7.2.3 Quality Control and Contamination Challenges in Recycled Powders

Quality control and contamination challenges in recycled powders have emerged as central issues in the sustainability of PM and AM (differences in powder characteristics are observed in Figs. 7.3 and 7.4). Recycled powders, though essential for reducing costs and environmental impact, undergo significant degradation during repeated use, necessitating rigorous quality control to maintain process reliability and mechanical integrity of the final components. According to Lanzutti and Marin

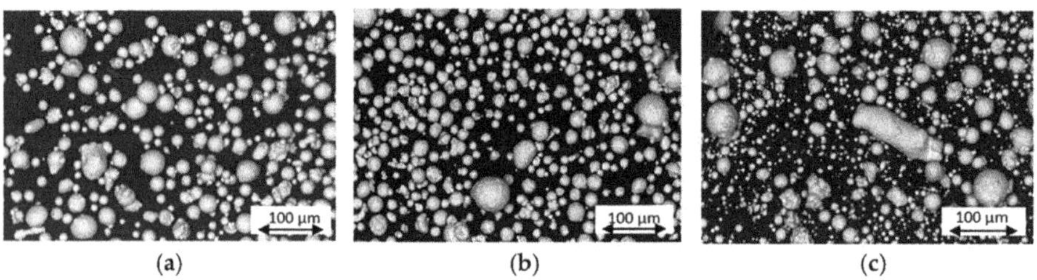

Fig. 7.3 Scanning electron microscopy (SEM) images of MS1 powder: (**a**) virgin, (**b**) used, (**c**) recycled [16]

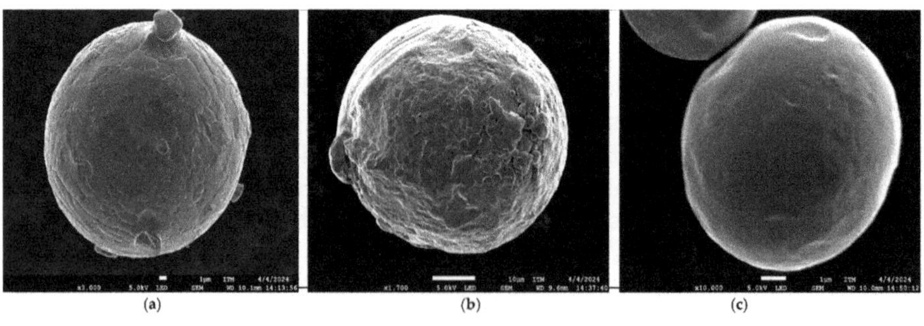

Fig. 7.4 SEM images of particles of MS1 powder: (**a**) virgin, (**b**) used, (**c**) recycled [16]

[27], metal powders used in AM exhibit property changes during successive recycling cycles, including oxidation, particle coarsening, and morphology alteration, all of which compromise powder flowability and packing density. Such degradation directly affects layer uniformity in processes like Laser Powder Bed Fusion (L-PBF) and Electron Beam Powder Bed Fusion (EB PBF), leading to porosity and inconsistent melting behavior. The presence of surface oxides and contaminant particles introduces further variability in energy absorption during laser processing, making consistent quality assurance indispensable throughout the recycling chain.

Powell et al. [28] described powder degradation as a multifactorial process governed by interactions among thermal exposure, oxidation kinetics, and mechanical handling during reuse. The absence of a universal end-of-life criterion was noted, as degradation depends on material type, process parameters, and environmental control. Instead, direct analysis of components produced from recycled powders remains the most reliable indicator of powder usability. This necessitates integrating feedback from mechanical testing, such as tensile strength and hardness, into powder quality protocols. Moisture and residual binder contamination were identified as major contributors to powder oxidation and agglomeration, particularly in stainless steel and titanium alloys. These factors result in reduced flowability and increased porosity in built parts, underscoring the need for strict humidity control and post-processing sieving as part of the quality assurance routine.

Soundarapandiyan et al. [29] provided a detailed mechanistic analysis of contamination in LPBF processes, focusing on laser spatter as a major source of chemical and physical nonuniformities in recycled powders. Spatter particles, typically larger and compositionally distinct from virgin powder, reentered the powder bed and modified both particle size distribution and surface chemistry. These spatters, enriched in oxides of manganese, silicon, and other elements, increase oxygen content in subsequent builds, especially in reactive alloys like titanium and aluminum. Such contamination not only deteriorates powder flow but also leads to lack-of-fusion defects and higher surface roughness in printed parts. Spatter control through optimized gas flow and chamber design, combined with preheating strategies to stabilize the melt pool, was suggested as an effective contamination mitigation technique.

7.2.4 Industrial Scale Recycling in Powder Metallurgy

Industrial scale recycling within PM has increasingly demonstrated its viability for achieving sustainable production through the integration of AM, direct recycling of metallic scrap, and material life cycle optimization. Rane and Date [30] described the conversion of finely divided ferrous scrap into functional components through PM and metal injection molding, indicating that solid-state recycling via PM provides higher recyclability indices as a result of increased densification and lower energy consumption (Fig. 7.5). Their process emphasized carbothermic reduction during

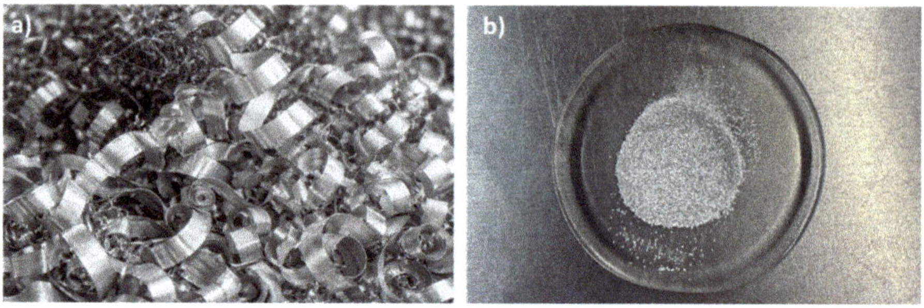

Fig. 7.5 (**a**) Metal chips from lathe machining turn into (**b**) powder metal after the milling process

sintering, converting iron oxides into dense metallic parts with minimal material losses, achieving a high recyclability index for porous structural components used in tribological applications.

Complementing this, Bulei et al. [14] focused on recycled aluminum alloys to produce metal matrix composites via stir casting. This approach provided a low cost and energy-efficient method for integrating reinforcing particulates uniformly within recycled aluminum matrices. Recycled aluminum alloys used as matrix materials in composites achieved high strength-to-weight ratios suitable for automotive and aerospace applications, supporting the use of solid-state recycling routes in the production of lightweight components.

Yamanoglu and coworkers [31] explored the recycling of Ti6Al4V support structures generated during selective laser melting (SLM) at a technological level. These support structures, typically considered waste, were consolidated by vacuum hot pressing to create dense materials exhibiting improved wear resistance. Their experiments compared spherical and irregular Ti6Al4V powders with recycled supports, reporting that samples from recycled supports achieved the highest wear performance. This illustrates the circular potential of AM by converting production residues into high performance materials while reducing waste generation and powder consumption.

The ecological implications of such recycling approaches are further reinforced by Wurst et al. [13], who conducted a multiple case life cycle assessment (LCA) on metal additive repair and refurbishment using powder bed fusion by laser beam (PBF LB). Material usage, particularly the amount of metallic powder consumed, was identified as the main factor influencing environmental impact. Additive repair and refurbishment were associated with extended product lifetimes and reduced environmental footprints relative to full material recycling or remanufacturing. Optimization of material flows and lower powder consumption emerged as effective means for improving ecological performance in metal additive repair chains.

Stieberova et al. [10] expanded on these concepts through a life cycle assessment comparing direct metal laser sintering (DMLS) molds to conventionally machined counterparts for zinc die casting. Despite higher powder production energy demands, DMLS molds demonstrated lower environmental impacts over their entire life cycle due to improved cooling efficiency, reduced cycle times, and extended mold lifespans. The cumulative energy demand and greenhouse gas emissions were significantly lower in the additive process during the use phase, proving that even energy-intensive AM can yield superior sustainability outcomes when evaluated holistically.

Beiker Kair and Sofos [32] examined the use of metallic AM in the automotive sector, with a focus on turbocharger components. Their sustainability analysis compared AM with casting and concluded that, while AM remains cost-intensive at low production volumes, it offers significant environmental and performance advantages. Reduced material waste, improved thermal management, and lower component weight were highlighted, along with projections that future reductions in metal powder costs and machine operating expenses would further support large-scale recycling within PM-based production systems.

Ford and Despeisse [33] reported a broader perspective by identifying AM as a cornerstone of sustainable industrial transformation. AM promotes material efficiency, local production, and extended product life cycles through repair and remanufacture strategies. Potential challenges include increased resource consumption associated with decentralized production and the need for robust recycling systems to manage metal powders and process byproducts. In this regard, Arrizubieta et al. [34] examined the environmental implications of metal powder handling and reuse in AM, with attention to occupational safety, waste management, and the development of standardized powder recycling protocols to limit health risks and maintain process efficiency. It was reported that repeated powder reuse can maintain material integrity up to a point, beyond which contamination and oxidation increase environmental impacts. Hence, efficient powder recovery and safe handling practices are integral to industrial sustainability in PM recycling systems.

7.3 Energy-Efficient Sintering and Green Binders

7.3.1 Fundamentals of Energy Consumption in Sintering Processes

Sintering and processing (Fig. 7.6) is a thermal step in PM that accounts for a substantial portion of the energy required for component manufacturing. Kruzhanov and Arnhold [35] reported that sintering, together with other furnace-based operations such as powder annealing, dominates the energy consumption profile of a typical PM facility, with an energy demand of approximately 1.8 kWh per kg of finished components. This value represents a substantial portion of the average total energy consumption of 5.0 kWh per kg for PM part production. Their experimental investigation involved a small belt furnace with a 0.3-m-wide conveyor belt, equipped to measure energy consumption across heating zones, cooling water power, and furnace surface temperature. The energy balance for this furnace, operating at 1120 °C with a parts throughput of 75 kg per hour, demonstrated that

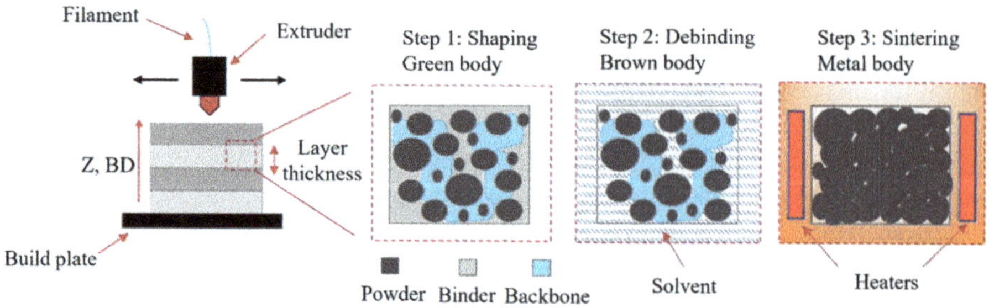

Fig. 7.6 Illustration of the processing steps to produce dense metal or ceramic parts [36]

less than 30% of the total energy input was used as productive heat to raise the temperature of the parts themselves. The majority of the energy was dissipated through radiation losses, heating of the conveyor belt and protective atmosphere, and cooling systems, resulting in a furnace efficiency of under 30%. This low efficiency is attributed to inherent system losses that are challenging to eliminate entirely though improvements in furnace design and loading practices are identified as key areas for potential savings.

Azevedo et al. [37] define energy efficiency as the ratio of the theoretical minimum specific energy consumption to the actual consumption observed in practice. For the sintering process, their calculated theoretical minimum energy requirement is 0.6 MJ/kg, while the actual energy consumption from efficient industrial practice is reported as 2.4 MJ/kg, yielding an energy efficiency of 26%. This agrees with the observations of Kruzhanov and Arnhold [35], indicating that sintering involves substantial energy losses. Furnace efficiency is constrained by the limited green strength of compacted components, which prevents dense stacking and requires the use of continuous furnaces with higher heat losses than batch systems. The energy consumption for sintering is also influenced by the generation of the protective atmosphere. Kruzhanov and Arnhold estimate that the energy associated with creating common sintering atmospheres, such as nitrogen-hydrogen mixtures or endothermic gas, plus the energy for the burn-off atmosphere, can add an average of 0.5 kWh per kg to the total energy footprint of the sintering step [35].

7.3.2 Low Temperature Sintering Routes

Low temperature sintering routes such as spark plasma sintering (SPS, illustrated in Fig 7.7), microwave sintering, and induction sintering have emerged as transformative techniques for energy-efficient PM processing. These technologies enable rapid densification at reduced temperatures, minimizing grain growth, oxidation, and energy consumption while enhancing the mechanical and functional properties of metallic and ceramic composites. In spark plasma sintering, the application of pulsed direct current and mechanical pressure facilitates localized Joule heating and plasma

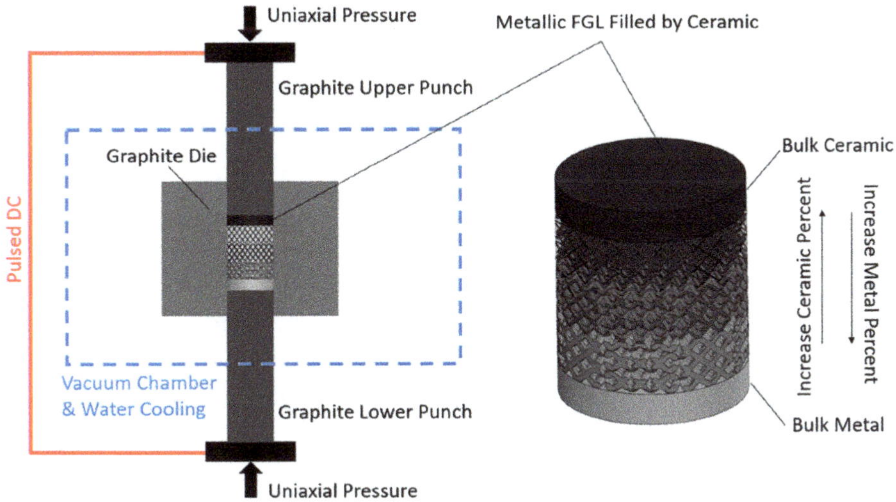

Fig. 7.7 Schematic of the spark plasma sintering process [38]

discharge between powder particles, promoting diffusion-driven bonding and high densification in very short times. Daoush et al. [17] demonstrated that SPS processing of CNTs-Si3N4/Cu nanocomposites produced almost full density structures with relative densities approaching 100% and a reduced coefficient of thermal expansion (1.6×10^{-5} °C^{-1}) compared to conventionally sintered counterparts, which reached only about 85% density. A sintering time of 1 min at 850 °C under 50 MPa limited grain coarsening and preserved the nanostructure, leading to improved thermal and mechanical stability for applications such as heat sinks and electronic packaging.

Similar advances in alloy systems have been reported for refractory and light metal alloys. Rahman and coworkers [39] used SPS to consolidate elemental powders of the niobium-based alloy C103 (Nb-10Hf-1Ti) at temperatures well below the solidus, achieving nearly full densification with only 1.3 vol.% residual porosity and significant mechanical improvements. The resulting alloy exhibited a yield strength of 394 MPa, tensile strength of 458 MPa, and 23% elongation, surpassing wrought and additively manufactured counterparts despite higher oxygen content (>2000 ppm). The plasma-assisted mechanism of SPS enhanced diffusion and homogenization of alloying elements while minimizing oxide inclusion effects. Moreover, the capacity of SPS to consolidate elemental powders directly into near-net-shape components without extensive post-processing demonstrated its promise for high temperature structural applications such as turbine blades and propulsion systems, where conventional sintering would require longer durations and higher thermal input.

Golmohammadi et al. [18] expanded the application of SPS to lightweight materials through the synthesis of an Mg-0.7Ca alloy, illustrating how short sintering cycles at 420 °C under 38 MPa for less than 15 min yield fully dense compacts with

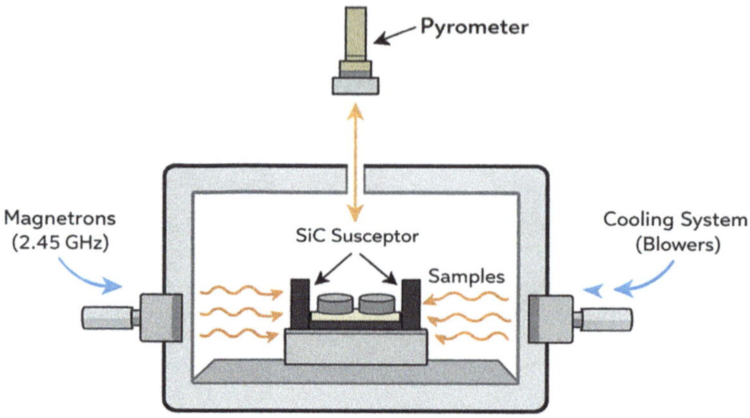

Fig. 7.8 Schematic diagram of a microwave sintering furnace

refined grain boundaries and tailored textures. The thermomechanical behavior of the alloy showed sensitivity to sintering time and binder type, with paraffin-assisted powders favoring densification and starch-assisted systems enhancing compressive strength through the formation of secondary oxide and carbon phases. The microstructural observations revealed that the field-assisted sintering effectively reduced porosity while preventing oxidation, a typical limitation in magnesium processing.

On the other hand, microwave-assisted sintering (illustrated in Fig. 7.8), has proven equally valuable in producing dense and defect-free metal matrix composites at lower temperatures and with shorter cycles. Manohar et al. [40] investigated AA7075/SiC/ZrC hybrid composites and demonstrated that microwave sintering enhanced tensile strength by up to 184.5% and compressive strength by 118% compared with conventionally sintered materials. The superior performance was attributed to uniform internal heating, reduced thermal gradients, and minimal pore formation (1.39% porosity versus 2.89% in conventional sintering). The absence of secondary phases such as Al_3Zr, observed in conventionally sintered samples, confirmed that rapid volumetric heating restricted undesirable intermetallic reactions. The finer grain size (4.99 μm) and homogeneous reinforcement distribution obtained by microwave energy promoted both dislocation and grain boundary strengthening mechanisms. Similar effects were reported in a related study on AA7075/B_4C/ZrC nanocomposites, where microwave sintering generated strong and clean matrix–reinforcement interfaces without detectable intermetallic compounds, doubling ductility from 6.12% to 12.24% compared to conventional processing. The mechanism involved dipole oscillation within the ceramic reinforcements, which served as internal heat sources, ensuring uniform temperature distribution and clean bonding interfaces throughout the composite structure.

Induction sintering offers another energy-efficient route for rapid densification, particularly effective for conductive powders. Jia and Gabbitas [41] applied induction sintering in combination with warm compaction to synthesize a near-β titanium alloy Ti-13 V-11Cr-3Al by blended elemental PM, achieving homogeneity after only 10 min of isothermal holding at 1300 °C. The alternating electromagnetic field induced localized heating at particle interfaces, promoting diffusion without excessive grain coarsening. Pore coarsening followed Ostwald ripening during extended hold times, whereas intermediate sintering durations produced a balance between density and compositional uniformity. The approach minimized porosity and produced mechanical properties consistent with fully dense wrought alloys, confirming the potential of field-assisted techniques in producing advanced β titanium materials efficiently.

7.3.3 Development and Characterization of Bio-Based and Non-Toxic Binders

Based on the research by Tsygoda et al. [42], the influence of organic binders, such as those illustrated in Fig. 7.9, and their decomposition products on the microstructure and thermoelectric properties of conductive materials was systematically investigated. The work addressed silicon nitride-based composites containing additions of transition metal carbides, specifically hafnium carbide (HfC) and zirconium carbide (ZrC). The binders examined were carboxymethyl cellulose (CMC), used in combination with a glycerin plasticizer, and natural rubber. System mass loss during processing depended on binder type, with samples containing resistive HfC produced using rubber binders showing an increase in mass loss with HfC content up to 27%, followed by a subsequent decrease. When CMC was employed as the binder, the mass loss of the system was approximately 1.3 to 1.6 times lower, while following similar trends relative to HfC concentration. The pyrolysis behavior of

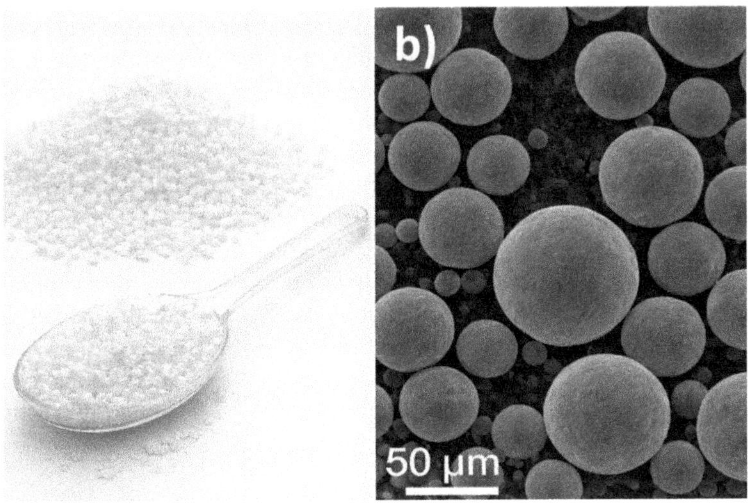

Fig. 7.9 (**a**) Binders used in (**b**) PM process

these binders was detailed, showing that CMC leaves a significant concentrated charred residue, whereas rubber enriches the material with carbon black. Furthermore, an admixture of glycerin in CMC was found to shift its decomposition temperature to a lower region and accelerate its decomposition, making such a binder promising for use in resistive materials manufactured using in situ reactions.

The decomposition pathways of these binders were linked to subsequent chemical reactions during high temperature consolidation. For the rubber binder, decomposition produced hydrocarbon gases, while CMC decomposition led to the release of carbon, hydrogen, and sodium. These decomposition products participated in gas phase reactions, such as the reduction of oxides and the formation of water and carbon monoxide. The presence of these gases facilitated in situ reactions, including the formation of silicon carbide (SiC) and silicon oxynitride (Si_2ON_2), which were dependent on the local atmosphere created by the binder pyrolysis. The binder species was found to influence the morphology of the resulting conductive cluster microstructure. Hafnium carbide particles formed in the presence of CMC and zirconium carbide particles formed with rubber exhibited a flake-like morphology aligned with the plane of the applied hot pressing force. In contrast, hafnium carbide particles obtained with rubber had a cylindrical shape elongated in the same direction, and zirconium carbide particles with CMC became more spherical [42].

The role of binders in PM processes is extensively covered by Vicenzi and colleagues [43], particularly in the Metal Injection Molding (MIM) technique. In MIM, organic binders, typically thermoplastic polymers and waxes, are mixed with metal powders to form a feedstock that can be injection molded. The primary function of the binder is to confer flowability for shaping, after which it must be removed through debinding processes such as thermal, solvent, or catalytic debinding. The complete removal of the binder is crucial, as insufficient removal can lead to defects like blistering or cracking in the final sintered part. Following debinding, the parts are sintered at high temperatures, achieving densities often above 96% of the theoretical value. The selection of binder components is critical, as they influence the rheology during molding, the efficiency of the debinding step, and the final dimensional stability and properties of the component.

Ashrafi et al. [44] described the use of magnesium stearate as a binder in the PM fabrication of aluminum matrix composites reinforced with Fe_3O_4 nanoparticles. The binder was introduced during ball milling to limit particle agglomeration and promote a homogeneous distribution of the nanofiller within the aluminum matrix. The mixed powders, containing the binder, were then compacted and sintered. The incorporation of the binder was a necessary step to refine the distribution of the reinforcement and to facilitate the compaction process before the thermal cycle. This approach resulted in composites with uniform microstructures and improved properties, demonstrating that the judicious use of a binder is a key technological factor in controlling the microstructure and, consequently, the physical and magnetic properties of the final material.

7.4 Circular Economy Models in Powder Metallurgy

7.4.1 *Principles of Circular Economy Applied to Powder Metallurgy*

The circular economy principles applied to PM involve the transformation of industrial waste into high value materials through the recovery, reprocessing, and reuse of metallic byproducts within metallurgical cycles. This approach is consistent with the sustainable production and consumption framework that encourages resource efficiency and environmental responsibility. The integration of circular economy models into PM has been demonstrated through several innovative practices that transform byproducts such as metallurgical slags, mill scale, and metallic chips into new functional materials, reducing raw material consumption and waste generation.

One of the most illustrative cases is the recycling of industrial metallic waste to produce high wear-resistant hardfacings using submerged arc welding (SAW). The Bendikiene group [21] developed a process in which alloyed steel shavings, cast iron chips, and debris from worn grinding wheels were combined with waste rich in carbide-forming elements like tungsten and vanadium to form hard coatings with improved wear resistance. The process utilized the SAW technique with optimized parameters, integrating waste metal powders into the weld zone either by surface spreading or by flux mixing. The resulting hardfacings achieved hardness values up to 63 HRC and exhibited wear resistance nearly twice that of conventional alloys, proving the viability of waste-based feedstock for producing durable metallurgical products.

Similarly, the research conducted by Sadenova et al. [45] extended these concepts to the reuse of metallurgical slags for the synthesis of ceramic catalyst carriers through PM. Metallurgical slag, with a composition similar to that of natural aluminosilicates, was shown to substitute conventional mineral raw materials in the formation of catalyst substrates. By combining zeolite, bentonite, and metallurgical slag, and processing them using PM techniques involving fine grinding, extrusion, and sintering, it was possible to synthesize ceramic materials with the desired structural and catalytic properties. The use of slags rich in iron and silicon oxides not only diverted industrial waste from landfills but also yielded materials suitable for environmental catalysis, supporting waste valorization within the metallurgical and chemical sectors. Analytical methods such as X-ray diffraction, thermogravimetry, and electron microscopy were applied to control the transformation of slag-based raw materials into stable ceramic matrices, supporting material quality in circular manufacturing systems.

In a related contribution to circular metallurgy, Tirado González and coworkers [11] investigated the valorization of mill scale, a byproduct of high temperature steel rolling, through PM to produce an oxide dispersion-strengthened iron compound termed "ironox." Partial reduction of calamine powders under hydrogen atmospheres was used to produce composite particles consisting of metallic iron and residual magnetite. This structure provided mechanical reinforcement due to phase coherency between the oxide and metallic matrix, leading to hardness values as high as 77.7 HRB. Furthermore, the incorporation of magnetite imparted ferrimagnetic behavior, expanding the potential functional range of the material for magnetic applications. The process demonstrated that PM can serve as a circular pathway for reintroducing byproducts into production cycles, enhancing both the mechanical and functional properties of recycled metals while conserving energy compared to conventional smelting processes.

7.4.2 Environmental and Economic Benefits of Scrap Recycling

The environmental and economic benefits of scrap recycling in PM are strongly linked to the circular economy paradigm, which emphasizes resource efficiency, waste minimization, and the valorization of secondary raw materials. Recycling metallic scrap (Fig. 7.10) goes beyond waste management and supports the development of more sustainable industrial systems. The group of Giurco [23] stressed that the global transition toward circularity will increasingly depend on the capacity to integrate mined and recycled materials into responsible supply chains, thereby reducing the reliance on virgin ores and the energy intensity of production. The reintroduction of scrap into PM production loops offers both environmental and economic advantages by lowering the cumulative energy demand and curbing greenhouse gas emissions that would otherwise arise from ore extraction and primary metal refining.

Recycling scrap metals, particularly aluminum and copper, can reduce energy consumption by up to 95% and 90%, respectively, compared to production from virgin ores, demonstrating a direct correlation between recycling efficiency and carbon footprint mitigation. Such reductions highlight the transformative potential of closed-loop material systems, where metal powders generated from scrap feedstocks substitute primary inputs without compromising performance standards. This

Fig. 7.10 Scrap metal

dynamic is also economically favorable since the volatility of raw material markets makes recycled metals a cost stabilizing alternative, mitigating supply chain risks associated with resource scarcity and geopolitical constraints. Moreover, Giurco et al. [23] underscored that improvements in separation technologies and advanced characterization tools are crucial for ensuring the purity of secondary metal streams, which directly influence their suitability for high performance PM applications.

From a life cycle perspective, Mecheter and collaborators [7] provided quantitative evidence of how additive and powder-based manufacturing processes, when coupled with effective recycling strategies, can achieve measurable environmental gains. Their life cycle assessment (LCA) of Direct Metal Laser Sintering (DMLS) and conventional CNC machining showed that electricity consumption and material waste are the dominant environmental burdens. However, when recycled metal powders are incorporated, both energy use and emissions decline substantially, aligning with the principles of dematerialization and design for circularity. Geometric complexity was shown to influence environmental performance, as optimized designs reduce waste during powder reuse cycles. Economically, while additive processes remain capital-intensive, the reuse of powders through recycling offsets part of the material cost and improves the overall life cycle cost structure, particularly for complex and low volume parts.

Oros Daraban et al. [8] described the operational aspects of recycling in metal AM, including powder recovery, reuse, and reconditioning as means of extending the service life of both materials and components. The production of metal powders from recycled feedstock was discussed in relation to technologies such as gas atomization, plasma atomization, and electrode induction melting gas atomization, which can convert recycled material into high-quality powders. These technologies maintain low oxygen content and high sphericity, parameters essential for consistent mechanical properties in PM products. The reuse of powders not only conserves raw resources but also significantly reduces waste generation and disposal costs. Furthermore, they reported that powder recycling contributes to sustainability performance by supporting multiple life cycles of use and enabling repairs or remanufacturing of defective or end-of-life components. Such practices, when integrated into industrial workflows, reinforce the 6R framework: reduce, reuse, recover, redesign, recycle, and remanufacture, central to circular economy strategies.

The environmental dimension of scrap recycling is therefore complemented by its economic viability. In PM processes, scrap-derived powders lower the dependency on primary raw materials, reducing procurement and transportation costs while enhancing resilience against material shortages. Life cycle costing analyses from Mecheter et al. [7] demonstrated that although AM incurs higher upfront investment, recycling loops can decrease material input costs over time, especially when combined with energy-efficient equipment and process optimization. Additionally, powder recovery systems reduce waste management expenses and enhance profitability in industries where high value metals such as titanium, nickel, and stainless steel dominate production.

7.4.3 *Future Trends: Digitalization, Traceability, and Additive Manufacturing Integration*

The integration of digitalization, traceability, and AM technologies represents the future direction of sustainable PM within circular economy (CE) models. In recent years, the convergence of these technologies has been instrumental in enhancing resource efficiency, extending product life cycles, and minimizing environmental impacts across all manufacturing stages. The emerging paradigm emphasizes intelligent, data-driven, and closed-loop production systems in which AM, artificial intelligence (AI), and digital monitoring frameworks jointly enable transparency, process optimization, and design innovation in metal-based production (see Fig. 7.11).

Digitalization plays a crucial role in establishing a foundation for the circularity of materials and products within PM systems. As identified by the group of Ponis and colleagues [9], Industry 4.0 technologies such as the Internet of Things (IoT), big data analytics, and cyber physical systems are essential enablers of circular manufacturing frameworks, supporting real-time process monitoring, automated decision-making, and optimized material flow management. The integration of digital platforms enables seamless communication between design, production, and end-of-life recovery phases, thereby reducing inefficiencies and promoting energy optimization. These advancements align with the triple bottom line (TBL)

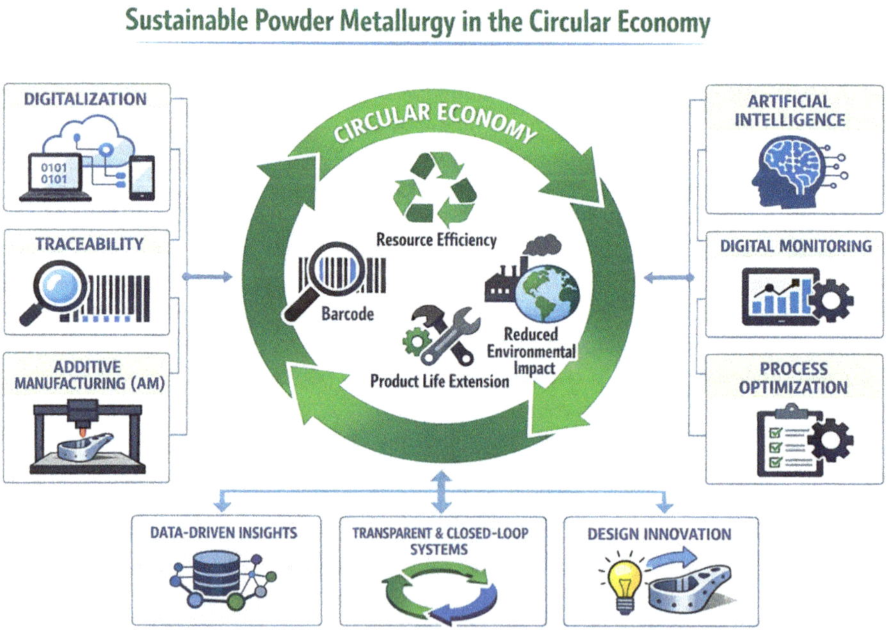

Fig. 7.11 Conceptual framework of sustainable PM within the circular economy, integrating digitalization, traceability, artificial intelligence, and process optimization

perspective discussed by Bahuriddin et al. (2025) [46], which highlights that a truly sustainable manufacturing framework must balance environmental, economic, and social dimensions through digital traceability and integrated assessment methodologies.

Traceability, as an extension of digitalization, provides the capacity to monitor and document the entire life cycle of metallic powders and AM components, ensuring material accountability and compliance with sustainability goals. According to Duve et al. [22], traceability mechanisms anchored in digital twins and standardized data frameworks enable the mapping of R strategies such as reuse, remanufacturing, and recycling across all manufacturing phases. These strategies support circular economy implementation by enabling the origin, composition, and transformation history of each metal powder batch to be digitally recorded. The digitization of the AM value chain allows for precise control over powder reuse, process parameter optimization, and waste minimization, which ultimately facilitates closed-loop resource management. Furthermore, the work of Leino et al. (2016) underscores that the integration of laser-based AM processes with digital feedback systems enables the repair, refurbishment, and remanufacturing of metallic components while minimizing environmental burdens through reduced raw material consumption and extended product lifetimes.

The integration of AM with digitalization introduces advanced monitoring and predictive maintenance capabilities, which are critical to sustaining circular production frameworks. The research group of Büschera [3] presented a trustworthy artificial intelligence framework for condition monitoring in metal AM systems that employs machine learning (ML) algorithms combined with Industrial Internet of Things (IIoT) infrastructures to detect early signs of component degradation. This approach enables predictive maintenance and process optimization, allowing equipment and material use to be increased while reducing downtime and waste. Explainable and uncertainty-aware artificial intelligence models have been shown to improve the reliability of automated systems, supporting the use of AM in circular and sustainable production.

Furthermore, AI-driven design optimization has emerged as a pivotal element of digitalized AM ecosystems. Abd-Aziz et al. [1] emphasized that artificial intelligence enables automated decision-making during component design, allowing for geometry optimization and the development of repairable and remanufacturable structures. This integration reduces the reliance on manual labor, increases process efficiency, and extends product life cycles by enabling the digital reconstruction and adaptive redesign of worn or obsolete parts. Similarly, Habeeb et al. [4] advanced this approach by developing hybrid AI optimization methods that combine artificial neural networks (ANNs) with genetic algorithms (GAs) to enhance design flexibility and facilitate sustainable repair and restoration using AM. These intelligent hybrid systems contribute to adaptive digital workflows where data collected during the AM process continuously informs the redesign and optimization of future components, reinforcing the feedback-driven nature of circular digital ecosystems.

In the broader context of sustainability assessment, Bahuriddin et al. [46] identified a pressing need for integrating life cycle assessment (LCA) and life cycle costing (LCC) approaches within digital AM environments to produce unified sustainability scores. Digital traceability systems can bridge this gap by enabling the automatic collection and evaluation of process data across design, manufacturing, use, and end-of-life stages. Such systems ensure that sustainability metrics are continuously updated and accessible, providing a foundation for data-driven decision-making in resource management. In this regard, the compatibility of AM with circular economy principles has been widely examined in relation to the development of sustainable industrial systems. Duve et al. [22] demonstrated that AM supports R-strategies such as repair, refurbish, and remanufacture through localized, flexible, and on-demand production enabled by digital connectivity. These processes not only conserve material and energy resources but also enable a transition from mass production to distributed manufacturing systems supported by digital networks. This decentralization promotes greater resilience and adaptability of supply chains, which are essential components of sustainable PM.

7.5 Conclusions

The chapter establishes sustainable PM as a closed-loop system in which metallic scrap, energy-efficient consolidation, and digitalized management converge to maintain material performance while reducing environmental burden. Mechanical milling and gas atomization transform machining chips, construction steel, aluminum scrap, and additive manufacturing residues into powders with refined microstructures and controlled morphology suitable for sintering and laser-based fabrication. Quality assurance practices, including sieving, blending, and plasma spheroidization, counteract oxidation, spatter contamination, and particle coarsening, preserving flowability and fusion behavior during repeated reuse cycles. In this regard, energy analyses demonstrate that conventional sintering is a dominant consumer within PM, motivating the adoption of spark plasma, microwave, and induction sintering, which achieve rapid densification, low porosity, and microstructural control at reduced thermal input. Circular economy models further extend sustainability through the valorization of metallurgical slags, mill scale, and recycled alloys into hardfacings, composites, and functional oxides. Moreover, digitalization, traceability, and artificial intelligence integrate these material and process loops by recording powder histories, optimizing reuse strategies, and enabling repair, remanufacture, and predictive maintenance.

References

1. Aziz NA et al (2021) Component design optimisation based on artificial intelligence in support of additive manufacturing repair and restoration: current status and future outlook for remanufacturing. J Clean Prod 296:126401

2. Batista CD et al (2021) From machining chips to raw material for powder metallurgy—a review. Materials 14:5432. https://doi.org/10.3390/ma14185432
3. Büscher J et al (2025) Trustworthy artificial intelligence for condition monitoring in metal additive manufacturing to facilitate a circular economy framework. Procedia CIRP 134:55–60
4. Habeeb HA et al (2023) Design optimization method based on artificial intelligence (hybrid method) for repair and restoration using additive manufacturing technology. Metals 13:490. https://doi.org/10.3390/met13030490
5. Ludger Bernsmann J-S, Schleifenbaum JH (2025) From construction for construction: additive manufacturing with gas-atomized recycled steel scrap. Circ Econ 4(3):100157
6. Ma K et al (2017) Environmental sustainability of laser metal deposition: the role of feedstock powder and feedstock utilization factor. Procedia Manuf 7:198–204
7. Mecheter A, Tarlochan F, Kucukvar M (2023) A review of conventional versus additive manufacturing for metals: life-cycle environmental and economic analysis. Sustainability 15: 12299. https://doi.org/10.3390/su151612299
8. Oros Daraban AE et al (2019) A deep look at metal additive manufacturing recycling and use tools for sustainability performance. Sustainability 11:5494. https://doi.org/10.3390/su11195494
9. Ponis S et al (2021) A systematic literature review on additive manufacturing in the context of circular economy. Sustainability 13:6007. https://doi.org/10.3390/su13116007
10. Stieberova B et al (2022) Life cycle assessment of metal products produced by additive manufacturing: a metal mold case study. ACS Sustain Chem Eng 10(16):5163–5174
11. Tirado González JG et al (2020) An innovative magnetic oxide dispersion-strengthened iron compound obtained from an industrial byproduct, with a view to circular economy. J Clean Prod 268:122362
12. Wei LK et al (2023) Producing metal powder from machining chips using ball milling process: a review. Materials 16:4635. https://doi.org/10.3390/ma16134635
13. Wurst J et al (2023) Assessment of the ecological impact of metal additive repair and refurbishment using powder bed fusion by laser beam based on a multiple case study. J Clean Prod 423:138630
14. Bulei C, Todor MP, Kiss I (2018) Metal matrix composites processing techniques using recycled aluminium alloy. IOP Conf Ser Mater Sci Eng 393(1):012089
15. Cacace S et al (2020) Using recycled material to produce gas-atomized metal powders for additive manufacturing processes. J Clean Prod 268:122218
16. Svozilova S et al (2024) Evaluation of recycled and reused metal powders for DMLS 3D printing. Materials 17(24):6184
17. Daoush WM et al (2025) Novel synthesis of CNTs–Si3N4/Cu nanocomposites: electroless deposition, powder metallurgy, spark plasma sintering, microstructure, and physical properties. Front Mater 12:1427381
18. Golmohammadi P et al (2025) Spark plasma sintering of a novel Mg-0.7Ca alloy: a comprehensive study. J Magnesium Alloys 13(3):1325–1340
19. Manohar G, Pandey KM, Maity SR (2021) Effect of microwave sintering on the microstructure and mechanical properties of AA7075/B4C/ZrC hybrid nano composite fabricated by powder metallurgy techniques. Ceram Int 47(23):32610–32618
20. Verma P, Saha R, Chaira D (2018) Waste steel scrap to nanostructured powder and superior compact through powder metallurgy: powder generation, processing and characterization. Powder Technol 326:159–167
21. Bendikiene R, Ciuplys A, Kavaliauskiene L (2019) Circular economy practice: from industrial metal waste to production of high wear resistant coatings. J Clean Prod 229:1225–1232
22. Duve M et al (2025) The role of metal additive manufacturing in a circular economy. In: Advances in automotive production technology—digital product development and manufacturing. Springer Nature, Cham

23. Giurco D et al (2014) Circular economy: questions for responsible minerals, additive manufacturing and recycling of metals. Resources 3:432–453. https://doi.org/10.3390/resources3020432
24. Gonçalves P et al (2023) Solid state recycling of Vanadis®8 steel scraps using powder metallurgy. Mater Res 26:e20230204
25. Kumar N, Bharti A (2021) Review on powder metallurgy: a novel technique for recycling and foaming of aluminium-based materials. Powder Metall Metal Ceram 60(1):52–59
26. Jäger S, Weber S, Röttger A (2021) Potential of the recycling of grinding sludge by various powder metallurgical processes. Procedia CIRP 104:893–899
27. Lanzutti A, Marin E (2024) The challenges and advances in recycling/re-using powder for metal 3D printing: a comprehensive review. Metals 14:886. https://doi.org/10.3390/met14080886
28. Powell D et al (2020) Understanding powder degradation in metal additive manufacturing to allow the upcycling of recycled powders. J Clean Prod 268:122077
29. Soundarapandiyan G et al (2021) A technical review of the challenges of powder recycling in the laser powder bed fusion additive manufacturing process. J Eng 2021(2):97–103
30. Rane K, P.P. (2018) Date recycling potential for finely divided ferrous metallic scrap using powder technology. Recycling 3:59. https://doi.org/10.3390/recycling3040059
31. Yamanoglu R, Bahador A, Kondoh K (2023) Support recycling in additive manufacturing: a case study for enhanced wear performance of Ti6Al4V alloy. Proc Inst Mech Eng Pt J J Eng Tribol 237(7):1603–1608
32. Beiker Kair A, Sofos K (2014) Additive manufacturing and production of metallic parts in automotive industry. KTH Royal Institute of Technology, Stockholm
33. Ford S, Despeisse M (2016) Additive manufacturing and sustainability: an exploratory study of the advantages and challenges. J Clean Prod 137:1573–1587
34. Arrizubieta JI et al (2020) Study of the environmental implications of using metal powder in additive manufacturing and its handling. Metals 10:261. https://doi.org/10.3390/met10020261
35. Kruzhanov V, Arnhold V (2012) Energy consumption in powder metallurgical manufacturing. Powder Metall 55(1):14–21
36. de Seijas MO et al (2023) Indirect induction sintering of metal parts produced through material extrusion additive manufacturing. Materials 16(2):885
37. Azevedo JMC, CabreraSerrenho A, Allwood JM (2018) Energy and material efficiency of steel powder metallurgy. Powder Technol 328:329–336
38. Rahmani R, Lopes SI, Prashanth KG (2023) Selective laser melting and spark plasma sintering: a perspective on functional biomaterials. J Funct Biomater 14(10):521
39. Jame MSR, Ravi Chandran KS (2025) Effect of spark plasma sintering on alloying, densification and mechanical properties of C103 niobium alloy. Mater Sci Eng A 932:148249
40. Manohar G et al (2025) Enhanced performance of AA7075/SiC/ZrC hybrid composite through microwave assisted powder metallurgy techniques. J Mater Eng Perform 34(5):4336–4349
41. Jia M, Gabbitas B (2015) Rapid synthesis of a near-β titanium alloy by blended elemental powder metallurgy (BEPM) with induction sintering. Metall Mater Trans A 46(10):4716–4729
42. Tsygoda V, Kyrylenko K, Petrovsky V (2019) The influence of organic binders and their decomposition products on the microstructure and thermoelectric properties of conductive materials based on si3n4 with transitional metal carbides additives. Technol Audit Prod Reserv 1(1(51)):18–25
43. Vicenzi B, Boz K, Aboussouan L (2020) Powder metallurgy in aerospace—fundamentals of pm processes and examples of applications. Acta Metall Slovaca 26:144–160
44. Ashrafi N et al (2022) Magnetic, electrical, and physical properties evolution in Fe3O4 Nanofiller reinforced Aluminium matrix composite produced by powder metallurgy method. Materials 15:4153

45. Sadenova MA, Utegenova ME, Klemeš JJ (2019) Synthesis of new materials based on metallurgical slags as a contribution to the circular economy. Clean Techn Environ Policy 21(10):2047–2059
46. Bahuriddin NS et al (2025) State of sustainability assessments of metal additive manufacturing: literature review and triple-bottom-line perspectives. Environ Sci Pollut Res 32(35): 20825–20843

Chapter 8
Future Directions and Disruptive Innovations in Powder Metallurgy and Mechanical Alloying

Abstract Mechanical alloying and advanced powder metallurgy define a nonequilibrium materials design space in which extreme plastic deformation, defect generation, and kinetically constrained diffusion enable access to metastable phases, extended solid solubility, and hierarchical nanostructures unattainable by equilibrium processing. High energy milling drives repeated cold welding and fracture, producing nanocrystalline, amorphous, and compositionally complex alloys whose properties are governed by defect density, grain boundary volume fraction, and controlled phase competition. Coupling mechanically alloyed powders with rapid densification routes such as spark plasma sintering, ultrafast Joule heating, or additive manufacturing preserves nanoscale features while achieving high relative density, enabling concurrent solid solution, precipitation, and dispersion strengthening. These principles underpin substantial gains in hardness, yield strength, radiation tolerance, hydrogen storage kinetics, magnetic efficiency, and superconducting current density across high entropy alloys, oxide dispersion-strengthened systems, nanocomposites, and functional intermetallics. Hybridization with additive manufacturing further enables spatially resolved composition, multimaterial architectures, and metamaterials exhibiting exceptional energy absorption and strength to weight ratios. Despite challenges associated with energy intensity, powder handling, and scalability, emerging data driven process control, environmentally conscious milling strategies, and integrated design frameworks position mechanically alloyed powder metallurgy as a foundational route for aerospace, energy, quantum, and extreme environment materials where performance is dictated by deliberate manipulation of nonequilibrium microstructural states.

8.1 Introduction

Future developments in powder metallurgy is increasingly driven by mechanical alloying, high energy milling, and powder-based coating technologies that enable alloy systems and microstructures unattainable through equilibrium processing. Mechanical alloying is described as a nonequilibrium processing route enabling

R. Pérez-Bustamante, *Fundamental, Practical and Disruptive Applications of Powder Metals*, Topics in Mining, Metallurgy and Materials Engineering,
https://doi.org/10.1007/978-3-032-27130-3_8

the formation of metastable, nanostructured, and compositionally complex alloys for structural, energy, and functional applications. High entropy and medium entropy alloys synthesized by staged and conventional milling demonstrate that phase selection, grain size, and strengthening mechanisms can be tuned through controlled energy input and elemental sequencing. Alloy systems such as AlCoCrFeNi, Fe40Mn20Cr20Ti10Al10, and CoCrFeNiTix show that milling time, titanium content, and processing route determine whether solid solution, χ, Laves, or σ phases dominate, with corresponding changes in hardness, compressive strength, and thermal stability [1–13].

High energy ball milling extends this processing route through the induction of amorphization, nanocrystallization, and solid solution formation in systems including FeCoTaB, CuFeAl, and refractory element alloys. Through repeated cold welding and fracturing, diffusion distances are reduced and defect densities increased, allowing alloying between elements with positive heats of mixing and the formation of nanoscale or amorphous structures [7, 14–18]. When combined with spark plasma sintering, such powders are consolidated into dense bulk materials while retaining nanostructural features and promoting in situ precipitation associated with enhanced strength and creep resistance, and the same alloyed powders are incorporated into powder-based coating technologies, including mechanically alloyed feedstocks for deposition, spark plasma-sintered sputtering targets, and laser cladding for surface engineering. Ni-Ti, CoCrFeNiTi, and Cu-Fe systems illustrate how milling parameters and powder homogeneity govern lamellar formation, corrosion behavior, hardness, and multifunctional responses in deposited layers. Hybridization with additive manufacturing further expands this capability, as mechanically alloyed Al-TiB_2 and Ti-modified 7075 powders achieve stable flowability, refined grain structures, and suppression of hot cracking during laser powder bed fusion [10, 19–21]. In this regard, this chapter describes how nonequilibrium powder processing combined with rapid consolidation and digital deposition enables the design of advanced alloys, coatings, and multimaterial systems in which properties are governed by engineered microstructures rather than by compositional constraints alone.

8.2 Advanced Processing Routes: Mechanical Alloying, High Energy Milling, and Coating Technologies

8.2.1 Mechanical Alloying as a Disruptive Tool for Novel Alloy Design

Mechanical alloying (MA, Fig. 8.1) has emerged as a disruptive tool for designing advanced metallic systems due to its capacity to produce metastable, nanostructured, and compositionally complex alloys beyond the thermodynamic limits of conventional processing. The integration of high energy milling into alloy development has

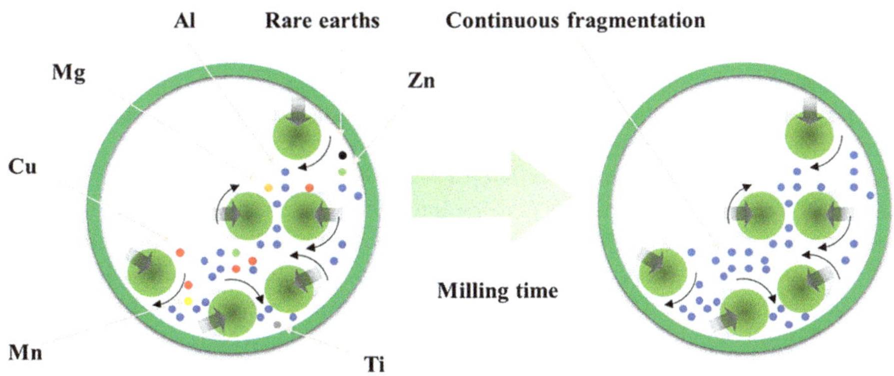

Fig. 8.1 Schematic illustration of the mechanical milling process, showing particle fragmentation and redistribution under repeated impacts

enabled the synthesis of novel phases, supersaturated solid solutions, and refined microstructures that significantly modify mechanical and functional responses. Recent studies demonstrate that MA facilitates compositional control, microstructural tailoring, and property optimization in complex concentrated alloys, high entropy alloys (HEAs), and oxide dispersion strengthened (ODS) steels, positioning it as a critical method for future materials innovation.

The group of Custodio [1] applied conventional and sequential high energy milling to AlCoCrFe and AlCoCrFeNi systems to produce equiatomic medium and HEA powders for coating feedstock design. The conventional route, where all elemental powders were milled simultaneously, yielded a dual-phase BCC and FCC structure in AlCoCrFeNi with phase fractions of 55% and 45%, respectively. In contrast, sequential milling, in which elements were introduced in stages, resulted in a single-phase BCC FeNiCoCrAl alloy after 24 h. The incorporation of Ni promoted FCC phase formation and refined the grain structure, enhancing hardness and mechanical strength. The systematic design-of-experiment approach linked milling parameters, thermodynamic factors, and microstructural evolution, emphasizing MA as an adaptable platform for producing feedstocks with tailored mechanical behavior and compatibility with advanced deposition methods such as cold spraying.

In addition, Jain et al. [3] reported the synthesis of a non-equiatomic Fe40Mn20Cr20Ti10Al10 alloy via MA and subsequent spark plasma sintering (SPS), demonstrating the method's capability for controlling complex phase equilibria. Extended milling (up to 40 h) produced a dual-phase structure consisting of a BCC matrix (a = 0.287 nm) and χ phase (a = 0.889 nm), with both phases contributing to enhanced mechanical stability. Differential scanning calorimetry and ex situ XRD analyses indicated that the alloy remained stable up to 400 °C and began forming FCC solid solution near 510 °C, confirming that mechanically alloyed powders can retain thermodynamic metastability. The SPSed specimens exhibited high hardness (~6.4 GPa) and compressive yield strength (~1900 MPa), reflecting strengthening from fine grains, dislocation entanglement, and nanoscale precipitates. This study established the synergy between MA and rapid sintering for developing dense nanostructured alloys with superior strength ductility balance suitable for elevated temperature service.

The influence of titanium content in mechanically alloyed and sintered CoCrFeNiTix high entropy alloys was examined by Kratochvíl et al. [2]. The systematic variation of Ti concentration ($x = 0.44$–2.15) caused a compositional transition from a single-phase FCC to an intermetallic-rich alloy containing Laves and σ phases. At 10–15 at.% Ti, solid solution strengthening dominated, but higher Ti levels induced secondary phase precipitation, boosting hardness (up to 968 HV1) and compressive yield strength (~2157 MPa). MA was shown to promote nonequilibrium solubility and the formation of metastable phases, allowing control over strengthening mechanisms. Compositional adjustment through mechanical alloying also modified wear behavior and thermal stability, with titanium-rich compositions exhibiting a transition from adhesive to abrasive wear associated with oxide debris formation.

Beyond HEAs, MA has also transformed the development of oxide dispersion strengthened steels for nuclear applications. Murphy et al. [22] reviewed disruptive manufacturing strategies for ODS steels, emphasizing that high energy milling ensures a homogeneous dispersion of nano oxides such as Y2Ti2O7 and Y2Ti2O5 within ferritic or martensitic matrices. These nano oxides serve as effective dislocation pinning sites, improve creep resistance, and act as helium traps under irradiation. Mechanically alloyed ODS steels exhibit nanoscale oxide dispersoids between 2 and 50 nm uniformly distributed in grains of approximately 100–200 nm. The MA step is therefore critical for generating stable, nanostructured dispersions that maintain strength at high temperatures and inhibit grain growth through Zener pinning. The combination of optimized PM and MA reduced processing time by up to 25% while increasing toughness by 30 to 50% relative to conventional ferritic martensitic steels [22].

8.2.2 High Energy Ball Milling for Nanostructured Powders

High energy ball milling (Figs. 8.2 and 8.3) has emerged as a highly versatile solid-state technique to produce nanostructured powders through repeated cold welding, fracturing, and rewelding of particles under extreme mechanical stress. This process enables alloying between immiscible elements and promotes the formation of metastable phases and amorphous structures unattainable under equilibrium conditions. Recent investigations have demonstrated that the control of milling parameters, alloy composition, and staging sequence profoundly influences microstructural evolution, phase transitions, and the resulting physical properties of the powders. In particular, the application of high energy MA has been central in developing nanocrystalline and amorphous alloys, high entropy materials, and multiphase catalytic powders.

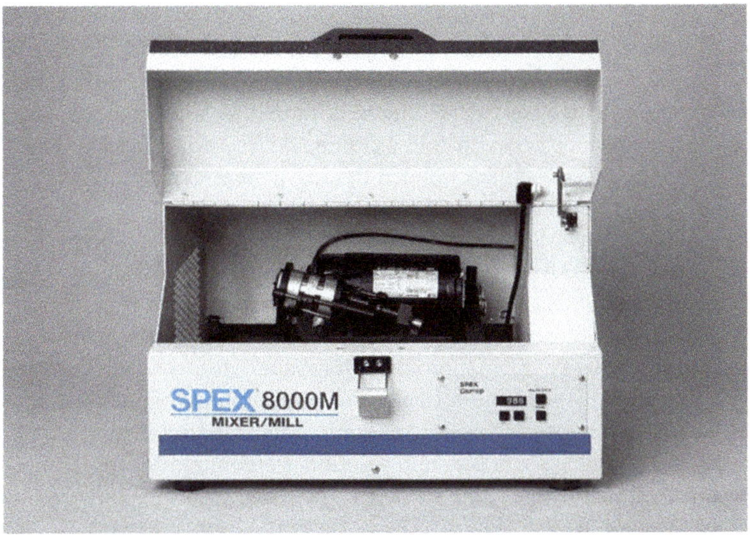

Fig. 8.2 High energy milling apparatus

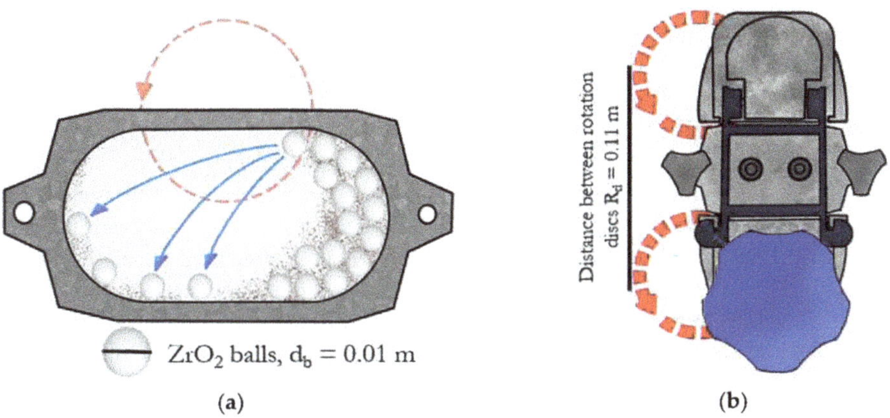

Fig. 8.3 Retsch E_{max} mill: (**a**) Vessel, (**b**) Clamping system [23]

The synthesis of nanocrystalline amorphous Fe-Co-Ta-B systems provides an illustrative example of the capacity of high energy milling to engineer fine-grained structures. Msetra et al. [18] reported the formation of a nanocrystalline Fe (Co) phase embedded within an amorphous matrix after 200 h of MA of $Fe_{31}Co_{31}B_{30}Ta_8$ powders under an argon atmosphere using a planetary ball mill at 350 rpm. Progressive milling induced a sequence of transformations from crystalline Fe, Co, and Ta to metastable Fe(Co), Co_2Ta, and Fe type borides, followed by amorphization. Scanning electron microscopy revealed a transition from irregular and coarse particles to uniformly fine powders with extended milling, while X-ray diffraction evidenced the coexistence of bcc and FCC phases and the eventual dominance of amorphous regions. The authors attributed the increased coercivity and irregular variation of magnetization to grain refinement and the growth of the amorphous fraction, as well as to the influence of nonmagnetic Ta and B atoms in the Fe Co lattice. The balance between cold welding and fracturing was shown to govern the equilibrium grain size, while extended milling increased defect density, promoting amorphization and magnetic hardening.

Similarly, the group of Tikhov [17] investigated a two-stage high energy ball milling approach for 33Al-45Cu-22Fe alloys, aiming to eliminate compositional segregation. A binary Cu-Fe mixture was milled first to form a metastable Cu (Fe) solid solution with an FCC structure, after which aluminum was introduced for further milling. X-ray diffraction and Mössbauer spectroscopy confirmed that this staged sequence avoided the formation of undesirable intermediate phases and promoted homogeneous elemental distribution. Transmission electron microscopy revealed the coexistence of Cu(Al) solid solution, Fe-containing amorphous phases, and a fine spinel phase after the second stage. The resulting powder exhibited mesoporosity and uniform microstructure, properties advantageous for catalytic applications. This two-stage process demonstrated that by decoupling alloying steps, solid-state diffusion can be better managed to prevent phase segregation and enhance atomic scale mixing, a principle transferable to other multicomponent systems. In this regard, the research group of Mitra [11] described theoretical and experimental approaches for understanding MA in the synthesis of high entropy materials, including alloys, metallic glasses, and ceramics. They described MA as a nonequilibrium top-down route enabling interdiffusion, amorphization, and nanostructuring across diverse multicomponent systems. The process parameters—rotational speed, ball-to-powder ratio, milling duration, process control agent, and atmosphere—govern energy transfer, defect generation, and thermodynamic pathways. Ultrafine grain sizes and high dislocation densities generated during mechanical alloying shorten diffusion paths, allowing alloying even between elements with large differences in melting point or atomic size. It also outlined how coupling MA with SPS preserves nanocrystallinity in bulk compacts and enhances strength via grain boundary strengthening. Case studies included mechanically alloyed CoCrFeNiTa and NbMoTaWVTi systems, in which nanostructuring and extended solid solubility yielded improved hardness and compressive strength.

8.2.3 Surface Coatings Via Powder-Based Techniques

Powder-based coating technologies, such as those illustrated in Fig. 8.4, represent a major area of modern surface engineering, integrating MA, SPS, and laser cladding to produce protective layers with enhanced mechanical and chemical performance. The Ni-Ti, CoCrFeNiTi, and Cu-Fe systems represent distinct but complementary strategies to exploit the thermodynamic and kinetic advantages of PM for the formation of high performance surface coatings. In the case of Ni-Ti coatings

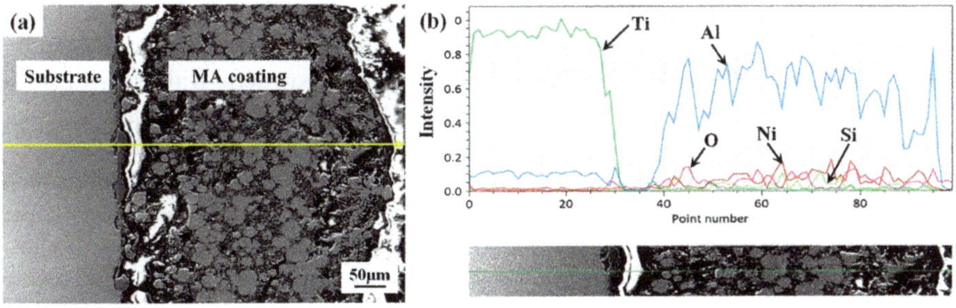

Fig. 8.4 Cross sectional characterization of 100-h oxidized MA coating: (**a**) microstructure, (**b**) EDS line scanning profile [24]

produced by MA, Chen et al. [21] demonstrated that the interplay between milling parameters and microstructural evolution governs coating adhesion, hardness, and corrosion resistance. Coatings synthesized at a rotational speed of 350 rpm and a milling duration of 13 h exhibited optimized lamellar microstructures and continuous alloyed layers with microhardness values approximately twice that of the 316L stainless steel substrate. The coating formation was dictated by cold welding and localized heat accumulation during repeated impacts, leading to dense inner lamellar regions and refined outer layers. Excessive milling, however, induced coating flaking due to work hardening and high interfacial stresses, revealing the narrow process window for achieving ideal thickness and homogeneity. A correlation was observed between milling-induced plastic deformation and corrosion behavior, with moderate milling times improving corrosion resistance through more uniform Ni-Ti solid solution formation, whereas extended milling increased defect density and promoted localized attack in saline environments.

A more complex compositional approach was followed in the development of CoxCrFeNiTi HEAs coatings designed for magnetron sputtering applications. Manea et al. [10] employed MA followed by SPS consolidation to fabricate equiatomic and low cobalt content targets. The powders achieved morphological uniformity after 30 h of milling, indicating completion of solid-state diffusion and homogenization among the principal elements. SPS at temperatures up to 950 °C produced dense, defect-free compacts with homogeneous elemental distributions and nearly stoichiometric compositions between target and deposited film. The subsequent magnetron sputtered coatings exhibited amorphous structures with nanohardness values exceeding 590 HV, surpassing those of traditional stainless steel substrates. Cobalt concentration influenced both hardness and ductility, with lower cobalt content slightly reducing hardness to about 570 HV while improving microstructural uniformity. The high configurational entropy of the CoCrFeNiTi system suppressed intermetallic segregation, enabling a stable single-phase structure that resisted oxidation and wear under operational conditions. Furthermore, the SPS-assisted route provided near-net-shape sputtering targets with high densification degrees and minimal carbon contamination, reflecting the synergy between MA and advanced sintering for coating precursor preparation.

Regarding these alloy systems, Deng et al. [25] explored the hybridization of MA and laser cladding for Cu-Fe immiscible coatings, a system inherently challenged by liquid phase separation. Mechanically alloyed Cu-Fe powders with varying iron content and milling duration produced fine, uniformly distributed particles that limited macrosegregation during rapid solidification. Laser cladding under optimized parameters (2 kW laser power, 200 mm/min scanning speed) resulted in α Fe particles dispersed within an ε Cu matrix, exhibiting gradient microstructures driven by Marangoni and Stokes convection mechanisms. Increased Fe content promoted higher hardness and electrical resistivity but reduced corrosion resistance due to intensified Fe-rich clustering. Conversely, longer ball milling durations improved powder homogeneity and enhanced corrosion performance by refining interfacial bonding and suppressing unmelted particle remnants. The resulting coatings demonstrated a balance of mechanical strength and electromagnetic performance, with the highest saturation magnetization (10.172 emu/g) and lowest coercivity (17.249 Oe) observed in moderately alloyed specimens, illustrating the multifunctional potential of immiscible systems processed via powder-based techniques.

8.2.4 *Hybrid Approaches: Combining Powder Metallurgy with Additive Manufacturing*

Hybrid approaches integrating PM with additive manufacturing (AM), as it is illustrated in Fig. 8.5, have emerged as a transformative route for engineering aluminum-based composites with enhanced microstructural control and mechanical performance. The research carried out by Parsons and Shaik [26] demonstrated that the MA of AlSi10Mg with TiB_2 enables the production of highly reinforced

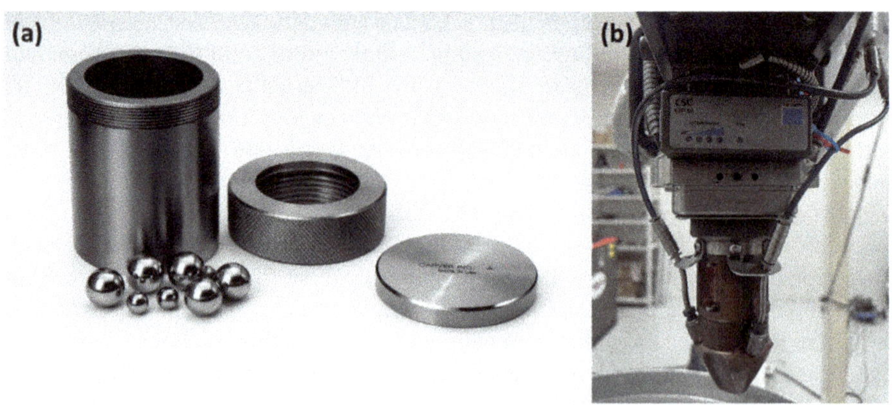

Fig. 8.5 (**a**) MA media used in the production of advanced powder metals for their processing via (**b**) laser cladding

aluminum composite powders optimized for laser powder bed fusion (LPBF). The ball milling process produced homogeneous, equiaxed particles containing up to 20 vol% TiB_2, exhibiting powder flowability comparable to gas-atomized AlSi10Mg. These powders successfully formed smooth, continuous single tracks under LPBF processing, confirming the stability of the melt pool and demonstrating that MA can effectively tailor the morphology and composition of feedstock powders for AM applications.

Wang et al. [20] expanded on this concept by employing a hybrid inoculation strategy that combined MA with conventional LPBF of 7075 aluminum alloy. Their approach utilized Ti powders mechanically alloyed with a fraction of 7075 alloy to form MA7075 Ti powders, which were subsequently blended with the bulk alloy powder. This dual-step method preserved powder sphericity and achieved homogeneous dispersion of Ti-rich particles, mitigating issues of poor powder flowability commonly associated with prolonged high energy milling. The result was a composite powder with stable rheological behavior and excellent printability across a broad processing window, underscoring the compatibility of mechanically alloyed hybrid feedstocks with commercial AM systems. Moreover, microstructural analysis revealed that the inclusion of MA7075-Ti particles induced a transition from coarse columnar grains to ultrafine equiaxed grains, achieving a 97.8% reduction in average grain size and complete suppression of hot cracking. The presence of $L1_2$ Al_3Ti nanoparticles formed during laser processing acted as potent heterogeneous nucleation sites, refining the grain structure and eliminating liquation cracking typically observed in high strength 7xxx series alloys. After T6 heat treatment, the Ti-modified 7075 alloy achieved yield and ultimate tensile strengths of 429 MPa and 513 MPa, respectively, with a uniform elongation of 12.8%, outperforming conventional LPBF-processed aluminum alloys in both strength and ductility.

8.2.5 Industrial and Energy-Related Applications

Industrial and energy-related applications of advanced powder metallurgy and mechanical alloying, as illustrated by examples from the automotive industry in Fig. 8.6, demonstrate how these processes enable improved performance and

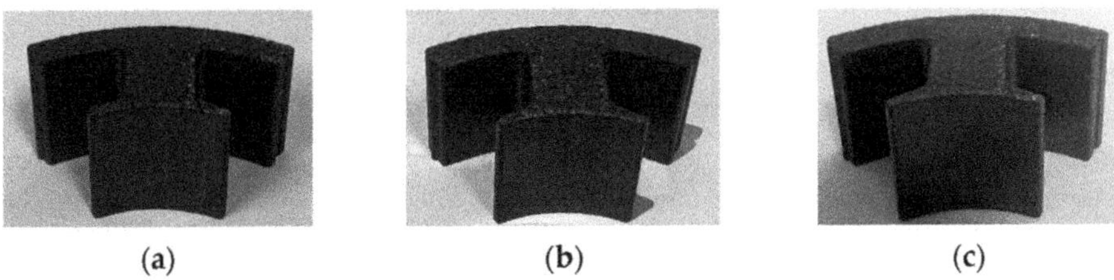

Fig. 8.6 Stator cores manufactured under optimal process conditions: (**a**) Sample 1; (**b**) Sample 2; (**c**) Sample 3 [28]

sustainability across multiple sectors through microstructural control, efficient resource utilization, and enhanced functional properties. In the automotive industry, PM has become a dominant method for manufacturing structural and functional parts, achieving up to 70% of total spare part production due to its ability to yield near-net-shape products with high precision and minimal material waste. Elkady and colleagues [27] highlighted that ferrous, aluminum, and titanium-based powders are widely employed to fabricate lightweight automotive components, reducing vehicle weight and consequently fuel consumption and emissions. Aluminum matrix composites synthesized through PM have been successfully used for brake systems, achieving both low density and improved wear resistance, while maintaining cost efficiency and environmental compatibility.

Energy-related innovations rely heavily on MA to develop intermetallic compounds with high hydrogen storage capacities. The group of Liu [29] discussed the synthesis of magnesium and titanium-based hydrides using MA as an alternative to traditional melting or chemical routes. These intermetallic systems, such as Mg-based and TiFe type alloys, provide a balance between volumetric hydrogen density, safety, and operating temperature. MA facilitates nanostructuring, creating defects and grain boundaries that enhance hydrogen diffusion kinetics, thereby improving absorption and desorption rates. The process also disrupts oxide layers that inhibit hydrogen uptake, ensuring reproducibility and stability under cycling conditions. Modified TiFe alloys obtained through alloying or severe plastic deformation exhibit rapid activation and improved resistance to poisoning by gas impurities, positioning them as promising candidates for renewable hydrogen energy systems.

Beyond hydrogen energy storage, MA contributes to the production of superalloys and HEAs tailored for industrial and energy-intensive environments. Altas [4] detailed that HEAs produced via MA exhibit superior strength, corrosion resistance, and high temperature stability compared to conventional alloys. These properties arise from their high configurational entropy and simple solid solution phases that enhance phase stability even under thermal and mechanical stress. Their low density, excellent oxidation resistance, and tunable magnetic or wear properties make them suitable for use in power generation turbines, chemical reactors, and structural components operating under extreme conditions. These materials offer longer service life and reduced maintenance requirements, addressing efficiency and sustainability goals in modern industry.

The adaptation of MA in aerospace sectors further demonstrates its industrial significance. Froes and Trindade [30] emphasized mechanochemical processing as a means to synthesize advanced materials like nanocrystalline titanium and aluminum alloys, which are essential for lightweight yet strong aerospace structures. Mechanochemical activation induces non solid-state reactions and generates metastable or amorphous phases with refined microstructures. These characteristics contribute to high strength-to-weight ratios, improved fatigue resistance, and enhanced oxidation tolerance, essential for components exposed to high thermal gradients. Moreover, the process allows alloying of difficult combinations of elements and the formation of nanograin structures that cannot be achieved through conventional melting techniques, expanding the material design space for aerospace energy systems and propulsion applications.

MA has also proven effective in synthesizing materials for magnetics, electronics, and energy conversion systems. Suryanarayana [15] described the capacity of MA to generate metastable structures, supersaturated solid solutions, amorphous alloys, and nanocomposites, allowing adjustment of mechanical and functional properties. Mechanically alloyed oxide dispersion strengthened superalloys, for example, provide improved high temperature creep strength and oxidation resistance in turbine engines. Similarly, nanocrystalline and amorphous alloys derived from this technique are increasingly used in soft magnetic cores for electrical and energy storage devices due to their low coercivity and high saturation magnetization. The scalability of MA, from gram level to industrial scale production, further underscores its utility in developing energy materials with consistent quality and performance.

In the context of electric vehicles, PM plays a central role in optimizing magnetic performance and energy conversion efficiency. Kang and Lee [31] demonstrated that soft magnetic composites (SMCs) fabricated through PM exhibit isotropic magnetic properties and minimal eddy current losses, making them ideal for motor stator cores. They reported optimal process parameters, such as compaction temperature and forming speed, through Taguchi statistical design to achieve uniform density and enhanced electrical properties. The use of materials like AncorLam HR and Fe 6.5 wt.% Si significantly reduced core losses while maintaining high permeability, directly contributing to higher motor efficiency and reliability in electric mobility applications. In addition, Kim and Lee [28] verified that stator cores produced under optimized PM conditions exhibit improved saturation flux density and bulk conductivity, with SEM analyses confirming microstructural uniformity across different compaction depths.

8.3 Nanostructured Materials Produced Via Powder Metallurgy

8.3.1 Bottom-Up and Top-Down Approaches for Nanostructured Powders

Under PM, nanostructured powders can be produced through two main strategies, known as bottom-up and top-down approaches. The bottom-up route relies on the assembly of atoms or molecules into nanostructured materials, while the top-down method involves the fragmentation of bulk materials into nanosized grains. MA of novel alloys such as NiAlFeCoCu presented in Fig. 8.7 represents a hybrid process that effectively integrates both approaches, as it promotes repeated cold welding, fracturing, and rewelding of particles, progressively reducing particle and crystallite size while enabling alloy formation at the atomic level. This dual nature of MA is extensively described by Suryanarayana et al. [16], who demonstrated that the

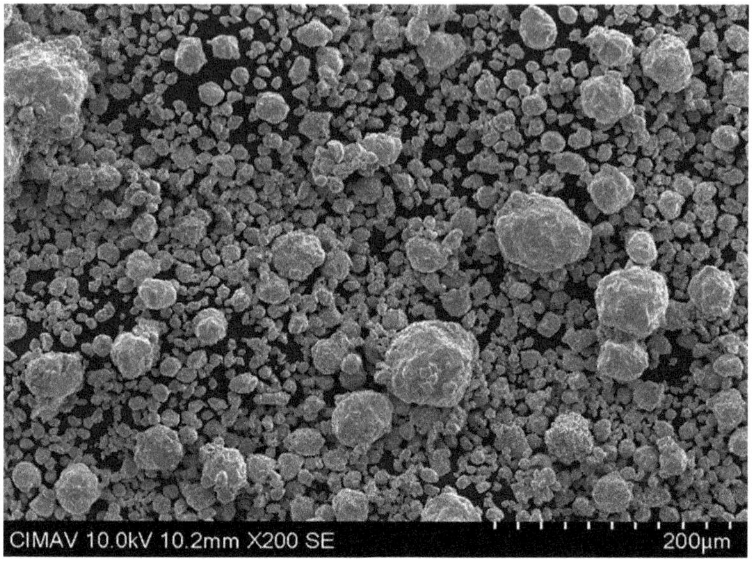

Fig. 8.7 Mechanically alloyed NiAlFeCoCu powders

reduction of diffusion distances, the introduction of high defect densities, and localized temperature rises facilitate the formation of supersaturated solid solutions, amorphous phases, and nanocomposites in a single processing sequence.

In bottom-up processing, atomic diffusion and solid-state reactions dominate the synthesis. As particles undergo repeated plastic deformation, atomic mobility is enhanced by defect generation and strain energy, leading to alloying and phase transformations even among immiscible elements. The group of Suryanarayana [16] emphasized that this phenomenon enables the production of alloys between metals with positive heats of mixing, which would otherwise be impossible through equilibrium routes. The bottom-up aspect is also evident in the formation of nanocrystalline phases during extended milling, where crystallite sizes can reach the 5–50 nm range. The dislocation pile-up model described in the same work correlates the minimum attainable grain size with material hardness and stacking fault energy, predicting that BCC metals achieve smaller crystallites due to their higher resistance to plastic deformation.

On the other hand, the top-down approach within MA arises from the mechanical fragmentation of large powder particles by high energy impacts. Koch and coworkers [32] demonstrated that mechanical attrition not only refines powders to nanoscale dimensions but also promotes the formation of metastable structures with superior mechanical and functional properties. In Mg97Y2Zn1 alloys, prolonged milling produced grains of 21 nm, which significantly enhanced hardness up to 2.13 GPa, representing the highest value reported for crystalline magnesium alloys at the time. This refinement process exemplifies a typical top-down behavior, in which mechanical energy is directly responsible for transforming coarse grains into

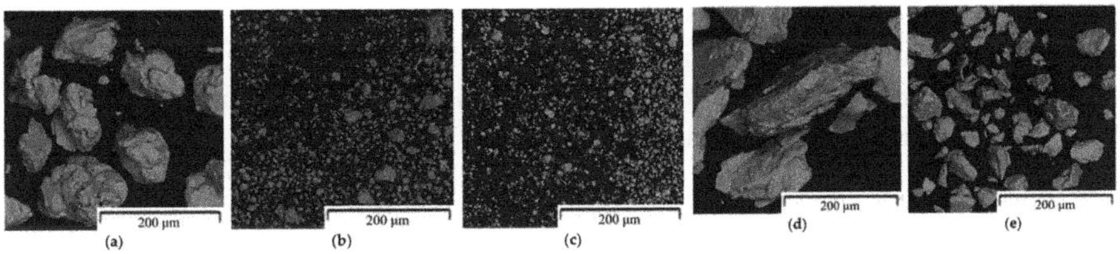

Fig. 8.8 Shape and morphology of the powders prepared by mechanical alloying: (**a**) Zn 1 Mg (1200 RPM, 60 min), (**b**) Zn 6 Mg (1200 RPM, 60 min), (**c**) Zn 16 Mg (1200 RPM, 60 min), (**d**) Zn 1 Mg (1200 RPM, 30 min), (**e**) Zn 1 Mg 0.08 g stearic acid (1200 RPM, 30 min) [23]

nanosized structures through repeated fracture during milling. The evolution of powder shape and morphology produced by mechanical alloying further reflects this mechanism, as observed in Zn-Mg systems processed under different compositions, milling times, and the presence of a process control agent. Variations in Mg content, milling duration, and the addition of stearic acid lead to distinct particle morphologies, indicating differences in fracture, cold welding, and refinement kinetics. Similarly, in C-Al-Zn and Al-Mg systems, in situ consolidation after cryogenic and room-temperature milling generated bulk nanocrystalline metals with strengths several times higher than those of their coarse-grained counterparts while retaining ductility, highlighting the benefits of controlled top-down processing combined with bottom-up atomic diffusion during milling (Fig. 8.8).

The relationship of both methodologies is further illustrated in the fabrication of oxide dispersion strengthened (ODS) molybdenum alloys [33]. During MA, dissolution of Y and Zr atoms into the Mo lattice occurs via a bottom-up mechanism, facilitated by the extreme local pressures exceeding 5 GPa that arise from collisions between ZrO_2 milling media and powder particles. This process dissolves interstitial atoms beyond their equilibrium solubility limits and promotes atomic scale mixing. Simultaneously, the top-down contribution is evident in the fragmentation and refinement of Mo particles from micrometer to submicrometer sizes as rotation speed increases. It was reported that at 250 rpm, a balance between cold welding and fracturing achieved crystallite sizes near 100 nm, which stabilized even with further increases in speed. Subsequent SPS acted as a continuation of the bottom-up process, allowing in situ precipitation of Y Zr O particles with average diameters of approximately 20 nm and homogeneous distribution within the 300 nm Mo grains, resulting in hardness values near 570 HV0.3.

8.3.2 *Advanced Sintering Strategies for Nanostructure Preservation*

Advanced sintering of mechanically alloyed powders relies on thermal histories that densify rapidly while limiting diffusion driven coarsening, so that nanoscale or ultrafine features remain stable in bulk form. In PM, the design rule is to minimize the time that compacted powders spend at temperatures where grain boundaries are mobile while still promoting neck growth, pore closure, and controlled phase evolution. Methodologies that combine nonequilibrium heating with rapid cooling enable densification before grain growth becomes dominant and allow phase selection from metastable precursors. The strategies below show how control of temperature, time, and chemistry preserves nanostructures while achieving high relative density: SPS of mechanically alloyed powders, ultrafast high temperature sintering with Joule heating, and laser-based AM of multimaterial architectures. MA followed by SPS demonstrates the balance between densification and nanostructure retention in a refractory HEA. In an NbMoTaWRe system, extended milling created metastable and supersaturated powders that exhibited a single BCC signature. Subsequent sintering between 1700 and 1900 °C produced a refined BCC matrix together with an FCC type dispersion of niobium tantalum carbide along boundaries. The carbide formed by reaction of carbon introduced during milling and minor tungsten carbide debris with metastable niobium and tantalum during the thermal cycle. The consolidated alloy combined high compactness, high strength, plastic strain, and near gigapascal hardness through densification, grain refinement, precipitation, and solid solution strengthening [14].

Zinc magnesium alloys benefit from the same framework at far lower temperatures. MA with stearic acid as a process control agent prevented cold welding, increased solid solubility above equilibrium, and produced powders that transformed into Mg2Zn11 and MgZn2 during consolidation. SPS at about 300 °C for 10 min under argon and 19.1 MPa, with a heating rate near 100 °C/min, yielded dense compacts while suppressing coarsening. The Zn-1 Mg alloy reached an average grain size of approximately 629 nm and showed a favorable balance between compressive strength and ductility among the compositions, associated with a refined matrix and controlled intermetallic dispersion. Milling speed and duration were quantified in terms of their influence on particle size and phase fractions, providing guidance for powder selection and preservation of the nanostructure during densification [23].

Ultrafast high temperature sintering pushes rapid densification further by sandwiching a green pellet between carbon heaters that deliver uniform Joule heating in seconds. Typical processing cycles reach temperatures of approximately 1000 to 3000 °C within 10–30 s, with heating and cooling rates on the order of 10^3–10^4 °C min^{-1}. Because the sample experiences seconds at temperature, densification, alloying, and reactive sintering proceed while abnormal grain growth and phase separation are strongly curtailed. The configuration limits current flow through the sample and operates without applied pressure, making it suitable for complex geometries produced by additive manufacturing [34]. AM conditions can also be adjusted to preserve nanostructures through localized melt pools and rapid solidification. In laser powder bed fusion of a multimaterial Ti6Al4V and CuNi2SiCr system, bidirectional scanning generated narrow copper alloy tracks within a titanium matrix with metallurgical bonding, limited intermixing, occasional porosity, and microhardness values of approximately 356 HV for the titanium regions and about 128 HV for the copper alloy regions, reflecting the influence of thermal confinement and high cooling rates on microstructural control [35].

8.3.3 Powder Metallurgy-Based Fabrication of Multimaterials and Metamaterials

PM combined with AM (Fig. 8.9) represents a transformative route for the fabrication of multimaterials and metamaterials, merging microstructural control with the design flexibility of layerwise deposition. The integration of PM into additive processes such as selective laser melting (SLM) or laser powder bed fusion (L-PBF) allows the development of hybrid systems that achieve functional gradation, tunable porosity, and enhanced mechanical behavior. Recent research has demonstrated that these combined approaches enable the production of advanced meta architectures with both high structural efficiency and energy absorption capacity.

The development of metallic multimaterials fabricated by combining AM and PM has been thoroughly explored recently by the group of Yadav [37]. Ti6Al4V honeycomb precursors were manufactured by L-PBF and subsequently filled with various metallic reinforcements through SPS. Reinforcements, including Mg, Al, Fe, Ni, and Cu, were consolidated within the Ti6Al4V lattice under controlled thermal and mechanical conditions, yielding complex multimaterial systems. The composites exhibited stable microstructural interfaces without interfacial reactions, particularly in Ti6Al4V Mg Al and Ti6Al4V Mg Al Fe systems, where reinforcement phases were homogeneously distributed. XRD and SEM analyses confirmed the coexistence of α' Ti martensitic phases with reinforcement phases such as Mg and Al, indicating strong structural integrity after sintering. This approach demonstrated the feasibility of producing multimaterial architectures with spatially varying

Fig. 8.9 The sample of Mg-10Zn-0.8Ca-0.5Zr alloy powder processed by SLM; two macro-images of the same sample [36]

compositions and mechanical gradients, allowing the tailoring of stiffness and hardness profiles for multifunctional applications.

Kayacan et al. [38] observed hybrid selective laser melting and powder metallurgy routes for metamaterial fabrication, in which the external shell and lattice framework were produced by selective laser melting and the internal volumes were filled and sintered using PM. Using 316L stainless steel, they achieved intricate lattice geometries combined with porous sintered interiors. This integration resulted in substantial reductions in both build time and material consumption while improving toughness by approximately 80%. Microstructural evaluation revealed well-sintered inner regions and mechanically continuous interfaces between SLM-processed and PM-derived zones. The hybrid parts displayed significant weight reductions, up to 83%, without severe compromise of mechanical strength. Compression testing indicated an enhanced energy absorption capability due to the coexistence of dense and porous regions, which promoted controlled deformation during loading.

Complementary research on LPBF-produced AlSi10Mg metamaterials by the laboratory group of Gao [39] further reports the potential of PM-related design principles in metamaterial fabrication. In LPBF, powder metallurgical characteristics such as particle size distribution, powder flowability, and controlled layer deposition govern structural precision. Plate-based FCC architectures incorporating engineered perforations enable residual powder removal while preserving high stiffness and strength. The designed metamaterials achieved peak compressive strengths up to 300% higher and energy absorption capacities up to 400% greater than those of conventional truss based or gyroid counterparts at similar relative densities. The controlled pore distribution and plate orientation minimized stress concentrations, promoting stable layer-by-layer deformation rather than catastrophic collapse.

8.3.4 Limitations, Scalability, and Environmental Considerations in Nanostructured Powder Metallurgy

The limitations, scalability, and environmental implications of nanostructured materials synthesized through PM and MA constitute critical factors influencing their transition from laboratory to industrial applications. Although MA has enabled the production of nanostructured alloys with superior mechanical and functional properties, such as enhanced strength, corrosion resistance, and biocompatibility, several constraints persist regarding process control, reproducibility, and sustainability. These challenges are evident when comparing literature across different alloy systems, including Fe-Mn, Cu/W/Co, Mg-Zn, Ca-Zr, TiS_2, and Si-$CoSi_2$ C, all of which underscore that process scalability and ecological considerations are inseparable from material performance.

From a processing standpoint, MA remains an energy-intensive and time-consuming route that relies on high energy ball milling to induce solid-state reactions. This limitation is accentuated in systems that require precise phase stability and nanocrystalline control. In biodegradable Fe-Mn-based nanostructured alloys, the MA time of 5 h at 300 rpm with a 10:1 ball-to-powder ratio produced uniform nanocrystalline solid solutions; however, further stress-relief treatments were necessary to improve compressibility and reduce defect density [40]. Such additional processing steps illustrate the delicate balance between achieving high densification and maintaining nanostructure stability. Moreover, scalability challenges emerge due to the sensitivity of powder characteristics to milling parameters; any deviation can lead to heterogeneity, contamination from the milling media, or excessive oxidation.

Similarly, the Mg-xZn-Zr-Ca biodegradable alloys demonstrate how composition and milling conditions directly affect powder morphology and uniformity. Raducanu et al. [36] achieved nearly spherical powders with minimal porosity when using 10 wt.% Zn and optimized milling speed and duration; nevertheless, the strong reactivity of Mg with oxygen and the risk of powder agglomeration during long milling cycles represent significant barriers for upscaling this technology. These processing issues influence the suitability of powders for AM, particularly selective laser melting, which requires controlled particle size distribution and surface chemistry. Additionally, despite the feasibility of combining MA and SLM for biomedical components, the industrial scalability of such integrated processing routes is limited by high cost, energy consumption, and the necessity for inert environments to prevent oxidation.

As nanostructured materials are increasingly adopted in Industry 4.0 applications, the environmental footprint of powder metallurgy is shaped by synthesis routes and the recycling of input materials, as described by Phani and Narsimhachary [41]. Wet ball milling, as introduced by Veluswamy et al. (2021), exemplifies an approach to mitigate energy and environmental costs while retaining scalability. The wet process produced two-dimensional TiS_2 nanoplatelets with reduced thermal conductivity and enhanced thermoelectric performance at 100 °C, reaching a figure of merit of 0.35. The use of a liquid medium improved homogeneity, limited oxidation, and allowed for more efficient heat dissipation during milling.

Scalability was also a central consideration in the synthesis of porous Si-$CoSi_2$-C composites developed for lithium-ion batteries. The research group of Seo [42] demonstrated that a two-step high energy mechanical milling followed by chemical etching yielded a mass-producible composite with superior electrochemical performance and improved cycle life. The in situ formation of $CoSi_2$ not only increased electrical conductivity but also provided mechanical buffering against Si expansion, illustrating a synergy between process design and material functionality. The authors emphasized that the process is cost-effective and scalable, suggesting that MA combined with chemical post-processing can achieve industrial viability when optimized for throughput and environmental safety. Nevertheless, environmental and health-related challenges persist across all MA-derived materials. The handling of fine nanostructured powders introduces risks of airborne particle release and contamination, while the use of organic process control agents or reactive metals necessitates careful waste management. For biodegradable alloys such as Fe-Mn and

Mg-Zn systems, biocompatibility does not inherently guarantee environmental neutrality during production or disposal, especially when using solvents or high energy equipment. Future process optimization should thus focus on reducing emissions, employing recyclable media, and integrating energy recovery systems within milling operations.

8.4 Emerging Markets: Space Exploration, Renewable Energy, and Quantum Technologies

8.4.1 PM for Lightweight, High Strength Aerospace Components

PM has become a cornerstone technology for producing lightweight, high-strength materials suited for next-generation aerospace applications (Fig. 8.10), as it allows for precise microstructural control and the integration of novel reinforcement systems. In emerging areas such as space exploration, where performance, durability, and mass efficiency define mission feasibility, powder metallurgy routes are being adapted through the use of nanoparticle reinforcement, HEA incorporation, and microgravity processing. These approaches enable the design of sustainable, high performance materials for extraterrestrial and atmospheric aerospace systems.

Agureev et al. (2015) described aluminum-based composites reinforced with nanoscale Al_2O_3 and ZrO_2 particles produced by conventional powder metallurgy. Microhardness and tensile strength increased by more than a factor of two relative to unreinforced aluminum at nanoparticle contents up to 0.15 vol%, associated with grain boundary stabilization and reduced diffusion creep induced by the nanoparticles. The experimental data underscored that nanoparticle positioning,

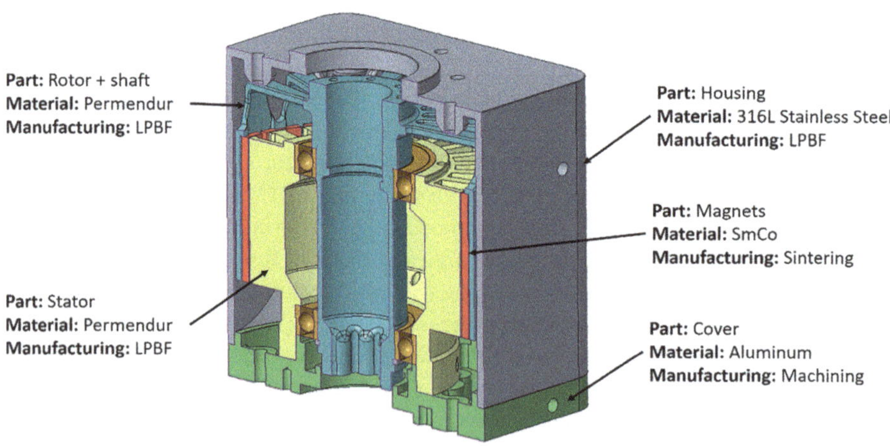

Fig. 8.10 Electric aerospace actuator manufactured by laser powder bed fusion [43]

either within grains or along grain boundaries, critically influenced crack propagation, subgrain formation, and vacancy annihilation, resulting in materials with enhanced creep and fracture resistance. Controlled nanoparticle dispersion provides a basis for lightweight yet mechanically robust alloys in spacecraft and satellite structures, where low density and structural integrity are required.

German et al. (2015) examined the sintering of tungsten heavy alloys under microgravity, with implications for in-space repair and component fabrication. Their microgravity experiments revealed that conventional sintering principles fail in reduced gravity: densification was hindered by the absence of pore buoyancy, leading instead to coalescence of large voids up to 200 μm, which distorted the structure and limited strength. Extraterrestrial PM systems therefore require artificial compaction or hybrid field-assisted techniques to address the role of gravity in densification mechanisms. Empirical data and modeling provide a basis for designing autonomous in situ resource utilization systems capable of producing high density aerospace components on the Moon or Mars [44].

The incorporation of HEAs into powder metallurgical processing marks a paradigm shift toward multifunctional and lightweight aerospace materials. Cui et al. [6] established a data-driven framework for designing lightweight high entropy alloys (LHEAs) optimized for strength-to-weight performance. Their material genetic engineering approach, employing high-throughput computation and machine learning, identified correlations between atomic size differences, mixing enthalpy, and valence electron concentration that predict solid solution formation in HEAs. These criteria, particularly $\delta \leq 6.6\%$ and $\Omega \geq 1.1$, enable the targeted design of solid solutions that maintain low density while preserving structural stability at elevated temperatures, aligning perfectly with aerospace needs for adaptive and dynamic materials.

Xu and coworkers [13] produced lightweight Al-Cr-Fe-Mn-Ti high entropy alloy coatings by laser cladding. The resulting BCC and $L2_1$ dual-phase microstructures exhibited improved oxidation resistance and thermal stability in the temperature range from 400 to 800 °C. Molecular dynamics and CALPHAD analyses confirmed that aluminum and titanium clustering within the lattice reduced alloy density without compromising bonding integrity, while controlled Ti addition minimized vacancy concentration and improved oxidation kinetics. The coatings exhibited lower oxidation activation energies (133–155 kJ/mol) and strong metallurgical bonding to Ni-based substrates, suggesting that HEA-based claddings can extend component life in propulsion systems and reentry vehicles.

On the other hand, the group of Verma [19] investigated aluminum matrix composites reinforced with CoMoMnNiV high entropy alloy particles produced by stir squeeze casting followed by heat treatment. A reinforcement level of 6 wt % provided the highest mechanical response, with an ultimate tensile strength of 287.6 MPa and a hardness of 117.2 HRB. Microstructural examination revealed uniform particle dispersion and improved interfacial wettability compared to ceramic reinforcements, leading to superior load transfer and reduced crack propagation. Heat treatment improved particle-matrix cohesion, illustrating the adaptability of hybrid powder metallurgy and casting approaches for aerospace-grade materials that require wear resistance and fatigue durability.

In addition, Akinwande et al. [45] embedded a lightweight $Al_{30}Li_{20}Mg_{10}Be_{20}Ti_{10}Cu_{10}$ HEA into AA7068 aluminum through microwave-assisted PM. Their results showed that a 10 wt% addition achieved optimal dispersion and the coexistence of BCC and FCC phases, enhancing strain hardening sensitivity and reducing necking susceptibility. The composite exhibited both superior hardness and improved ductility, with the HEA phase acting as a ductile metallic reinforcement rather than a brittle inclusion. This approach resolves the classical trade-off between strength and formability that often constrains powder-processed aerospace alloys.

8.4.2 *High Temperature and Radiation-Resistant Powder Metallurgy Alloys for Space Missions*

PM alloys such as the NbTaTiZrHfV0.5 (Fig. 8.11) with high temperature and radiation resistance are used to support the reliability of spacecraft exposed to cosmic radiation, atomic oxygen erosion, and thermal extremes. These alloys, developed through advanced PM and AM techniques, exhibit microstructural precision and stability that conventional casting or forging processes cannot achieve. Recent investigations have particularly focused on titanium-based systems,

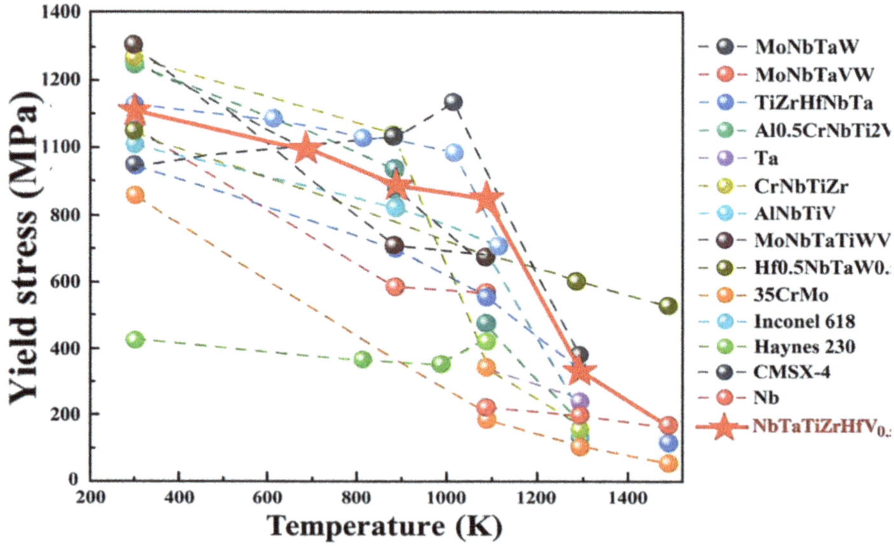

Fig. 8.11 Tensile strength versus the testing temperatures of NbTaTiZrHfV0.5 with those of high temperature metals and alloys, including conventional RHEAs, Ni-based superalloys and refractory metals (Nb) [47]

refractory high entropy alloys (RHEAs), and boron nitride reinforced coatings due to their high temperature strength retention and superior radiation tolerance.

Titanium alloys produced via PM, such as Ti 6Al-4Zr-0.5Mo-0.6Si, demonstrate exceptional microstructural refinement and mechanical stability under sustained thermal exposure. The controlled formation of $(TiZr)_6Si_3$ silicides within the α/β matrix enhances both solid solution and precipitation strengthening mechanisms, allowing the alloy to maintain a yield strength of approximately 510 MPa and an elongation of 13% at 700 °C. These fine silicide dispersions prevent grain coarsening and suppress creep deformation during long-duration, high temperature service. Furthermore, the homogeneous microstructure obtained through sintering and hot rolling improves oxidation resistance by limiting diffusion pathways along grain boundaries, making this alloy suitable for long-term applications in high radiation or high vacuum environments such as lunar bases or planetary landers [46].

PM processing of RHEAs has broadened the range of compositions available for space propulsion and shielding applications. Mechanical alloying and spark plasma sintering of multicomponent Nb-Mo-Ta-W systems have produced dense, fine-grained microstructures with high strength and oxidation resistance above 1200 °C. These RHEAs maintain yield strengths above 1000 MPa at 600 °C and exhibit minimal grain growth up to 1600 °C due to their sluggish diffusion kinetics and lattice distortion effects. The inclusion of nanoscale oxide dispersions, such as Y_2O_3 or Al_2O_3, further enhances creep resistance and defect recovery, limiting void swelling under neutron and ion irradiation. Such alloys represent an advance in the design of multifunctional materials that combine mechanical endurance and radiation stability, particularly for high heat flux components in nuclear propulsion systems [9].

In addition to bulk alloys, composite coatings incorporating hexagonal boron nitride (hBN) have been demonstrated to significantly improve radiation shielding performance. Experiments performed aboard the International Space Station under the MISSE 17 mission exposed Ti/2 vol.% hBN coatings, fabricated using Atmospheric and Vacuum Plasma Spray (APS and VPS) methods, to 171 days of continuous low Earth orbit radiation. The VPS coatings maintained structural integrity with no evidence of microcracking or delamination while exhibiting a 56% increase in hardness and a 26% rise in elastic modulus after exposure. X-ray diffraction confirmed the formation of TiO, TiO_2, and TiN phases, indicating radiation-assisted surface hardening, while boron retained its atomic configuration despite neutron transmutation. The combination of titanium oxidation resistance and the neutron absorption cross section of hBN, approximately 760 barns, resulted in a multifunctional barrier against radiation, thermal stress, and abrasive regolith particles [48]. A broader assessment of radiation-resistant materials for space missions has emphasized the necessity of layered and composite shielding systems that combine metals with ceramics or polymers to achieve optimal protection-to-mass ratios. Titanium, aluminum, and magnesium alloys are favored as base materials for their low density, while tungsten and tantalum layers provide primary protection against high energy particles. Recent reviews have shown that Ti-BN and Ti-SiC multilayer composites exhibit lower secondary radiation production compared to traditional metallic shields. Moreover, long-term irradiation and thermal cycling tests have validated the mechanical reliability of these composites, confirming their potential for deployment in both low Earth orbit and deep space habitats [48].

At the frontier of alloy manufacturing, AM has emerged as a transformative method for producing PM-based alloys with intricate geometries and controlled microstructures optimized for extreme environments. The development of alloys such as GRX 810, an oxide dispersion strengthened NiCoCr alloy, and HR-1, a hydrogen-resistant Fe-Ni-Cr alloy, illustrates this trend. When produced by laser powder bed fusion and directed energy deposition, these materials exhibit improved creep strength, higher thermal conductivity, and increased tolerance to radiation-induced defects. GRX 810, for example, achieves nanoscale Y_2O_3 dispersion that reinforces grain boundaries and prevents void formation under proton and neutron irradiation. The integration of computational modeling through Integrated Computational Materials Engineering (ICME) further accelerates the optimization of alloy chemistry and processing parameters, reducing the time required to achieve flight certification [49].

8.4.3 Powder Metallurgy in Renewable Energy

PM has emerged as a key fabrication route in renewable energy systems, particularly for the advancement of solid oxide fuel cells (SOFCs), hydrogen storage materials, and high performance battery technologies. In the field of fuel cells, PM-based processes have enabled the creation of metal-supported solid oxide fuel cells (MS SOFCs) that overcome the mechanical fragility and cost limitations of traditional ceramic-supported systems. Haydn et al. [50] demonstrated that oxide dispersion strengthened Fe-Cr alloys fabricated via powder metallurgical processing serve as highly durable, porous substrates capable of operating at temperatures up to 850 °C under both oxidizing and reducing conditions, thus offering superior mechanical strength and oxidation resistance suitable for mobile SOFC applications. Through the control of sintering parameters and compaction pressure, these metallic supports maintain sufficient porosity for gas transport while providing a stable mechanical backbone for thin-film electrolytes and electrodes deposited by vapor or plasma techniques. The resulting architectures improve thermal shock resistance and allow for lower temperature operation, aligning with the efficiency and reliability requirements of next-generation renewable power systems.

The group of Topcu [51] further explored PM for SOFC metallic interconnects using alloys such as Crofer 22 APU, stainless steel 316 L, and Inconel 600, highlighting the impact of compaction pressure and sintering temperature on thermal expansion and microhardness properties. Crofer 22 APU exhibited a thermal expansion coefficient suitable for integration with ceramic electrolytes, providing dimensional stability across thermal cycles. Together with the analysis of metallic substrates for solid oxide fuel cells by Ju, this supports the use of PM to adjust structural and chemical compatibility in multilayer fuel cell assemblies while retaining scalability at low cost (Ju, 2019). These improvements contribute to the longevity and performance of distributed renewable energy generators based on solid oxide fuel cell technology.

The relevance of PM extends to hydrogen related electrochemical systems, where it contributes to the development of advanced anode materials. Huang and coworkers [52] studied Mg Al Pb anode alloys fabricated through PM for seawater-activated batteries and demonstrated that controlled sintering and solution treatment modify the dispersion of Mg_2Pb and $Mg_{17}Al_{12}$ intermetallic phases, improving both corrosion resistance and discharge uniformity (Huang et al., 2020). Refined microstructures were associated with higher electrochemical efficiency and longer operational stability relative to cast alloys, reflecting the ability of powder metallurgy to control phase distribution and surface chemistry. These characteristics are applicable to hydrogen evolution and storage systems, where uniform phase interfaces influence reaction kinetics and corrosion behavior, and also extend to lithium-based energy storage, as illustrated by the fabrication of porous copper scaffolds for lithium metal anodes using powder metallurgy routes [53]. By infiltrating molten lithium into the sintered Cu framework and subsequently reacting it with Freon R134a, the resulting LiF-coated Li Cu composite anodes exhibited enhanced rate performance and cycling stability. The porous Cu scaffolds reduced local current density and stabilized the solid electrolyte interphase (SEI), effectively suppressing dendrite formation. This simple and scalable process provides a low-cost route for producing air-stable lithium metal anodes, advancing the practical deployment of high energy density batteries.

Kim et al. [54] applied MA to synthesize FeS_2 powders for Na/FeS_2 batteries, achieving an optimal discharge capacity of 447 mAh g^{-1} with 30 h of milling. MA reduced the FeS_2 particle size and improved phase homogeneity, resulting in enhanced electrochemical performance and reversibility at room temperature. PM routes have been applied to sodium-based systems to reduce production costs and support their use as alternatives to lithium-ion batteries for large-scale energy storage. Mechanical milling has also been used to produce Fe-Si alloy anodes for lithium-ion batteries, where controlled high energy milling refined the silicon phase and promoted the formation of alpha $FeSi_2$, yielding materials with balanced electrical conductivity, structural integrity, and electrochemical stability [55]. The Fe-Si alloy milled for 12 h exhibited the best cycle performance, attributed to minimized cracking and suppressed inactive β-$FeSi_2$ formation. Such mechanically engineered nanocomposites address silicon's intrinsic volume expansion problem and enhance the durability of next-generation lithium-ion batteries.

The research group of Welch [56] in 2024 developed a physics-based model describing the ball milling process for electrode material synthesis, relating operational parameters such as milling frequency, ball-to-powder ratio, and the energy threshold for amorphization to phase evolution and alloying kinetics. The model clarified how mechanical breakdown and atomic diffusion jointly influence phase formation, providing essential predictive tools for process optimization in electrode design. Yang et al. [57] discussed progress in ball milling-based silicon anodes, highlighting the industrial advantages of this top-down approach relative to more complex bottom-up synthesis routes. Chemical synthesis routes provide high laboratory scale performance, whereas ball milling offers simplicity, cost efficiency, and scalability for large-scale electrode production, particularly in electric vehicle applications. High speed mechanical milling followed by annealing has been used to produce uniform, cube-shaped microstructures in $Li_{5.3}PS_{4.3}ClBr_{0.7}$ argyrodite solid electrolytes [58]. The resulting material exhibited significantly reduced porosity and enhanced resistance to dendrite penetration, achieving a critical current density of 3.8 mA cm^{-2} and stable cycling over 1000 cycles in Li/NCM all-solid-state cells.

8.4.4 Powder Metallurgy in Superconductors

The use of PM in superconducting materials has enabled new approaches for improving performance in technologies such as space exploration, renewable energy systems, and quantum devices. Superconductors produced by these routes allow microstructural control of flux pinning, current-carrying capacity, and the upper magnetic field limit that defines their operational range. In NbTi-based systems, the adoption of oxide dispersion strengthening has proven highly effective in forming artificial pinning centers that significantly improve magnetic performance. Mousavi et al. [59] produced nanostructured NbTi–$Y_2Ti_2O_7$ composites by mechanically alloying Nb, Ti, and Y_2O_3 nanopowders followed by annealing, yielding finely dispersed $Y_2Ti_2O_7$ nanoparticles smaller than 5 nm. These nanoscale dispersions improved flux pinning relative to conventional alpha Ti precipitates, leading to current densities of 5.2 kA mm^{-2} at 4.2 K and 5 T, exceeding those obtained in conventionally processed NbTi wires.

In Nb_3Sn superconductors, internal oxidation techniques have achieved remarkable grain refinement and flux pinning enhancement through the formation of ZrO_2 nanoparticles. The group of Buta [60] in 2021 investigated Nb-Ta-Zr alloys, where the internal oxidation of Zr produced nanometric ZrO_2 precipitates that inhibited grain growth during Nb_3Sn formation. The resulting microstructure exhibited grains with aspect ratios between 1.5 and 1.7 and up to a 40% reduction in grain boundary intercept distance compared with reference Nb-Ta alloys. These structural modifications produced record breaking upper critical fields of 29.2 T at 4.2 K and improved Jc values above 10 to 15 T magnetic fields. Oxidized and non-oxidized zirconium were associated with different effects, with non-oxidized Zr modifying superconducting electron scattering and oxidized Zr contributing to Nb_3Sn grain refinement. Nanoscale control of secondary oxide phases achieved through powder metallurgy and internal oxidation influences flux pinning behavior and magnetic field limits in high field applications such as accelerator magnets.

The synergy between PM and AM further expands the possibilities for producing superconductors with intricate architectures and tailored microstructures. Ersoz et al. [61] described how additive manufacturing techniques, including laser powder bed fusion, electron beam melting, and extrusion-based processes, can produce complex superconducting geometries beyond the geometric and compositional constraints of conventional wire and tape fabrication. AM allows in situ alloying and fine control of microstructural features, enabling optimization of pinning center distributions and phase purity. AM allows microstructural parameters to be adjusted during fabrication, which is particularly relevant for superconducting alloys where current-carrying capacity depends on the distribution of defects and secondary phases. This integration of AM with PM-based feedstocks could lead to multifunctional superconducting components that merge structural and electrical optimization in a single step, aligning with the future requirements of energy-efficient systems and quantum technologies.

The exploration of superconductivity in HEAs introduces another frontier in PM-based material design. Sajjanar and Archana [12] analyzed HEAs such as Ta-Nb-Hf-Zr-Ti, which exhibit tunable superconducting transition temperatures driven by strong electron-phonon coupling and atomic disorder. These materials, synthesized through arc melting and MA, combine mechanical robustness with superconducting properties that are resistant to microstructural inhomogeneities. Their BCC single-phase structures display high configurational entropy that stabilizes uniform lattices, reducing phase segregation and enabling controllable Tc values. Density functional theory studies showed that disorder broadens electronic bands, creating opportunities for engineering superconductors with tailored transport properties and enhanced flux pinning. The combination of disorder-driven superconductivity and solid solution strengthening makes high entropy alloys suitable for quantum logic circuits and for structural applications that require both mechanical strength and superconducting behavior.

Iron-based superconductors also benefit from PM and MA approaches that control particle morphology and grain connectivity. Duchenko et al. [62] in 2025 carried out an interesting research related to $CaKFe_4As_4$ compounds and showed that varying the milling energy during high energy ball milling critically determines powder morphology, particle aggregation, and superconducting current pathways. Using milder mechanochemical conditions and substituting metallic calcium with calcium hydride enhanced homogeneity and minimized unwanted secondary phases, yielding denser and more uniform polycrystalline bulks. Reducing the transferred milling energy from 2.3 to 0.1 MJ g^{-1} limited excessive agglomeration and improved intergrain connectivity, leading to more effective current transport in polycrystalline samples. In iron-based superconductors, as in NbTi and Nb_3Sn systems, adjustment of powder metallurgy parameters such as milling energy, precursor chemistry, and thermal treatment governs superconducting behavior.

8.5 Conclusions

The chapter establishes mechanical alloying and high energy milling as enabling technologies for future powder metallurgy, allowing the synthesis of metastable, nanostructured, and compositionally complex alloys that extend far beyond equilibrium phase fields. Through controlled milling sequences, phase constitution, grain size, and defect density are tuned in high entropy, refractory, and oxide dispersion strengthened systems, producing powders that retain thermodynamic metastability and respond favorably to rapid sintering routes. Spark plasma sintering and laser-based consolidation preserve nanoscale features while promoting densification and beneficial in situ reactions. Powder-based coating technologies translate these engineered feedstocks into functional surfaces, where lamellar architectures, amorphous or solid solution phases, and controlled interfacial bonding govern hardness, corrosion resistance, magnetic behavior, and wear. The integration of mechanical alloying with additive manufacturing enables hybrid feedstocks that maintain flowability while inducing grain refinement, eliminating hot cracking, and improving strength and ductility in printed components. These nonequilibrium powder routes define a materials design paradigm in which microstructural architecture, rather than melting-based chemistry, dictates performance and enables disruptive innovations in powder metallurgy.

References

1. Custodio AGD, Saha GC, Aranas C (2024) Mechanical alloying of AlCoCrFe and AlCoCrFeNi: design and experimental evaluation of medium- and high-entropy alloy particles for cold spraying. Powder Technol 437:119556
2. Kratochvíl P et al (2025) Understanding the influence of Ti content on mechanically alloyed and sintered CoCrFeNiTix high entropy alloy. J Mater Res Technol 35:7371–7383
3. Jain H et al (2025) Structure and properties of Fe40Mn20Cr20Ti10Al10 high-entropy alloy fabricated by mechanical alloying and spark plasma sintering. J Mater Sci 60(3):1705–1725
4. Altas E (2024) Mechanical alloying of super alloys like high-entropy alloys and smart materials. In: Rajendrachari S (ed) Mechanically alloyed novel materials: processing, applications, and properties. Springer Nature, Singapore, pp 363–378
5. Balaji V et al (2024) Development of light-weight high entropy alloys for aerospace applications: perspective of artificial intelligence. In: Innovations in Mechanical Engineering III. Springer Nature, Cham
6. Cui K et al (2024) Data driving design of high-entropy alloys for lightweight and dynamic applications. Sci China Phys Mech Astron 67(2):227101
7. Daryoush S, Mirzadeh H, Ataie A (2021) Nanostructured high-entropy alloys by mechanical alloying: a review of principles and magnetic properties. J Ultrafine Grained Nanostruct Mater 54(1):112–120
8. Feng R et al (2021) High-throughput design of high-performance lightweight high-entropy alloys. Nat Commun 12(1):4329
9. Ivannikov A, Yusupov V (2024) Recent development in powder metallurgy of high-entropy alloys for high-temperature applications: brief review. Izvestiya Ferr Metall 67:509–519
10. Manea CA et al (2023) CoxCrFeNiTi high-entropy alloys prepared via mechanical alloying and spark plasma sintering for magnetron sputtering coatings. Materials 16:6386. https://doi.org/10.3390/ma16196386
11. Mitra R, Bajpai A, Biswas K (2025) A critical review on mechanical alloying of high-entropy materials. High Entropy Alloys Mater 3(1):1–40
12. Sajjanar SM, Archana (2024) High-entropy alloys in solid-state physics: electronic structure, mechanical properties, and superconductivity. J Sci Res Technol 3(1):1–17
13. Xu J et al (2024) New lightweight high-entropy alloy coatings: design concept, experimental characterization, and high-temperature oxidation behaviors. Surf Coat Technol 491:131154

14. Gu T et al (2022) Effect of mechanical alloying and sintering behavior on the microstructure and properties of NbMoTaWRe refractory high entropy alloy. Met Mater Int 28(11):2571–2582
15. Suryanarayana C (2019) Mechanical alloying: a novel technique to synthesize advanced materials. Research 2019:4219812
16. Suryanarayana C, Al-Joubori AA, Wang Z (2022) Nanostructured materials and nanocomposites by mechanical alloying: an overview. Met Mater Int 28(1):41–53
17. Tikhov S et al (2022) Elimination of composition segregation in 33Al–45Cu–22Fe (at.%) powder by two-stage high-energy mechanical alloying. Materials 15:2087. https://doi.org/10.3390/ma15062087
18. Msetra Z et al (2023) Structural, microstructural, and magnetic properties of nanocrystalline-amorphous Fe–co–ta–B alloy processed by high-energy mechanical alloying. J Mater Res Technol 26:8934
19. Verma PK et al (2025) Microstructure, mechanical, and wear characteristics of heat-treated aerospace-grade aluminium composite reinforced with HEA particles. Mater Chem Phys 331:130195
20. Wang T et al (2025) A novel inoculation strategy combined with mechanical alloying for preparing crack-free 7075 alloy in laser powder bed fusion. J Alloys Compd 1040:183364
21. Chen C et al (2024) Effects of process parameters on microstructure and properties of Ni–Ti coatings synthesized by mechanical alloying method. Results Surf Interfaces 14:100198
22. Murphy C et al (2025) Disruptive manufacturing of oxide dispersion-strengthened steels for nuclear applications: advances, challenges, and future prospects. PRO 13:1572. https://doi.org/10.3390/pr13051572
23. Nečas D et al (2022) Advanced zinc–magnesium alloys prepared by mechanical alloying and spark plasma sintering. Materials 15:5272. https://doi.org/10.3390/ma15155272
24. Xie H et al (2025) Oxidation-resistant Ni-AlSi12 composite coating with strong adhesion on Ti-6Al-4V alloy substrate via mechanical alloying and subsequent laser cladding. Coatings 15(11):1329
25. Deng C et al (2025) Microstructure and properties of Cu-Fe immiscible coatings fabricated via combined mechanical alloying and laser cladding. Materials 18:4436. https://doi.org/10.3390/ma18194436
26. Parsons EM, Shaik SZ (2022) Additive manufacturing of aluminum metal matrix composites: mechanical alloying of composite powders and single track consolidation with laser powder bed fusion. Addit Manuf 50:102450
27. Elkady O (2023) Recent advances in production of automotive industry's spare parts by powder metallurgy. Int J Mater Technol Innov
28. Kim J, Lee S (2024) A study on powder metallurgy process for x electric vehicle stator core. Metals 14:858. https://doi.org/10.3390/met14080858
29. Liu Y, Chabane D, Elkedim O (2021) Intermetallic compounds synthesized by mechanical alloying for solid-state hydrogen storage: a review. Energies 14:5758. https://doi.org/10.3390/en14185758
30. Froes FH, Trindade B (2004) The mechanochemical processing of aerospace metals. J Mater Process Technol 153–154:472–475
31. Kang S, Lee S (2024) A Study on the optimal powder metallurgy process to obtain suitable material properties of soft magnetic composite materials for electric vehicles. Metals 14:815. https://doi.org/10.3390/met14070815
32. Koch CC et al (2010) Nanostructured materials by mechanical alloying: new results on property enhancement. J Mater Sci 45(17):4725–4732
33. Yao L et al (2022) Preparation of nanostructural oxide dispersion strengthened (ODS) Mo alloy by mechanical alloying and spark plasma sintering, and its characterization. Int J Refract Met Hard Mater 105:105822
34. Wang C et al (2021) Rapid synthesis and sintering of metals from powders. Adv Sci 8(12):2004229

35. Marques A et al (2024) 3D multi-material laser powder bed fusion: Ti6Al4V–CuNi2SiCr parts for aerospace applications. Prog Addit Manuf 9(2):391–400
36. Raducanu D et al (2022) Mechanical alloying process applied for obtaining a new biodegradable Mg-xZn-Zr-Ca alloy. Metals 12:132. https://doi.org/10.3390/met12010132
37. Yadav MK et al (2025) Metallic multimaterials fabricated by combining additive manufacturing and powder metallurgy. J Compos Sci 9:80. https://doi.org/10.3390/jcs9020080
38. Kayacan MY, Yılmaz MS, Üzün A (2025) Mechanical characterization of meta-materials manufactured by a novel hybrid SLM technique utilizing powder metallurgy. Met Mater Int 31(6):1541–1556
39. Gao C et al (2024) Mechanical properties and energy absorption capabilities of plate-based AlSi10Mg metamaterials produced by laser powder bed fusion. J Mater Res Technol 30:3851–3862
40. Ammar HR, Sivasankaran S, Alaboodi AS (2021) investigation of the microstructure and compressibility of biodegradable Fe-Mn-Cu/W/Co nanostructured alloy powders synthesized by mechanical alloying. Materials 14:3088. https://doi.org/10.3390/ma14113088
41. Phani MK, Narsimhachary D (2024) Present and the future of mechanically alloyed materials. In: Rajendrachari S (ed) Mechanically alloyed novel materials: processing, applications, and properties. Springer Nature Singapore, Singapore, pp 459–469
42. Seo H et al (2021) Scalable synthesis and electrochemical properties of porous Si-CoSi2-C composites as an anode for Li-ion batteries. Materials 14:5397
43. Lizarribar B et al (2023) Electric aerospace actuator manufactured by laser powder bed fusion. Aerospace 10(9):813
44. Gereman R et al (2005) Critical learning from microgravity sintering of tungsten alloys: implications for extraterrestial fabrication and repair
45. Akinwande AA, Kumar MS, Romanovski V (2023) Mechanical strengthening and micromechanical modelling of newly fabricated 7068-aluminium/lightweight-high-entropy-alloy as aero-material. Mater Chem Phys 309:128344
46. Guo R et al (2020) Microstructure and mechanical properties of powder metallurgy high temperature titanium alloy with high Si content. Mater Sci Eng A 777:138993
47. Liu Z et al (2023) High-temperature mechanical properties of NbTaHfTiZrV05 refractory high-entropy alloys. Entropy 25(8):1124
48. Kunneparambil Sukumaran A et al (2025) Radiation-resistant Ti/BN coatings: insights from 171 days exposure to space radiation and atomic oxygen in low orbit. NPJ Mater Degrad 9(1): 93
49. Gradl P et al (2023) Advancement of extreme environment additively manufactured alloys for next generation space propulsion applications. Acta Astronaut 211:483–497
50. Haydn M et al (2013) Development of metal supported solid oxide fuel cells based on powder metallurgical manufacturing route. Powder Metall 56(5):382–387
51. Topcu A et al (2021) Thermal expansion, porosity, and microhardness properties of solid oxide fuel cell metallic interconnects manufactured by powder metallurgy approach. Turk J Electromech Energy 6(3):123–130
52. Huang J et al (2022) Effects of intermetallic phases on electrochemical properties of powder metallurgy Mg-6%Al-5%Pb anode alloy used for seawater activated battery. Mater Res Express 9(6):066504
53. Qin K et al (2020) A powder metallurgic approach toward high-performance lithium metal anodes. Small 16(24):2000794
54. Kim TB et al (2008) Electrochemical characteristics of Na/FeS2 battery by mechanical alloying. J Alloys Compd 449(1):304–307
55. Lee K et al (2022) Properties of Fe–Si alloy anode for lithium-ion battery synthesized using mechanical milling. Materials 15:1873. https://doi.org/10.3390/ma15051873
56. Welch C, Cho K, Srinivasan V (2024) Modeling analysis of ball-milling process for battery-electrode synthesis. Chem Mater 36:6748

57. Yang H et al (2023) Recent advances in ball-milling-based silicon anodes for lithium-ion batteries. Energies 16:3099. https://doi.org/10.3390/en16073099
58. Liu Y et al (2024) Inhibiting dendrites by uniformizing microstructure of superionic lithium Argyrodites for all-solid-state lithium metal batteries. Adv Energy Mater 14(31):2400783
59. Mousavi T et al (2021) New nanoscale artificial pinning centres for NbTi superconductors. Mater Des 198:109285
60. Buta F et al (2021) Very high upper critical fields and enhanced critical current densities in Nb3Sn superconductors based on Nb–Ta–Zr alloys and internal oxidation. J Phys Mater 4(2): 025003
61. Ersoz TT, Mohamed AE-MA, Attallah MM (2025) Additive manufacturing of superconductors: opportunities and challenges. Mater Res Bull 189:113448
62. Duchenko A et al (2025) Influence of milling energy and precursors on CaKFe4As4 polycrystalline superconductor morphology. Crystals 15:276. https://doi.org/10.3390/cryst15030276

Chapter 9
Real-World Innovations by Leading Companies

Abstract Industrial deployment of powder metallurgy and metal additive manufacturing reflects the maturation of feedstock engineering, process control, and digital integration into production-grade technologies. Powder- and wire-based routes exhibit distinct thermophysical regimes that govern melt pool stability, microstructural refinement, residual stress evolution, and defect sensitivity, with optimized parameter windows enabling relative densities exceeding 99% and mechanical performance competitive with wrought alloys. Fine wire deposition and hybrid energy inputs narrow the historical resolution gap between wire and powder systems while improving material utilization and scalability. Case-driven implementations demonstrate substantial reductions in component mass, lead time, and manufacturing cost through topology aware design and hybrid additive–subtractive workflows. Parallel advances in in situ monitoring, simulation-driven process optimization, and digital twins establish predictive control over quality and certification, enabling repeatable fabrication of safety-critical components across automotive, aerospace, energy, and medical sectors.

9.1 Introduction

Metal additive manufacturing is presented in this chapter as a powder and wire-based extension of powder metallurgy in which feedstock morphology, energy coupling, and thermal gradients govern microstructural evolution, defect formation, and component reliability [1–13]. Powder bed fusion systems selectively melt fine metallic powders under controlled atmospheres, producing extremely high cooling rates and microstructures capable of achieving relative densities above 99.8% in demanding systems such as tungsten heavy alloys when laser exposure, hatch spacing, and layer thickness are precisely regulated [11, 12, 14–19]. These rapid solidification conditions generate refined W-Ni-Fe phase distributions and elevated hardness, yet they simultaneously introduce residual stresses, lack-of-fusion porosity, and cracking when energy input and scanning strategies are not balanced. Optical and ultrasonic inspection methods therefore become integral to quality assurance, as defect

R. Pérez-Bustamante, *Fundamental, Practical and Disruptive Applications of Powder Metals*, Topics in Mining, Metallurgy and Materials Engineering,
https://doi.org/10.1007/978-3-032-27130-3_9

populations are closely coupled to melt pool dynamics and thermal gradients [15, 19].

Wire-based directed energy deposition represents a parallel materials route in which continuous metallic wires are fed into localized heat sources, enabling nearly 100% material utilization and higher deposition rates. The metallurgical response is dictated by droplet transfer, bead geometry, and thermal management. Advances in fine wire laser deposition and hybrid laser Joule heating have reduced surface roughness to the micrometer scale and produced densities exceeding 99%, approaching the microstructural uniformity of powder-based systems while avoiding powder oxidation and handling losses. In this regard, industrial implementation is further shaped by hybrid manufacturing strategies that integrate additive deposition with subtractive machining [2, 4, 6, 7, 9, 10, 12]. The Meltio wire laser DED platform exemplifies this approach through localized material placement, iterative parameter optimization, and digital modeling, enabling weight optimized, mechanically robust components for motorsport and repair applications. Similarly, laser powder bed fusion platforms deployed by Renishaw demonstrate the translation of refined feedstock control and process monitoring into automotive heat exchangers and lattice-based medical implants with regulated porosity and mechanical compatibility [20–28].

Across these systems, feedstock engineering, in situ monitoring, and digital twin platforms couple microstructural evolution with process stability, supporting certification and industrial scalability, while coordinated control of energy input, material delivery, and real-time data governs the reproducible fabrication of high performance powder metallurgy-derived components for structural and biomedical applications.

9.2 Powder Metallurgy Products and Technologies in Industrial Applications

9.2.1 Scientific Foundations of Metal Additive Manufacturing Technologies (Wire Vs. Powder)

Metal additive manufacturing (AM) technologies have evolved along two feedstock-based routes, namely powder-based and wire-based processes (Fig. 9.1). Each approach is governed by distinct mechanisms of energy coupling, material delivery, and thermal gradients, resulting in characteristic microstructural evolution. The resulting metallurgical integrity, mechanical response, and defect population are associated with feedstock morphology, processing parameters, and heat-transfer conditions.

Powder-based technologies, particularly laser powder bed fusion (L-PBF), rely on the selective melting of fine metallic powders under controlled atmospheres. In this process, a laser scans each layer of powder, inducing localized melting and rapid

Fig. 9.1 Robotic wire-laser metal deposition of an SS316L flange using the Meltio Engine, enabling support-free fabrication through radial deposition strategies. (Courtesy of Meltio)

solidification that produce extremely fine microstructures. The group of Verdonik [18] investigated this approach in tungsten heavy alloys and demonstrated that full densification above 99.8% theoretical density can be achieved by optimizing laser exposure time and energy input to control melting of both the tungsten and the Ni-Fe matrix. The fine microstructure produced under high cooling rates was associated with elevated hardness levels, exceeding those of liquid phase-sintered tungsten heavy alloys. This microstructural refinement in powder-based AM was governed by controlled energy distribution and melt pool dynamics.

However, powder-based systems also introduce challenges related to defect formation and residual stress accumulation. Chicote et al. [15] highlighted the significance of defect detection in L-PBF processes, noting that porosity, lack of fusion, and cracking can occur due to incomplete melting or suboptimal scanning strategies. Both in-line optical monitoring and off-line ultrasonic inspection resolved defects down to 0.3 mm, indicating that quality assurance in powder bed systems is governed by the integration of advanced nondestructive evaluation methods. Complementarily, Magana Carranza et al. [17] analyzed residual stress generation during L-PBF, showing that steep thermal gradients cause significant tensile and compressive stresses in successive layers. These residual forces depend strongly on laser power, scanning strategy, and point distance. The meander scanning pattern was found to minimize residual forces, emphasizing the thermomechanical coupling between process design and internal stress evolution.

The complexity of powder-based manufacturing is further reflected in its dependence on laser material interactions and process parameters. As noted by the group of Verdonik [18], the controlled melting of tungsten powders required low to moderate laser powers to prevent overheating of the Ni-Fe binder phase, while maintaining small hatch distances and layer thicknesses to promote remelting. These parameters directly determine the morphology of the microstructure, especially the distribution of the W-rich and matrix phases. Thus, powder bed fusion is fundamentally constrained by the trade-off between energy input and microstructural stability, where excessive power leads to keyholing or evaporation and insufficient power results in porosity and poor bonding. In contrast, wire-based AM technologies offer a different set of scientific and technological foundations centered on directed energy deposition (DED) principles. Wire feed systems eliminate issues of powder handling, improve material utilization to nearly 100%, and permit higher deposition rates. Wire feed AM systems are associated with higher deposition rates and lower processing costs, whereas reduced geometric fidelity and increased surface roughness have been reported relative to powder-based techniques [5]. The underlying mechanisms involve continuous feeding of a metallic wire into a localized heat source, where process control of wire melting, droplet transfer, and bead formation dictates the structural integrity of the final part.

In 2019, Shaikh et al. [12] developed a fine wire-based laser metal deposition (FW-LMD) technique that addresses the accuracy limitations of traditional wire processes. Using a 100 μm stainless steel wire and a pulsed Nd:YAG laser, they achieved layer thicknesses of 40–80 μm with surface roughness as low as 8–16 μm. The process optimization, particularly through a lateral overlap of 60–70%, resulted in dense morphologies without porosity. The microstructure consisted predominantly of austenite, with minor ferrite localized at bead junctions as resolved by EBSD and X-ray diffraction. The mechanical response remained uniform, with tensile strengths in the range of 700–750 MPa. Refined control of laser pulse energy combined with fine wire feedstock enabled the fabrication of precision components with quality levels comparable to powder-based methods, while avoiding powder oxidation and material losses.

The influence of reduced power input on wire-based processes was explored by the group of Li [10], who proposed a laser Joule wire feed direct metal deposition system operating at only 50 W. The combination of laser and Joule heating significantly decreased the step effect typical of wire AM, producing stainless steel parts with surface roughness below 5 μm and even as low as 0.4 μm after electrolyte plasma treatment. Their results showed a relative density of 99.37% and homogeneous Vickers hardness, indicating that stable thermal gradients and controlled heat input enable microstructural uniformity even at low energy levels. The use of compound heat sources exemplifies how thermal management remains a critical scientific foundation of all wire-based processes. Additional studies in this regard, such as that of Huang et al. [7] on laser wire feed manufacturing of aluminum alloys, further clarify the thermophysical relationships between process parameters and geometric stability. Through analysis of wire feeding direction, angle, laser power, traverse speed, and wire feed rate, the research revealed that front feeding produced

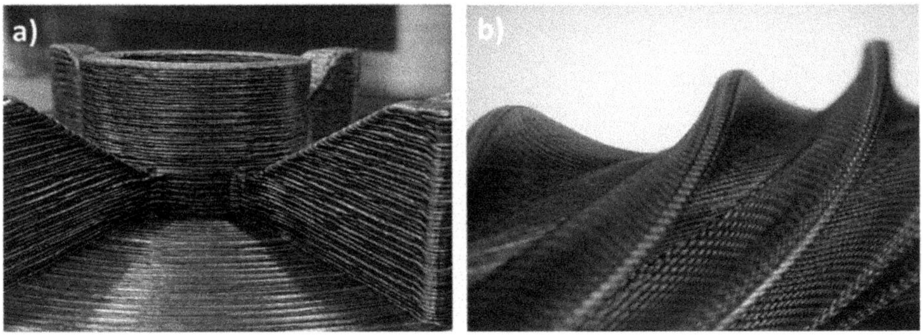

Fig. 9.2 Comparison between the resolution from (**a**) wire arc additive manufacturing and (**b**) Meltio's DED process. (Courtesy of Meltio)

superior clad quality due to improved melt pool stability and reduced spatter formation. High speed imaging confirmed that stable droplet transfer and controlled heat flow were crucial to consistent deposition. The ability to form defect-free aluminum structures with reduced residual stress highlighted the versatility of wire feed AM when process parameters are carefully optimized.

Wire-based methods such the presented in Fig. 9.2 have also been applied to reinforcement and hybrid manufacturing applications. Studies such as those by Wu et al. [13] on laser wire reinforcement blanks demonstrated that increased laser power improves bonding strength but also affects bending performance due to localized overheating. Complementary research on single bead deposits of Inconel 718 has demonstrated that process geometry, microstructure, and mechanical performance are strongly governed by the stability of the laser–wire interaction. Excessive heat input or unstable arc conditions lead to spatter formation, lack of fusion, and oxidation during wire arc additive manufacturing. These defects highlight the necessity of real-time process monitoring and closed-loop feedback to ensure stable and repeatable deposition [4].

9.2.2 Case Study: Meltio Wire Laser Directed Energy Deposition Systems for Hybrid and Repair Applications

The Meltio wire laser Directed Energy Deposition (DED) system represents a significant innovation within hybrid manufacturing, integrating additive and subtractive processes to achieve superior mechanical performance and geometric flexibility in critical industrial components. The Meltio M450 platform, as evaluated by Hirudi 3D Intelligence, was applied under processing conditions representative of automotive motorsport requirements. This environment was intentionally selected because it imposes stringent constraints on weight, mechanical strength, and geometric precision. The chosen component, a stub axle for a prototype hillclimb racing car (Fig. 9.3), exemplifies a mechanically intensive part where stiffness,

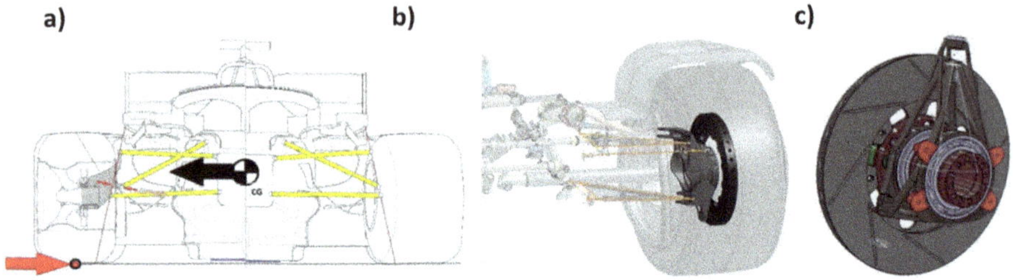

Fig. 9.3 (**a**) Racing car for mountain climbing. (Courtesy of Meltio); (**b**) Stub axle sub-assembly and (**c**) Mechanical design. (Courtesy of Meltio)

load-bearing capability, and dimensional accuracy are crucial for operational safety and performance [20].

The feasibility of hybrid manufacturing for functional components under industrial processing conditions was assessed through the application of the Meltio wire laser system to the production of lightweight structures with controlled load-bearing capability. Traditional fabrication approaches, including machining and mechanically welded assemblies, were used as comparative baselines. By integrating the wire laser DED process with precision machining, the researchers aimed to explore the potential of this hybrid method to reduce production time, material waste, and overall cost while improving geometric design freedom and metallurgical integrity. This aligns with Meltio's overarching goal of promoting AM as a viable industrial alternative for small batch, high performance components that require design flexibility without compromising on strength or quality [20].

The design process for the stub axle began by assessing the boundary conditions and manufacturing limitations inherent to the Meltio M450 system. The working volume of 150 × 200 × 450 mm defined the maximum part dimensions, while the design guide prescribed optimal values for geometrical parameters such as overhang angles and minimum wall thicknesses to ensure consistent material deposition. This initial step established the foundation for adapting the part design to the capabilities of the Meltio process. The design approach centered on decomposing the component into functional geometrical regions responsible for either load absorption or load transmission. This partitioning enabled selective material allocation to regions subjected to higher mechanical loading, resulting in an improved strength-to-weight ratio. The Meltio system enabled localized material deposition in regions of structural demand, resulting in material efficiency comparable to architectures typically achieved by powder-bed fusion processes [20].

Following the initial design phase, a series of iterative design and manufacturing cycles were conducted to refine both the digital model and the deposition parameters. Each iteration incrementally increased the size and complexity of the printed fragments, ensuring that each geometric feature met the required standards of accuracy and mechanical integrity. These iterations enabled the validation of previously assumed process constraints such as layer thickness, cantilever limits, and deposition paths. Moreover, this iterative methodology facilitated a learning-based calibration

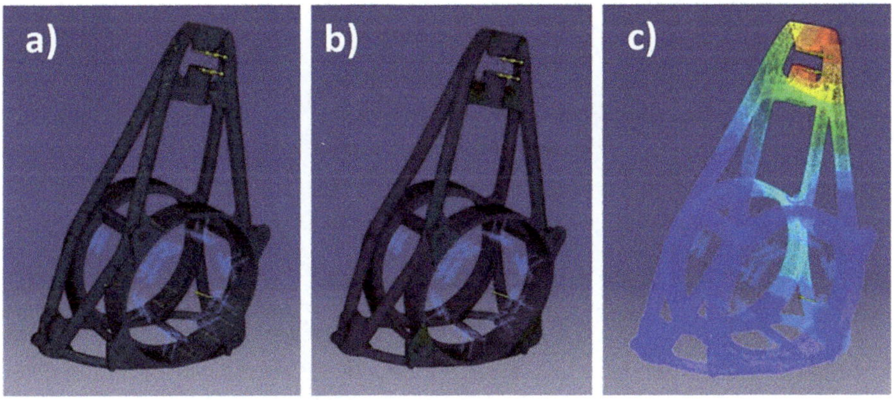

Fig. 9.4 (**a**) Lateral load case. Mesh size 0.5 mm (**b**) Lateral load stresses (**c**) Lateral load displacements. (Courtesy of Meltio)

of manufacturing variables, progressively improving print quality and dimensional stability. To minimize support material and post-processing needs, auxiliary elements were incorporated into the CAD model itself, allowing for the creation of self-supporting structures in areas such as the bearing housings and upper ball joint mountings. This design optimization reduced both material consumption and machining time, one of the key advantages of hybrid manufacturing in complex component fabrication [20].

Process characterization was conducted to determine parameter sets associated with stable deposition and uniform mechanical response across different regions of the component. Several factors influenced this optimization, including surface area size, overhang angles, and path overlaps. By segmenting the part into multiple manufacturing fragments, each fabricated with tailored deposition parameters, the researchers successfully controlled localized overheating and minimized residual stress accumulation. This granular process control allowed the team to reach the threshold of manufacturability concerning cantilever angles, wall thicknesses, and minimal cross section dimensions. Thus, the design team could precisely define the operational limits of the Meltio M450 system, integrating these findings into the final production design. The final manufacturing phase was executed only after empirical validation of all these parameters, demonstrating a methodical integration between experimental evidence and digital modeling [20].

Finite Element Analysis (FEA, Fig. 9.4) simulations were conducted at each iteration to validate the mechanical behavior of the evolving design under realistic operational loads. The primary load cases considered were lateral and longitudinal braking forces, representing the most critical mechanical conditions experienced by the stub axle in racing applications. These simulations utilized material properties corresponding to ER70S6 mild steel, selected for its high weldability and machinability. Before applying operational loads, a mesh convergence study was performed to determine an element size that adequately captured the complex geometry while

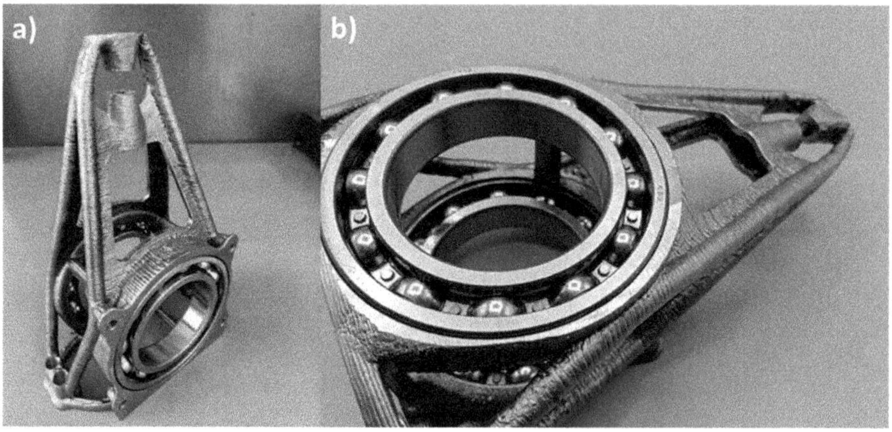

Fig. 9.5 Machined part with mounted bearings. (Courtesy of Meltio)

maintaining computational efficiency. The simulation results revealed that under the lateral load case, which simulated a 2G lateral acceleration, the component exhibited maximum stress values of approximately 367 MPa and displacements nearly 1.25 mm. For the longitudinal braking case, corresponding to a 2.5G deceleration, stresses peaked at 325 MPa with maximum displacements of 0.75 mm. In both loading cases, the computed stresses remained below the yield strength of 400 MPa, indicating structural adequacy of the hybrid-manufactured component. The numerical model excluded auxiliary elements such as bearings and fasteners, so the effective stiffness under assembled conditions is expected to exceed the predicted values [20].

The machining phase followed the completion of AM and was primarily confined to regions requiring high precision interfaces, such as the bearing housings and the bolted connection points. Since these critical surfaces needed tight dimensional tolerances to accommodate commercial components, post-processing was essential. However, the hybrid strategy significantly reduced the volume of material to be machined compared to conventional fabrication (Fig. 9.5). The Meltio process produced near-net-shape geometries that required minimal subtractive finishing. The final machined part demonstrated exceptional dimensional accuracy and surface quality, validating the synergy between additive deposition and conventional machining. Photographic evidence of the finished stub axle and its subassemblies confirmed the system's capacity to produce complex and mechanically robust parts suitable for demanding applications such as high performance automotive systems [20].

The hybrid manufacturing route exhibited measurable advantages over machining-only processing in terms of performance and production cost. The total weight of the stub axle was reduced by 62.5% relative to the fully machined counterpart, translating to substantial benefits in terms of vehicle dynamics and energy efficiency. Additionally, the hybrid process achieved a total cost reduction from 780 € to approximately 501.54 €, largely due to the lower raw material consumption and reduced machining time. The production lead time was reduced from 1.5 weeks to 1 week, reflecting the capacity of the Meltio wire laser system to increase manufacturing throughput and shorten delivery times for specialized components. The comparative analysis indicated that the raw material cost for AM was substantially lower (11.54 € versus 230 €), whereas the transformation cost remained competitive due to process efficiency and reduced labor input. Under these conditions, the Meltio DED system was associated with a cost-effective and scalable route for the production of geometrically complex, low volume components requiring controlled dimensional and mechanical performance [20].

Geometrical and metallurgical inspection of the component indicated stable processing conditions during deposition. Metallographic examination and visual analysis showed no macroscopic defects or layer-to-layer inconsistencies, while the uniformity of the deposited material and the continuity across interlayer boundaries were associated with effective metallurgical bonding, a prerequisite for fatigue resistance under cyclic loading in motorsport applications. Under optimized processing conditions, wire laser DED exhibited structural integrity comparable to that of selective laser melting. The compatibility of this process with hybrid manufacturing routes enabled material addition and subsequent machining for component restoration and geometric modification, with localized deposition minimizing material consumption and repair cost. The integration of additive deposition with CNC finishing further supported its applicability in maintenance and remanufacturing environments [20].

9.2.3 Case Study: Renishaw Laser Powder Bed Fusion Platforms Applied to the Production of Automotive and Medical Components

Renishaw L-PBF platforms have been applied in both automotive (Fig. 9.6) and medical (Fig. 9.7) manufacturing environments through collaborations with HiETA Technologies, Irish Manufacturing Research (IMR), and Croom Precision Medical (CPM). These implementations addressed sector-specific constraints related to thermal management, biocompatibility, and regulatory traceability, while maintaining dimensional control, process repeatability, and geometric flexibility. Under these conditions, the L-PBF systems supported the transition of additive manufacturing from laboratory scale prototyping to repeatable production of end-use components [23].

The collaboration between Renishaw and HiETA Technologies exemplifies the industrial application of L-PBF in the automotive sector. The objective of HiETA was to move from small-scale prototyping toward low volume, commercial production of metal components, particularly for heat management systems. The company, specializing in lightweight, thermally efficient designs, had previously faced

Fig. 9.6 Cross section of SLAMMiT recuperator. (Courtesy of Renishaw)

manufacturing challenges due to the complex geometries and thin-walled structures required in heat exchangers and microturbine recuperators. Traditional fabrication methods, which involved welding thin sheets, were not only time-consuming but also added unwanted weight and compromised performance. The deployment of the Renishaw AM250 followed by the RenAM 500 M enabled the fabrication of thin-walled Inconel structures with wall thicknesses down to 150 μm and leak-tight performance verified through joint testing at the Renishaw Staffordshire facility and the Bristol and Bath Science Park site. These trials generated process window data used to establish parameter sets and design guidelines for additively manufactured heat-transfer components [23].

The technological evolution continued with the investment of HiETA in the RenAM 500 M, a system featuring a 500-W laser and automated powder handling. These features not only enhanced consistency but also reduced operator involvement and cycle time. The transition from the AM250 to the RenAM 500 M allowed HiETA to reduce build times for complex components from 17 days to 80 h, marking a significant milestone in process scalability. Subsequent projects such as the MiTRE cuboid heat exchanger and the annular recuperator for electric vehicle range extenders validated the commercial feasibility of L-PBF. The additively manufactured components exhibited up to 40% mass reduction and 30% volume

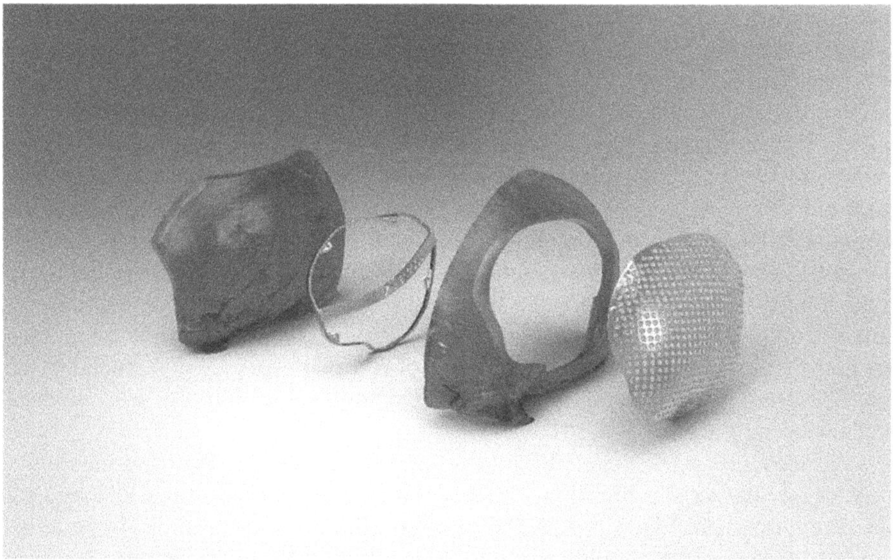

Fig. 9.7 Patient-specific cranial implant and cutting guide, additively manufactured in titanium, shown alongside the surgical fit and final implanted position for precise cranial reconstruction. (Courtesy of Renishaw)

reduction relative to conventionally produced counterparts, while maintaining equivalent thermal and pressure performance. Under these conditions, the L-PBF process was applicable to automotive thermal management components in which reduced mass is associated with improved energy efficiency and lower emissions [23].

In the medical sector, Renishaw L-PBF platforms have been applied to the fabrication of orthopedic and spinal implant components. The collaboration between Renishaw, IMR, and nTopology demonstrated the use of L-PBF for manufacturing spinal implants that closely replicate the mechanical properties of human bone. At IMR, the RenAM 500 M platform in combination with nTopology generative design tools was used to fabricate cervical spine implants incorporating lattice architectures with controlled topology. These porous architectures promoted osseointegration while maintaining load-bearing strength under physiological conditions. The design process was data-driven, incorporating patient-specific loading conditions and the mechanical properties of diseased bone tissue. IMR and nTopology jointly developed models that enabled geometric optimization at the unit cell level, aligning the implant stiffness with that of cortical bone to minimize stress shielding. Renishaw's L-PBF system was essential in achieving these designs with high dimensional accuracy and material homogeneity, using Ti 6Al 4V ELI powder that met ASTM F136 and F3302 standards [26].

Validation was performed through destructive testing of witness coupons produced alongside the implants to verify chemical composition, microstructure, and mechanical response. Mechanical characterization followed ISO 13314 for porous metallic structures, whereas interfacial adhesion between porous and dense regions was evaluated according to ASTM 1104 and ASTM 1147. Under these conditions, the RenAM 500 M platform produced components meeting medical device requirements for strength, porosity, and surface condition with limited post-processing. Process refinements implemented through joint operation between Renishaw and IMR reduced post-processing of critical features by approximately one order of magnitude, supporting a more direct transition from digital design to finished implant. The partnership also emphasized the importance of software-hardware compatibility; nTopology and Renishaw worked jointly to ensure seamless translation of complex lattice geometries into build-ready files. This integration minimized the potential for data loss or geometric distortion during conversion, a crucial factor for ensuring reproducibility and regulatory compliance in medical manufacturing [26].

Croom Precision Medical (CPM) incorporated Renishaw L-PBF and metrology systems into the serial production of orthopedic implants within a regulated medical device environment. The RenAM 500 M platform was used for the additive fabrication of titanium components, with QuantAM software applied for build preparation and process definition. Under these conditions, process traceability and dimensional control were supported by in-line metrology, enabling the acquisition of layer-resolved data on deposition quality and process stability. This approach shifted quality assurance from extensive off-line testing toward in-process verification, allowing component acceptance to be based on build data rather than destructive evaluation [26].

One of the notable applications of CPM was the production of acetabular cups with gyroid lattice structures. This design was previously infeasible using traditional machining due to its geometric complexity. The lattice not only improved mechanical compatibility with bone but also reduced implant weight and enhanced osseointegration potential. The produced components demonstrated high density and optimal Young's modulus, confirming their suitability for long-term orthopedic applications. Moreover, the integrated metrology solutions of Renishaw, including the OSP60 probe with SPRINT technology and the PH10 coordinate measuring machine (CMM) probe, allowed CPM to conduct precision inspections at various manufacturing stages. This level of control was pivotal in maintaining ISO 13485 certification and ensuring compliance with medical device regulations. The measurable outcomes of the adoption of CPM for L-PBF included not only improved product consistency but also expanded international market access, particularly in East Asia and North America, where the company's AM capabilities provided a competitive advantage [26].

9.3 Key Innovations in Industrial Powder Metallurgy Products

9.3.1 Scientific Advances in Feedstock Development, Process Control, and Microstructure Engineering

Advances in powder and wire feedstock design, process control, and microstructural engineering have altered the performance envelope and industrial applicability of metallic AM systems (Fig. 9.8). Feedstocks based on refractory metals, titanium alloys, nickel superalloys, and maraging steels have been developed to achieve high relative density, controlled phase distributions, and stable mechanical response. These developments are governed by the relationships between feedstock morphology, deposition strategy, and solidification behavior, enabling microstructural tailoring for components subjected to demanding service conditions.

The microstructural characterization of refractory powder materials such as molybdenum and molybdenum rhenium alloys has revealed the potential of controlled alloying to stabilize grains and enhance high temperature performance. In plasma-spheroidized Mo-Re powders, scanning and transmission electron microscopy analyses have shown the diffusion of rhenium within molybdenum grains, leading to a refined and thermally stable morphology [29]. The addition of rhenium not only alters the dislocation behavior and grain boundary chemistry but also mitigates brittleness through solid solution softening mechanisms. Such microstructural improvements in feedstock design illustrate how alloying and spheroidization processes can produce powders with superior flowability, packing density, and sintering behavior, directly impacting the quality of L-PBF builds. The resulting materials exhibit enhanced resistance to recrystallization and creep, which are essential for aerospace and high temperature components.

Fig. 9.8 SLM technology used in the production of high performance components. (Courtesy of Renishaw)

Advances in feedstock preparation have been complemented by innovations in wire-based AM, particularly the evolution of the wire arc additive manufacturing (WAAM) and laser wire additive manufacturing (LWAM) processes. Research carried out on these technologies highlight the advantages of wire feedstocks in achieving high deposition rates and low material wastage while maintaining excellent metallurgical bonding [1, 3]. The WAAM process benefits from precise control of heat input through gas tungsten arc, plasma, or metal inert gas systems, enabling the production of large-scale structural components with minimized residual stress when optimized process parameters are employed. The laser wire feed process, by contrast, offers improved spatial precision and lower thermal distortion, leading to better dimensional accuracy and refined grain structures. Abuabiah et al. [1] emphasized that LWAM provides near-net-shape parts with high metallurgical quality and reduced porosity, given the synchronized control of laser power and wire feeding rate. Such control over the melt pool dynamics and deposition path planning has been central to modern feedstock and process integration strategies.

Further progress has emerged from hybrid systems that integrate wire and powder-based directed energy deposition (DED) processes. A sequential hybridization of wire and powder-based DED was applied to Inconel 718, enabling high deposition efficiency in combination with fine structural resolution within a single build [11]. Under controlled cooling rates, interfacial bonding between wire- and powder-deposited layers resulted in defect-free joints associated with increased tensile strength and extended fatigue life. Electron backscatter diffraction showed grain refinement at the wire powder interface, associated with recrystallization-driven strengthening. The combined use of distinct feedstock forms under tailored process control enabled concurrent improvement in interfacial integrity and deposition efficiency.

The application of advanced alloys such as high molybdenum maraging steels fabricated via laser AM represents another milestone in feedstock engineering. Xu et al. [30] demonstrated that direct aging treatments at controlled temperatures can trigger precipitation strengthening through the formation of Fe_2Mo and nanoscale spherical phases, resulting in tensile strengths approaching 2000 MPa with favorable ductility. The preservation of a predominantly martensitic structure before and after aging, coupled with the reverse transformation of austenite, highlighted the stability and reproducibility of the laser-processed microstructure. Such precise microstructural control through process parameter optimization and thermal treatment exemplifies how AM enables tailored strengthening mechanisms previously unattainable with conventional metallurgy.

Microstructural control was central to the performance of L-PBF-processed titanium alloys in lattice architectures. The relationship between lattice geometry, porosity, and compressive response in Ti-6Al-4V structures was evaluated, showing that topology governed defect distribution and apparent density, with diamond lattices exhibiting higher strength as a consequence of increased solid volume fraction [19]. Surface roughness measurements further indicated that uniform strut spacing and reduced staircase effects limited fatigue-crack initiation. Under these conditions, feedstock quality and process stability were associated with predictable mechanical response in architected cellular materials.

Process control in L-PBF has been addressed through sensor integration and defect detection methodologies. Multimodal sensing combined with selective signal processing enabled identification of lack-of-fusion porosity, keyhole-induced voids, and spatter during laser melting. Advanced monitoring systems combining acoustic emission, photodiode feedback, and thermal imaging provide real-time data for adaptive control algorithms that stabilize the melt pool and prevent cumulative defects. Monitoring performance is coupled to feedstock uniformity, as variations in powder size distribution and particle morphology influence signal stability and defect detection thresholds. The application of control algorithms provided a link between feedstock preparation and microstructural reproducibility [16, 31, 32].

9.4 Future Industry Needs and Challenges

9.4.1 Scientific Perspective: Characterization, In Situ Monitoring, and Certification Requirements

Metal AM has undergone extensive advancement over the last decade, driving new paradigms in process control, part qualification, and material integrity assessment. The integration of comprehensive characterization techniques with in situ monitoring systems and rigorous certification frameworks has become essential to meet the performance and safety standards required in critical sectors such as aerospace, biomedical, and energy industries (see an aerospace component presented in Figs. 9.9 and 9.10). Characterization of metal AM materials involves understanding

Fig. 9.9 Front view of the hypersonic propulsion system under experimental development, illustrating structural integration and associated instrumentation for validation testing. (Courtesy of Renishaw)

Fig. 9.10 Metal additive manufactured components arranged for series production, illustrating the transition of AM technologies toward industrial scale manufacturing. (Courtesy of Renishaw)

the interplay among process parameters, microstructure, and resulting mechanical properties, which remain significantly influenced by factors like heat input, cooling rate, and layer orientation. Armstrong and colleagues [14] emphasized that process-induced variations, including residual stresses, porosity, and microstructural heterogeneities, directly determine the functional reliability of additively manufactured metallic components. Localized thermal gradients generated by laser or electron-beam energy sources govern melt pool behavior and require quantitative control to ensure reproducible component quality. Variations in layerwise deposition and solidification response remain a primary source of build-to-build variability in mechanical performance, supporting the use of integrated multiscale characterization approaches linking microstructural, thermal, and mechanical fields.

In situ monitoring systems have been applied to enable real-time control and defect mitigation during processing. Optical pyrometry, high speed imaging, and acoustic emission sensing were used to track melt pool temperature, layer thickness, and powder distribution [33]. These sensors enable immediate feedback loops for controlling process deviations during powder bed fusion (PBF) and directed energy deposition (DED) processes. However, the authors noted that despite these technological advances, full correlation between sensor signals and resulting part integrity remains incomplete, primarily due to the high dimensionality of the data and the lack of universally accepted calibration standards. Fan et al. [31] advanced this concept by proposing an autonomous additive manufacturing (AAM) framework in which artificial intelligence (AI) and digital twin systems collaborate to interpret multi-modal data from diverse sensors. Sensor-driven feedback architectures enabled autonomous adjustment of laser power, scan velocity, and environmental conditions, resulting in adaptive stabilization of the process window. AI-assisted in situ monitoring was associated with certification-oriented AM platforms in which consistency is maintained through data-driven control models that evolve with accumulated process data.

Accurate microstructural and mechanical characterization remains a cornerstone for material certification. Topology-optimized structures produced by metal additive manufacturing were evaluated with respect to manufacturability and mechanical stability through structure-property relationships [34]. X-ray computed tomography, scanning electron microscopy, and tensile testing were applied to characterize defect populations, anisotropy, and material response under service-relevant loading. The integration of computational methods with experimental validation provides a comprehensive understanding of how topology-optimized geometries respond to thermal stresses and load conditions. This synergy between computational prediction and physical characterization is critical in developing certification guidelines that ensure functional performance without overconservatism. Hybrid inspection protocols combining nondestructive techniques, including ultrasonic and eddy current methods, with post-build destructive validation were used to correlate defect signatures with mechanical failure modes.

The biomedical sector illustrates the importance of precise characterization and monitoring for regulatory compliance. In 2025, the group of Razzaq [35] examined the use of AM in bone implants, emphasizing that material characterization must encompass both mechanical and biological performance criteria. Micro-computed tomography and surface profilometry were used to quantify pore morphology, interconnectivity, and surface roughness, which govern osseointegration and load transfer behavior. In situ monitoring during fabrication of patient-specific implants provided early quality verification, ensuring that critical dimensions and porosity remained within clinically defined limits. Regulatory agencies such as the U.S. Food and Drug Administration (FDA) require extensive data on biocompatibility, mechanical testing, and long-term performance validation before approval of additively manufactured implants. Process certification protocols therefore included traceability of digital design data, processing parameters, and machine calibration records in addition to physical inspection, supporting regulatory compliance while enabling controlled evolution of implant architectures and material functionality.

Characterization and certification are complicated by the increasing use of multimaterial and compositionally graded architectures. The concept of digital metallurgy enables spatial control of alloy chemistry through computationally defined compositions and tessellated deposition paths [32]. These heterogeneous systems require multiscale characterization capable of resolving compositional gradients, phase evolution, and interfacial integrity across chemically dissimilar regions. In situ monitoring of multimaterial builds was based on machine learning algorithms trained to distinguish expected microstructural transitions from defect signatures using high speed spectroscopy, thermography, and image-based pattern recognition to regulate solidification in real time. The presence of spatially varying compositions within a single component necessitated revised qualification procedures for property assessment, motivating the use of dynamic certification models incorporating computational simulation and AI-based property prediction to verify compliance with industrial standards.

Industrial scale characterization and monitoring of metal AM have incorporated data analytics in combination with materials science to support certification workflows [36]. Machine learning models linked to process parameter databases enabled predictive identification of porosity, delamination, and related defects prior to their manifestation in the fabricated component. These approaches were aligned with the development of industry-wide qualification criteria designed to balance production throughput with repeatability and traceability. High entropy alloys were identified as candidate feedstocks for additive processing owing to their mechanical response and compositional flexibility although their use requires advanced phase characterization and mechanical evaluation under dynamic loading. Under these conditions, standardization of in situ monitoring and its formal linkage to certification protocols were associated with reproducible qualification of additively manufactured components across industrial sectors.

The coupling between characterization, monitoring, and certification is accentuated in hybrid and autonomous AM platforms. Cyber physical architectures integrating sensor streams with simulation data and historical performance records enable hierarchical decision processes based on data fusion, anomaly detection, and real-time optimization of processing parameters [31]. These developments mark a transition from intelligent additive manufacturing (IAM) to AAM, wherein machines can self-diagnose and self-correct based on learned knowledge. This capability not only reduces human intervention but also enhances process consistency and supports certification through continuous self-validation. A multi-layered architecture comprising knowledge, generative, operational, and cognitive layers was defined to support refinement of predictive models and verification of quality compliance.

In situ monitoring is central to process qualification for aerospace and defense components. Lack of fusion, residual stress accumulation, and surface roughness were identified as recurrent features in high strength alloys processed for aerospace applications [14, 33]. To mitigate these, thermographic imaging, layerwise optical scanning, and acoustic emission analysis have been employed to detect deviations during buildup These nondestructive monitoring tools are integrated with computational models that simulate melt pool behavior and thermal gradients. The combination of empirical and simulated data aids in developing robust process windows, ensuring that each manufactured component meets the stringent mechanical and dimensional tolerances mandated by certification authorities. Furthermore, the group of Armstrong [14] pointed out that hybrid inspection strategies combining X-ray diffraction and microhardness mapping provide deeper insight into residual stress distribution, a critical factor influencing fatigue life in structural components.

Digital metallurgy and topology optimization have been recognized as mutually reinforcing approaches that strengthen the predictive power of characterization and monitoring systems by enabling the design of geometrically complex, mechanically efficient structures as well as functionally graded materials with spatially tailored properties, both of which require validation through in situ data acquired during fabrication and subsequently used to inform post-processing steps such as heat treatment and surface finishing that directly influence certification outcomes; in parallel, the convergence of machine learning, advanced sensor technologies, and multiphysics modeling has enabled additive manufacturing systems to correlate process signatures with actual manufacturing states, thereby distinguishing transient noise from meaningful deviations, a capability that is particularly critical in biomedical applications where tight control of porosity and surface texture is required, reducing process variability and the reliance on extensive post-production testing while improving traceability and supporting virtual validation prior to physical inspection; at the same time, the need to harmonize international regulatory frameworks has been emphasized through the development of digital thread architectures that provide fully traceable records of all stages of the process, from design and machine calibration to in situ monitoring and post-processing, linking manufacturing data to enterprise-level quality management systems in accordance with Industry 4.0 principles [14, 31, 32, 34–36].

9.4.2 Case Study: The Meltio Approach to Scalability, Sustainability, and Industrial Adoption

Meltio has demonstrated a forward-looking strategy for achieving scalability, sustainability, and industrial adoption through its collaboration with Ansys in the study titled "Improving Printability of Large Directed Energy Deposition Components with Simulation." This partnership exemplifies how simulation-based process optimization can transform AM into a scalable and industrially viable technology (Fig. 9.11). The Meltio wire laser metal deposition (WLMD) process was coupled with finite element simulation tools such as Ansys Additive Suite to address distortion, residual stress, and thermal management in large-scale metal AM. This integration supported assessment of process scalability and the establishment of production and qualification pathways for AM-directed energy deposition under industrial operating conditions [21].

The research collaboration focused on understanding the thermomechanical behavior of AM-DED processes during the production of a nickel-based turbine blade, a component chosen for its geometric complexity and high sensitivity to thermal distortions. DED technologies are inherently scalable, allowing for the fabrication and repair of large components beyond the size limits of powder bed fusion systems. Meltio WLMD technology employs multiple laser sources and dual wire feeding within a compact deposition head, enabling elevated deposition rates and geometric flexibility. This configuration supports industrial scalability through reduced equipment footprint and increased deposition volume, while cumulative heat input, residual stress development, and part distortion govern build reproducibility and long-term process stability [21].

A simulation-based engineering workflow using the AM-DED module of the Ansys Additive Suite was applied to address thermal accumulation, stress evolution, and geometric distortion (Fig. 9.12). The virtual process model predicted

Fig. 9.11 Turbine blade printed (DED) on the base. (Courtesy of Meltio)

temperature fields, residual stress development, and part deformation prior to physical fabrication. Integration of real G-code data enabled replication of layer activation sequences, laser power profiles, and material feed rates corresponding to the experimental build. This level of digital fidelity was key to validating process scalability, as it enabled researchers to predict and mitigate potential printability problems in larger components, such as warping near anchor points and excessive tip deformations due to heat accumulation. Such predictive capabilities directly contribute to sustainability by reducing the number of failed prints and the associated waste of energy and raw materials [21].

Real-time thermal measurements were coupled with simulation outputs to calibrate thermal coefficients and improve predictive accuracy. The turbine blade printing process took approximately 4 h and 53 min, involving 114 layers and consuming 183 m of metallic wire. Throughout this process, temperature sensors recorded thermal gradients that were later correlated with simulated temperature fields. These correlations confirmed the reliability of simulation-based predictions, particularly in identifying overheating zones near the blade tip. Overheating was shown to induce nonuniform microstructures, porosity, and mechanical

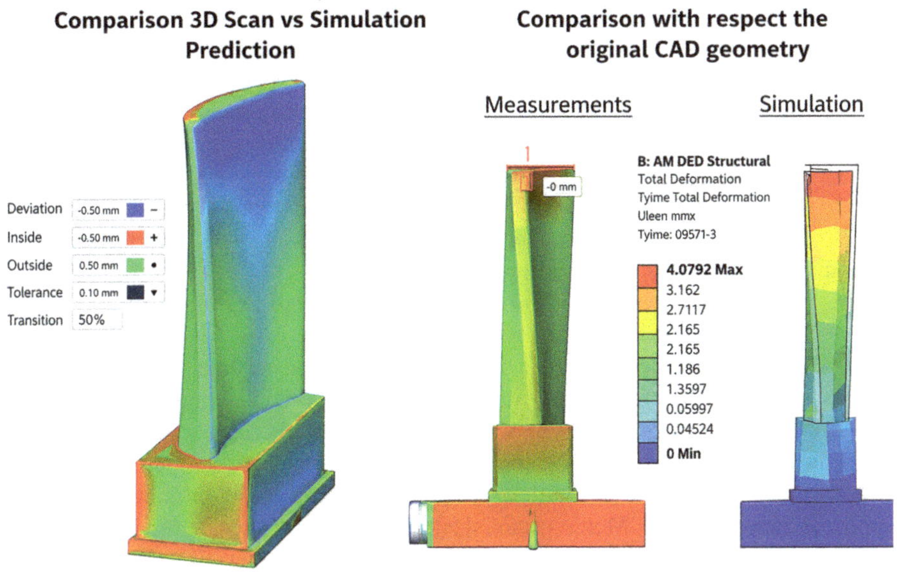

Fig. 9.12 (**a**) Comparison between the 3D scan and the simulation prediction; (**b**) comparison between the measurements, simulation results, and nominal geometry. (Courtesy of Meltio)

anisotropy—defects detrimental to part performance and durability. By enabling engineers to foresee these effects, simulation technology allows the optimization of process parameters, such as dwell times, cooling intervals, and deposition speeds, before production begins. This predictive control not only improves print quality but also enhances resource efficiency by minimizing rework and material wastage [21].

The emphasis of Meltio on scalability is also evident in its commitment to using simulation as a means of optimizing large-scale builds. The turbine blade case demonstrated that even a deformation as small as 4.0 mm at the blade tip could have functional consequences for components subjected to high mechanical and thermal loads. By accurately predicting this deformation, the simulation allowed Meltio engineers to redesign the printing sequence or modify thermal control strategies in future builds. This iterative digital feedback loop forms the cornerstone of scalable AM production, allowing companies to transition from prototyping to mass production with greater confidence in part reliability. Moreover, the scalability of systems of Meltio extends beyond component size to encompass production flexibility; the same hardware and simulation tools can be applied across different materials, geometries, and industries without requiring extensive reconfiguration [21].

Sustainability considerations were incorporated into the simulation-driven workflow, as conventional metal processing is associated with extensive material removal, energy consumption, and waste generation. In contrast, the wire laser metal deposition route operates under near-net-shape conditions, limiting raw material losses. When combined with simulation-driven process optimization, it further reduces the environmental footprint by eliminating the need for excessive trial-and-error iterations during production setup. By simulating the effects of different cooling rates and layer sequencing, engineers can identify process settings that achieve optimal thermal management without unnecessary energy expenditure. The ability to predict and prevent overheating not only extends component lifespan but also ensures a more consistent and repeatable production cycle, both of which are essential for sustainable industrial manufacturing [21].

The partnership between Meltio and Ansys exemplifies how digital twins (Fig. 9.13) can be leveraged for industrial adoption of AM. The Ansys Additive Suite AM-DED module generated a digital twin of the turbine blade deposition process using G-code inputs and material feed parameters representative of the physical build. This digital twin was capable of predicting both thermal and mechanical responses throughout the build, providing actionable insights that can be used to modify future print strategies. By validating these simulations against empirical measurements, Meltio established a robust workflow where virtual testing complements physical experimentation. Such integration significantly accelerates industrial adoption by reducing the uncertainties and risks associated with transitioning new manufacturing technologies into production environments [21].

The simulation-based workflow also supported quality assurance and regulatory compliance, which govern industrial adoption. The ability to simulate stress distributions and deformations allows manufacturers to guarantee mechanical integrity before a part is even printed. This requirement is pronounced in aerospace, energy, and defense sectors, where certification protocols impose rigorous validation of manufacturing processes. By offering a scientifically validated method for predicting and mitigating defects, Meltio strengthens the case for broader industrial deployment of AM-DED technologies. The digital record of simulated and experimental data also facilitates traceability, another essential requirement for industrial certification and quality management systems [21].

Meltio approach combined process sophistication with operational usability through the application of Ansys Additive Suite, Mechanical, and Discovery modules across successive stages of process development, with an integrated simulation wizard enabling efficient definition of boundary conditions and evaluation of process outputs. Simulation-derived data were used to refine printing parameters, including cooling intervals and deposition path strategies, to limit thermal accumulation at the blade tip, resulting in improved dimensional stability and extended service life of components. Within a data-driven and adaptive manufacturing context, this simulation-based methodology supported Industry 4.0 integration through coupling of digital process models with physical deposition, enabling regulation of productivity and resource efficiency under industrial operating conditions through a closed-loop alignment of virtual and physical process states [21].

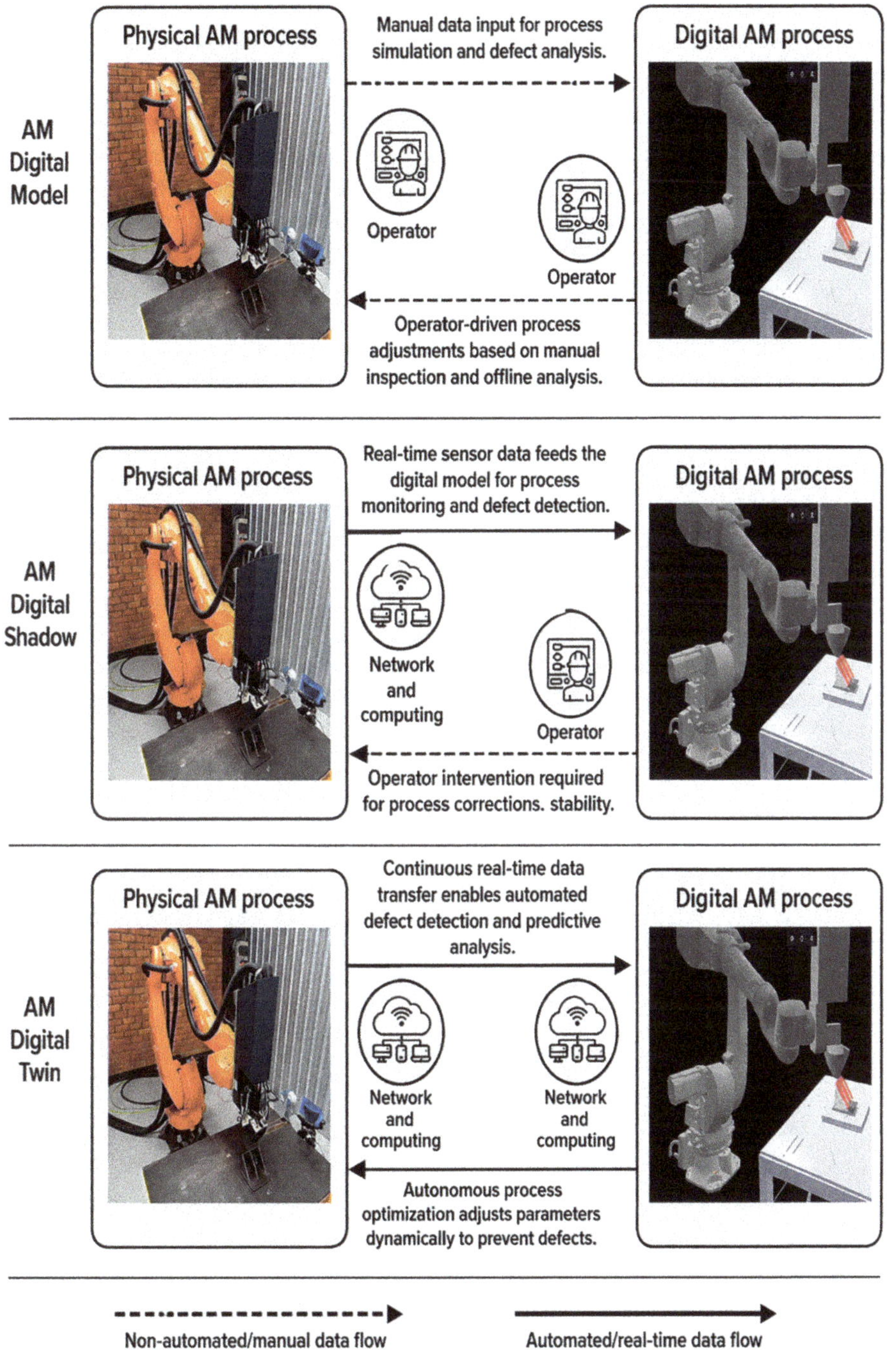

Fig. 9.13 Digital twin categories for AM processes with focuses on process monitoring and defect detection [2]

9.5 Conclusions

The chapter establishes metal additive manufacturing as an industrially mature extension of powder metallurgy in which powder and wire feedstocks are coupled with precise thermal control, in situ monitoring, and digital modeling to deliver reproducible microstructures and mechanical reliability. Powder bed fusion achieves extreme microstructural refinement and near-theoretical density through controlled melt pool dynamics, while advanced inspection and scanning strategies mitigate porosity and residual stress. Wire-based directed energy deposition complements this capability by providing high material utilization, scalable deposition, and refined surface quality through fine wire control and hybrid heating. Moreover, hybrid manufacturing platforms, exemplified by the Meltio wire laser DED system, demonstrate how localized deposition combined with machining yields lightweight, cost-efficient, and mechanically validated components under demanding service conditions. Laser powder bed fusion platforms deployed by Renishaw extend these principles into automotive thermal management and patient-specific medical implants, where feedstock quality, lattice design, and in-process metrology ensure dimensional accuracy, biocompatibility, and regulatory compliance. It is important to note that digital twins, sensor-driven feedback, and simulation-based process optimization integrate feedstock behavior with thermal and mechanical predictions, reducing empirical iteration and material loss while supporting certification oriented production, defining a data-driven, materials-centric manufacturing paradigm for powder metallurgy and metal additive manufacturing.

References

1. Abuabiah M et al (2023) Advancements in laser wire-feed metal additive manufacturing: a brief review. Materials 16:2030. https://doi.org/10.3390/ma16052030
2. Alvares AJ, Rodriguez E, Figueroa B (2025) Digital-twin-enabled process monitoring for a robotic additive manufacturing cell using wire-based laser metal deposition. Processes 13(8): 2335
3. Chaturvedi M et al (2021) Wire arc additive manufacturing: review on recent findings and challenges in industrial applications and materials characterization. Metals 11:939. https://doi.org/10.3390/met11060939
4. Derguini M et al (2025) Investigation of process-geometry-microstructure-properties relationships in Inconel 718 single-bead deposits using wire based laser metal deposition. Mater Res Proc 54:39–48
5. Ding D et al (2015) Wire-feed additive manufacturing of metal components: technologies, developments and future interests. Int J Adv Manuf Technol 81(1):465–481
6. Fulco E, Sorgente D (2024) Wire laser additively reinforced blanks: effect of the laser power on the bending strength of a single layer reinforcement. J Mater Res Technol 33:1276–1285
7. Huang W et al (2021) Laser wire-feed metal additive manufacturing of the Al alloy. Opt Laser Technol 134:106627
8. Jain S, Murtaza Q, Singh P (2025) Microstructural and mechanical characterization of wire arc additive manufactured stainless steel 316L. Arch Metall Mater 70:119–126

9. Kyvelou P et al (2025) Mechanical properties and microstructure of wire laser metal deposited austenitic stainless steel. Mater Des 250:113558
10. Li B et al (2021) Low-roughness-surface additive manufacturing of metal-wire feeding with small power. Materials 14:4265. https://doi.org/10.3390/ma14154265
11. Mudakavi D et al (2024) Sequential hybridisation of wire and powder-based additive manufacturing of Inconel 718: mechanical and microstructural characterization. Mater Sci Eng A 903:146639
12. Shaikh MO et al (2020) Additive manufacturing using fine wire-based laser metal deposition. Rapid Prototyp J 26(3):473–483
13. Wu B et al (2018) A review of the wire arc additive manufacturing of metals: properties, defects and quality improvement. J Manuf Process 35:127–139
14. Armstrong M, Mehrabi H, Naveed N (2022) An overview of modern metal additive manufacturing technology. J Manuf Process 84:1001–1029
15. Chicote B et al (2022) Defect detection in L-PBF metal additive manufacturing using on-line and off-line inspection processes. Procedia CIRP 111:351–354
16. Guillen D, Wahlquist S, Ali A (2024) Critical review of L-PBF metal print defects detection: roles of selective sensing technology. Appl Sci 14:6718. https://doi.org/10.3390/app14156718
17. Magana-Carranza R, Sutcliffe CJ, Patterson EA (2021) The effect of processing parameters and material properties on residual forces induced in laser powder bed fusion (L-PBF). Addit Manuf 46:102192
18. Verdonik TW et al (2023) Investigation of powder bed laser fusion additive manufacturing of 93W, 5.6Ni, 1.4Fe tungsten heavy alloys. Int J Refract Met Hard Mater 111:106055
19. Zluhan B et al (2025) Design, defect analysis, compressive strength and surface texture characterization of laser powder bed fusion processed Ti6Al4V lattice structures. J Mater Res Technol 35:2914–2933
20. Hirudi DI (2025) Application case study development with Meltio system. Meltio: www.meltio3d.com, Linares, pp 1–22
21. Meltio (2024) Improving printability of large directed energy deposition components with simulation. Meltio: Case Study together with Ansys, pp 1–9
22. Renishaw (2016) Case of study: metal 3D printing pushes the boundaries in Moto2™ through defiant innovation. Renishaw: www.renishaw.com, pp 1–4
23. Renishaw (2017) Case of study: exchanging metal 3D printing solutions with HiETA. Renishaw: www.renishaw.com, pp 1–4
24. Renishaw (2019) Case of study: streamlining additive manufacturing for spinal implants. Renishaw: www.renishaw.com, pp 1–4
25. Renishaw (2019) Case of study: material innovation through metal additive manufacturing: a case study with Uniform Wares and Betatype. Renishaw: www.renishaw.com, pp 1–3
26. Renishaw (2020) Case of study: additive manufacturing for serial production of orthopaedic implants. Renishaw: www.renishaw.com, pp 1–3
27. Renishaw (2020) Case of study: RAM3D partners with Renishaw to establish its high-quality volume manufacturing capabilities. Renishaw: www.renishaw.com, pp 1–4
28. Renishaw (2024) Case of study: rapid prototyping expert uses Renishaw AM systems to develop hypersonic technology for aerospace. Renishaw: www.renishaw.com, pp 1–7
29. Czech A et al (2025) Microstructure characterization of refractory powder materials for L-PBF method. Arch Metall Mater 70:897–904
30. Xu TZ et al (2024) Development and characterization of a novel maraging steel fabricated by laser additive manufacturing. Mater Sci Eng A 891:145975
31. Fan H et al (2025) New era towards autonomous additive manufacturing: a review of recent trends and future perspectives. Int J Extreme Manuf 7(3):032006
32. Ali S et al (2025) Additive manufactured enabled digital metallurgy processes, challenges and future prospects. Appl Mater Today 42:102580

33. Vafadar A et al (2021) Advances in metal additive manufacturing: a review of common processes, industrial applications, and current challenges. Appl Sci 11:1213. https://doi.org/10.3390/app11031213
34. Ibhadode O et al (2023) Topology optimization for metal additive manufacturing: current trends, challenges, and future outlook. Virtual Phys Prototyp 18(1):e2181192
35. Razzaq MH et al (2025) Additive manufacturing for biomedical bone implants: shaping the future of bones. Mater Sci Eng R Rep 163:100931
36. Zafar MQ et al (2025) A review on metal additive manufacturing—types, applications and future trends. Recent Progr Mater 07(01):006

Chapter 10
Scientific Foundations and Emerging Research

Abstract Advances in powder metallurgy increasingly arise from the convergence of high-resolution characterization, multiscale modeling, and data-driven design, enabling quantitative control over structure formation far from equilibrium. Diffraction, electron microscopy, and operando techniques resolve phase stability, defect density, and transient transformations across mechanical alloying, sintering, and additive manufacturing, linking lattice strain, grain refinement, and interfacial chemistry to functional response. In situ X-ray and neutron methods capture melt pool dynamics, residual stress evolution, and phase kinetics in real time, while machine learning frameworks integrate multimodal signals to classify process regimes and enable predictive control. Classical diffusion and grain growth models are extended by atomistic and mesoscale simulations that reconcile thermodynamic driving forces with kinetic constraints, explaining metastability, pinning, and coarsening resistance. These foundations underpin the rational design of high entropy alloys, MAX phases, amorphous metals, and hierarchical composites, where deliberate manipulation of defects, interfaces, and length-scale coupling yields exceptional combinations of strength, stability, conductivity, and functionality inaccessible through equilibrium processing routes.

10.1 Introduction

Progress in powder metallurgy increasingly depends on the capacity to resolve, predict, and control structural evolution across multiple length and time scales [1–8]. This chapter positions advanced characterization, theoretical modeling, and data-driven design as mutually reinforcing pillars for understanding how powders transform into functional materials under mechanical alloying, sintering, and additive manufacturing. Diffraction-based methods establish the crystallographic foundation of this framework. X-ray diffraction provides quantitative access to phase constitution, lattice strain, and crystallite size in systems ranging from Cu-graphene nanocomposites to multicomponent high entropy alloys, where peak broadening

R. Pérez-Bustamante, *Fundamental, Practical and Disruptive Applications of Powder Metals*, Topics in Mining, Metallurgy and Materials Engineering,
https://doi.org/10.1007/978-3-032-27130-3_10

and shifts reveal the effects of severe plastic deformation and solute partitioning during milling [4, 5, 9–16].

Microscopy extends the analysis to morphological and mesoscale regimes, with scanning electron microscopy resolving powder sphericity, agglomeration, and interfacial bonding in sintered and additively manufactured materials, through which particle size and shape govern melt pool geometry, grain morphology, and defect formation. Observations of columnar to equiaxed grain transitions in laser-processed stainless steel and dispersion quality in metal ceramic composites establish direct links between process parameters and microstructural outcomes. Automated image analysis and deep learning segmentation further formalize this relationship by converting SEM data into objective particle size and shape distributions that determine powder bed quality and flowability [7, 9, 17, 18].

The dynamic character of powder-based processing requires in situ and operando characterization capable of resolving transient phenomena beyond the reach of ex situ analysis. Time-resolved X-ray and neutron diffraction and imaging provide direct observation of phase transformations, melt pool dynamics, and residual stress development during laser processing, under which thermal cycling suppresses metastable phases in Ti-6Al-4V and controlled oxidation modifies spatter behavior. These physical measurements are increasingly integrated with machine learning frameworks that classify process regimes and correlate thermal and acoustic signals with defect formation in real time [19–21].

Alongside experimental observation, classical diffusion theory, grain growth models, and multiscale simulations relate atomic mobility, defect density, and second phase pinning to densification and coarsening. Predictive control in powder metallurgy is therefore governed by quantitative coupling between crystallography, morphology, and kinetic processes through integrated experimental and computational descriptions [17, 18, 22–27].

10.2 Advances in Powder Characterization Techniques

10.2.1 X-Ray Diffraction

X-ray diffraction (XRD, illustrated in Fig. 10.1) represents one of the most reliable and widely applied techniques for phase identification, crystallographic analysis, and structural quantification in metallic and composite systems developed by powder metallurgy (PM). The application of XRD provides essential information about phase formation, lattice parameters, crystallite size, and residual strain, which are fundamental for understanding the structure and property correlations of materials synthesized through mechanical alloying (MA) and sintering. In the case of graphene nanoplatelet-reinforced copper nanocomposites, the XRD analysis confirmed the predominant face-centered cubic (FCC) structure of copper and the absence of secondary phases or detectable graphene peaks, attributed to the low concentration and weak scattering of carbon atoms compared with copper. The

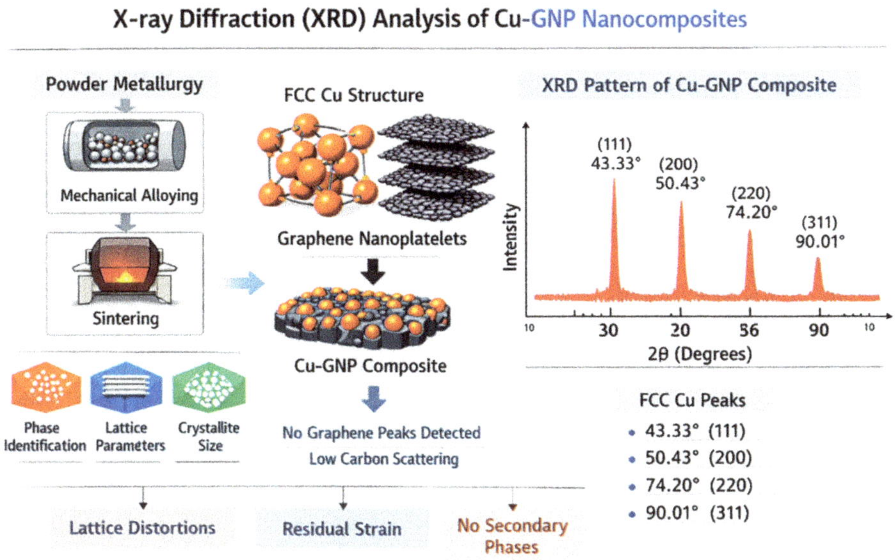

Fig. 10.1 X-ray diffraction (XRD) analysis of a metal matrix nanocomposite, illustrating phase identification, lattice characteristics, and the absence of secondary phases

typical diffraction peaks observed at 43.33°, 50.43°, 74.20°, and 90.01°, corresponding to the (111), (200), (220), and (311) planes, respectively, verify the FCC Cu lattice. The negligible shift in diffraction angles and the absence of new peaks in Cu-Graphene Nanoplatelets (GNPs) composites further indicate that the incorporation of GNPs did not alter the basic crystal structure of copper but contributed to subtle lattice distortions and strain due to interfacial interactions between Cu and GNPs [5, 7].

For FeCoNiCr-HEAs synthesized by PM, XRD was critical for tracking the evolution of phase constitution as a function of copper content. The diffraction patterns revealed a progressive transformation from a single FCC solid solution to a dual-phase system comprising an FCC matrix and a Cu-rich FCC phase. This dual-phase configuration arises from the thermodynamic balance imposed by the high mixing entropy and positive enthalpy of mixing between the alloying elements, particularly with the increasing concentration of copper. The refinement and broadening of diffraction peaks observed with higher Cu addition correspond to increased lattice strain and decreased crystallite size, both consequences of severe plastic deformation during mechanical milling. Quantitative estimation of structural parameters performed using the Scherrer and Williamson Hall methods, demonstrated a consistent reduction in crystallite size and an increase in lattice strain with copper concentration. The crystallite size decreased from approximately 22.24 nm at 5 wt% Cu to 21.24 nm at 20 wt% Cu, while the lattice strain rose from 0.10983% to 0.11468%, indicating enhanced lattice distortion and dislocation density [5]. Such structural refinement is critical for strengthening mechanisms in HEAs, as it promotes grain boundary strengthening and impedes dislocation motion.

In contrast, for the CoCrFeMnNi HEA prepared by gas atomization and subsequent high energy mechanical milling, XRD patterns confirmed the persistence of a single-phase FCC solid solution throughout varying milling times, signifying the thermodynamic stability of this configuration. As milling duration increased from 0 to 240 min, the full width at half maximum (FWHM) of diffraction peaks broadened significantly, reflecting grain refinement and increased microstrain induced by repeated fracturing and cold welding of powder particles during milling. Quantitative XRD analysis performed via the Williamson Hall method revealed a sharp decrease in crystallite size from 39.7 to 6.56 nm and a concurrent increase in lattice strain from 0.25% to 1.48%, emphasizing the impact of milling energy and duration on structural evolution [1].

A notable advancement in the quantitative analysis of XRD data was the integration of artificial neural network (ANN) modeling to predict crystallite size and lattice strain in CoCrFeMnNi HEAs. The ANN model, trained using experimental data obtained from XRD line profile analysis and Williamson–Hall plots, achieved prediction accuracies of 96.25% and 93.43% for crystallite size and lattice strain, respectively. This methodological innovation underscores the potential of combining experimental diffraction data with computational intelligence to accelerate materials design and microstructural optimization. Such predictive modeling also compensates for the limitations of traditional XRD analysis, where measurement errors and instrumental broadening can obscure fine structural details [1].

The comparative analysis of XRD patterns among these systems, Cu-GNP nanocomposites, FeCoNiCrCux HEAs, and CoCrFeMnNi HEAs, demonstrates the versatility of diffraction-based techniques in elucidating fundamental aspects of structure formation in PM materials. In the Cu-GNP composites, XRD primarily served as a qualitative confirmation of phase purity and the absence of oxide or carbide contaminants, while in HEAs, it functioned as a quantitative tool to monitor phase evolution and defect generation. The peak shifts and intensity variations observed in FeCoNiCrCux HEAs with increasing Cu addition directly correlated with enhanced lattice parameters and increased solid solution strain, which align with the calculated increases in dislocation density. Complementary microstructural examination by scanning electron microscopy (SEM) showed alloy homogeneity and the absence of contamination, supporting the phase analysis derived from XRD [5].

10.2.2 Scanning Electron Microscopy

SEM has proven indispensable for analyzing the morphology, microstructure, and surface defects of metallic and composite powders, as well as their sintered or additively manufactured counterparts. Its application across PM and additive manufacturing has enabled researchers to establish direct correlations between powder characteristics, processing conditions, and the resultant microstructural evolution that dictates mechanical and thermal performance. In the fabrication of aluminum matrix composites reinforced with boron carbide, SEM has been utilized to reveal the dispersion of B_4C particles within the aluminum matrix, confirming the homogeneity achieved through solid-state PM and the quality of interfacial bonding. Brillon and colleagues [4] reported SEM micrographs showing well-dispersed B_4C particles without significant agglomeration, which indicated effective mixing during the hot pressing stage at 620 °C. The micrographs also evidenced limited porosity and a smooth interfacial transition between Al and B_4C, attributed to minimal interdiffusion in solid-state sintering, leading to the formation of thin Al_3BC and AlB_2 phases at the interface. The morphological evaluation under SEM confirmed that the densification process reduced voids and improved load transfer across the matrix reinforcement interface, correlating with increased hardness and yield strength while reducing ductility due to the brittle behavior of B_4C-reinforced regions (see Fig. 10.2).

In laser additive manufacturing of 304 stainless steel, SEM was used to examine morphology as a function of laser scan speed, with variations in scan velocity associated with distinct solidification and surface features. Chai et al. [28] used SEM to characterize the microstructural morphology of cladded regions formed at different scanning speeds, revealing how thermal gradients and cooling rates dictate grain morphology. At lower scan speeds, the SEM images displayed coarser columnar grains aligned along the heat flow direction due to slower solidification and higher energy input, leading to larger molten pools. On the other hand, higher scan speeds were associated with finer and more equiaxed grain structures as the cooling rate increased and the heat-affected zone decreased. The columnar grain width decreased from 13.3 μm to 9.2 μm as the scan speed increased from 4 mm s^{-1} to 16 mm s^{-1}, with a concomitant increase in microhardness. The SEM characterization of the clad zone (CZ), heat affected zone (HAZ), and dilution zone (DZ) also confirmed the presence of directional solidification and microsegregation phenomena, which directly impacted hardness gradients within the build. SEM observations therefore validated that optimized scan parameters could refine microstructures and minimize surface defects, improving the overall integrity of the manufactured steel parts.

In powder bed fusion processes, SEM was used to characterize powder morphology, spreadability, and layer uniformity, with these attributes governing the consistency of the resulting component microstructure. In 2021, the group of Mussatto [29] performed detailed SEM analyses on three stainless steel powders with different morphological characteristics to investigate how particle sphericity, surface texture, and size distribution influence powder bed quality. The SEM micrographs revealed that powders with high sphericity and smooth surfaces exhibited superior flowability and packing density, producing more uniform and defect-free powder beds. Contrariwise, irregular or elongated particles generated voids and layer nonuniformities during spreading. SEM examination of the powder bed surface confirmed that higher spreading velocities increased surface roughness and particle segregation, while optimal conditions ($\leq$ 80 mm/s) minimized these effects. SEM imaging also

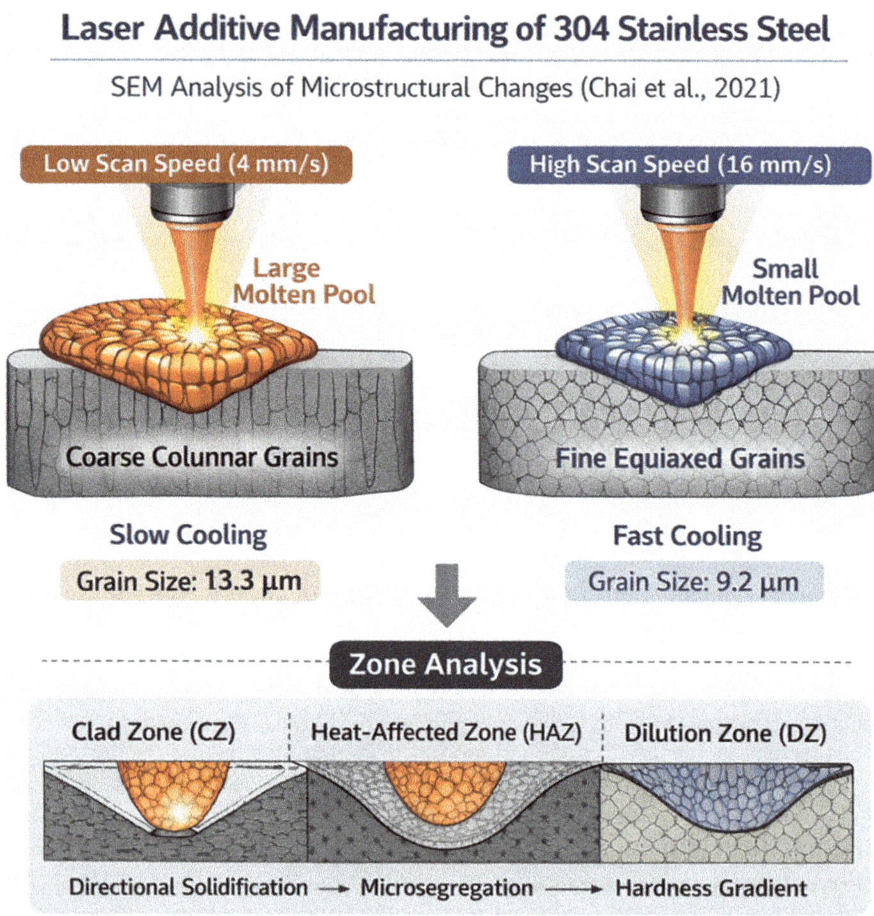

Fig. 10.2 SEM-based schematic illustrating the effect of laser scan speed on molten pool geometry, microstructural evolution, and hardness in laser additive manufacturing of 304 stainless steel, highlighting the transition from coarse columnar grains at low scan speeds to fine equiaxed grains at high scan speeds, as well as the characteristics of the CZ, HAZ, and DZ

identified the formation of scratches and particle clusters on powder layers under suboptimal spreading conditions, highlighting the strong dependence of bed uniformity on powder morphology and process control.

Additional microstructural information on powder-driven processes was obtained through SEM and electron backscatter diffraction, with powder size distribution governing melt pool morphology and solidification patterns in powder fusion additive manufacturing [30]. Their SEM observations revealed that fine powders led to the formation of deep, narrow melt pools with elongated columnar grains exceeding 100 μm in length, whereas coarse powders produced wider melt pools and fine equiaxed grains smaller than 10 μm. The contrast between these morphologies, observed through SEM imaging, provided direct visual confirmation of the counterintuitive phenomenon where coarse feedstock resulted in refined microstructures due to differences in preheat temperature and melt pool dynamics. The examination of melt pool cross sections under SEM confirmed a consistent increase in melt pool width and depth with increasing particle size, emphasizing that particle scale thermal effects could be exploited to achieve site specific microstructural control. Surface morphology analyses further indicated reduced porosity and smoother topographies in coarse powder builds, demonstrating that powder selection alone can serve as an effective lever for tailoring microstructural features without altering processing parameters.

10.2.3 Particle Size Distribution and Shape Analysis

The combination of MA, SEM, and artificial intelligence-based image analysis has provided a multidimensional approach to understanding these parameters. In the context of metal powder characterization, Bakas and coworkers in 2022 [16], developed an automated analytical tool utilizing deep learning algorithms such as Mask R-CNN to extract morphological and statistical information from SEM images of H13 steel powders. This system allows for objective evaluation of the powder morphology by detecting particle boundaries and classifying them into distinct categories, including spherical, elongated, and satellite forms. The methodology provides a rapid and reproducible alternative to traditional manual measurements, which are prone to subjectivity and limited sample representation. The software integrates instance segmentation and statistical post-processing to derive quantitative data on the size distribution, providing a comprehensive framework for assessing powder suitability for additive manufacturing applications (see Fig. 10.3).

In the MA domain, particle size evolution and morphological transformation are strongly influenced by milling conditions, material ductility, and reinforcement addition. Biyik [11] demonstrated how prolonged high energy ball milling of Ag-ZnO-Y_2O_3 composite powders results in a significant refinement of particle size, decreasing the average particle diameter from 18.646 μm to 1.077 μm after 25 h of milling. The initial stages of milling were dominated by cold welding events, leading to the formation of flaky and agglomerated structures. As milling progressed, fracturing mechanisms prevailed, leading to the formation of equiaxed and homogeneously distributed nanocrystalline powders. This evolution corresponds to a narrowing of the particle size distribution and increased morphological uniformity. The introduction of 1 wt % Y_2O_3 not only facilitated the refinement process by acting as a brittle phase but also improved the homogeneity of dispersion within the Ag matrix. The observed morphological transformations, verified through SEM and laser diffraction analysis, confirm that particle shape and size evolution during MA are highly dependent on the interplay between plastic deformation and fracture kinetics. The optimized milling duration of 25 h balanced refinement efficiency and minimized contamination, preserving the functional characteristics essential for electrical applications.

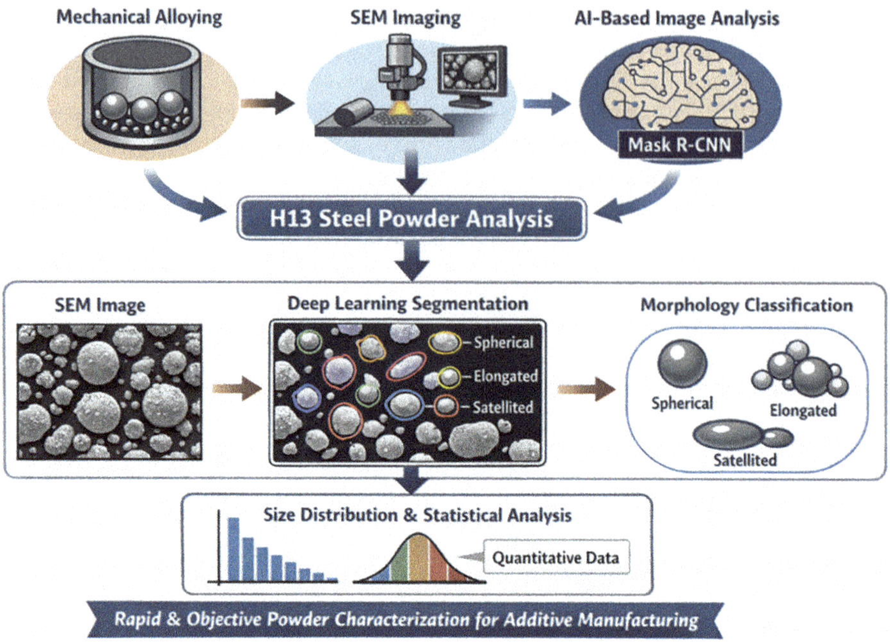

Fig. 10.3 Schematic 2D representation of the automated SEM-AI workflow for H13 steel powder characterization, including particle segmentation, morphology classification, and size distribution analysis for additive manufacturing

Similarly, Si et al. [22] investigated the influence of MA on the particle size and morphology of Fe-based amorphous powders initially produced by gas atomization. The as-received powders exhibited a lognormal size distribution with a median diameter (D_{50}) of 138.4 μm, which decreased to 24.22 μm after 9 h of ball milling. This refinement was accompanied by a transition from spherical to irregular polygonal morphologies, driven by repeated fracturing and re-welding events. The narrowing of the size distribution (ΔH) from 179.96 μm to 36.38 μm indicated a more homogeneous particle population, improving powder flowability and packing density. Importantly, XRD analysis revealed that the ball milling process not only reduced particle size but also increased the amorphous phase content, suggesting intensified interdiffusion and lattice strain. These structural modifications were obtained without degradation of thermal stability or soft magnetic behavior, with controlled mechanical alloying producing increased microstructural uniformity while preserving functional response. Morphological refinement induced by milling was associated with improved particle size control under conditions that maintained the targeted material characteristics.

10.2.4 In Situ, Operando, and Multimodal Characterization Techniques for Powder Evolution Monitoring

The continuous evolution of in situ, operando, and multimodal characterization methods has transformed the understanding of powder behavior during additive manufacturing and related synthesis processes. These methods enable direct observation of dynamic microstructural transformations, thermal gradients, and defect formation in real time, thereby providing fundamental insights into material evolution mechanisms that are inaccessible through ex situ analyses. The integration of X-ray, neutron, optical, acoustic, and thermal imaging modalities, combined with advanced data analytics, has established a new paradigm for real-time powder monitoring, ensuring process optimization, defect mitigation, and predictive modeling accuracy.

The operando XRD and X-ray imaging (XRI) techniques described by Singh and colleagues [7] have become central to understanding the transient behavior of powders and melt pools during metal additive manufacturing, particularly in laser-based processes such as Directed Energy Deposition (DED) and Powder Bed Fusion (PBF). Time-resolved diffraction and imaging enabled phase identification and quantification by operando XRD, while X-ray imaging was used to monitor melt pool dynamics and defect formation mechanisms. Real-time diffraction provided access to the solid-liquid-solid transition in Ti-6Al-4V, with the $\beta \rightarrow \alpha'$ transformation kinetics and texture evolution governed by the applied thermal cycles. These operando measurements showed that repeated laser rastering leads to the suppression of metastable β phase formation after multiple cycles, underscoring the role of intrinsic thermal cycling in defining the final microstructure of titanium alloys. Similar studies on Fe-based alloys using in situ diffraction validated classical solidification models and provided lattice strain evolution data under controlled cooling rates, which correlate with defect generation and residual stress accumulation in printed components (see Fig. 10.4).

Soundarapandiyan et al. [31] employed high speed synchrotron X-ray imaging to examine the influence of oxygen content in Ti6Al4V powders during Laser Powder Bed Fusion. Their study revealed that oxidized powders, characterized by surface TiO_2 and Al_2O_3 layers, modify the melt pool hydrodynamics and reduce spatter formation. Interestingly, controlled oxidation (0.40 wt% O) decreased defect density and improved surface finish, implying that intentional tuning of powder chemistry may enhance process stability. In situ characterization linked powder surface chemistry with melt pool dynamics and component integrity, with the oxidation state acting as a controllable processing variable rather than a constraint. Operando X-ray radiography provided sub-millisecond resolution of melt pool events, enabling a direct correlation between powder attributes and process response.

In situ and operando neutron diffraction approaches further extend this framework by enabling bulk-scale, non-destructive measurements of internal stresses and texture during additive manufacturing. The group of Gaudez [20] developed an L-PBF platform specifically adapted for neutron beamlines to investigate residual

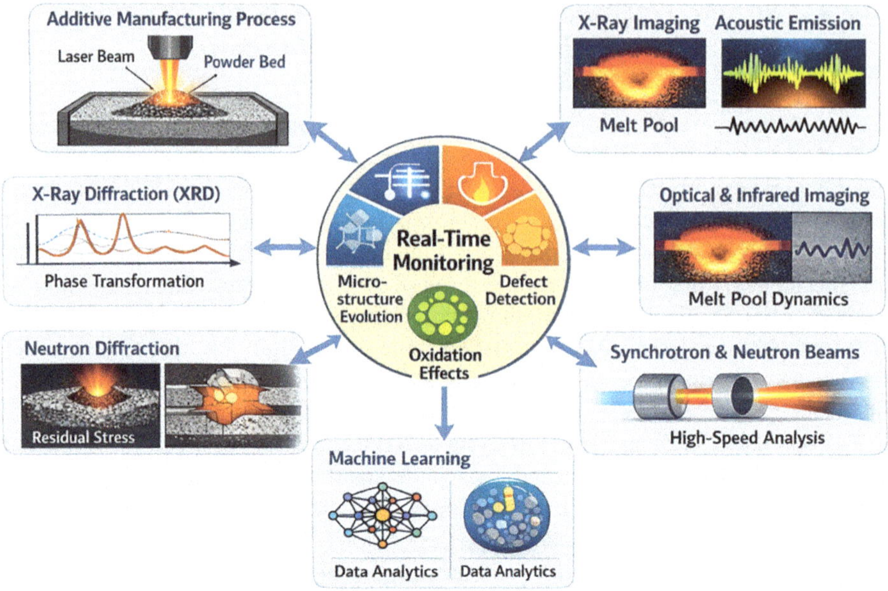

Fig. 10.4 2D schematic of in situ and operando powder evolution monitoring, integrating X-ray and neutron diffraction/imaging with optical and acoustic sensing for real-time tracking of melt pool behavior, phase transformations, and defect formation in additive manufacturing

stress formation in duplex stainless steel (2205). Their Bragg edge imaging and operando diffraction experiments revealed that stress redistribution predominantly occurs during the deposition of the first few layers and that crystallographic texture exerts a pronounced influence on the apparent stress values, introducing errors up to 96 MPa when neglected. Results indicated that orientation-dependent diffraction elastic constants are required for accurate stress modeling. Neutron-based strain mapping combined with real-time monitoring provided constraints for thermomechanical simulations, linking microscale texture evolution with the development of macroscopic residual stresses.

The integration of data-driven approaches into in situ characterization has also become a powerful strategy to interpret complex, multivariate sensor signals. Pandiyan et al. [32] introduced a deep learning framework that couples convolutional and recurrent neural networks to process heterogeneous in situ sensor data, back reflection, infrared emission, optical emission, and acoustic signals, synchronized with operando X-ray radiography during L-PBF. The neural model achieved near-perfect classification of process regimes, distinguishing conduction, keyhole, and lack-of-fusion modes across variable time scales. The authors demonstrated that operando X-ray radiography provides a ground truth for validating the learning algorithms, while the machine learning (ML) model enhances the interpretability and generalization of in situ monitoring systems. Such multimodal frameworks represent a crucial step toward autonomous process control, merging physical imaging and data analytics in a unified operando environment.

Thermal imaging and computed tomography (CT) have also been effectively integrated in operando environments to identify thermal anomalies and optimize structural performance. Lim and colleagues [6] developed an in situ/in operando methodology for topology optimization using synchronized thermal imaging, thermal analysis, and CT scanning. ML algorithms were used to identify cold spots associated with the balling effect during directed energy deposition and to predict the structural integrity of printed components in real time. The correlation between infrared thermal profiles and CT-detected porosity established a quantitative link between process thermal history and defect topology, enabling adaptive control of energy input and scan strategy.

Beyond additive manufacturing, operando characterization has found applications in monitoring powder-based electrochemical systems, where particle evolution directly determines functional performance. Ssendagire and colleagues [14] used operando optical microscopy to investigate the morphological evolution of lithium metal powder composite anodes containing silver nanoparticles. Silver nanoparticles were observed to alloy with lithium during electrochemical cycling, resulting in more homogeneous lithium nucleation and a reduction in dendritic growth. Operando imaging captured dynamic surface evolution and interfacial reactions in powder-based electrodes, with analogous behavior to metallic powder melting and solidification in additive manufacturing. Nanoscale transformations during cycling were resolved by operando techniques in reactive powder systems.

10.3 Theoretical Modeling of Powder Behavior

10.3.1 Classical Models of Densification

The classical models of densification provide the foundation for understanding diffusion mechanisms and sintering theory as applied to mechanically alloyed systems. In the case of nanostructured materials prepared through MA followed by spark plasma sintering (SPS), such as the CoSb3-xTex skutterudites investigated by Liu et al. [17], densification is governed primarily by diffusion-driven consolidation processes enhanced by fine grain structures. The MA step produces powders with nanometric grains and high defect densities, which increase diffusion fluxes during sintering. The SPS process promotes densification through electric field-assisted diffusion, providing high heating rates and short sintering times that reduce grain coarsening and retain the nanostructure. The combination of mechanical alloying and spark plasma sintering produced CoSb3-xTex alloys with an average grain size of approximately 160 nm, with diffusion-controlled densification exceeding that of conventionally melted materials, for which grain sizes are typically on the order of several micrometers (see Fig. 10.5).

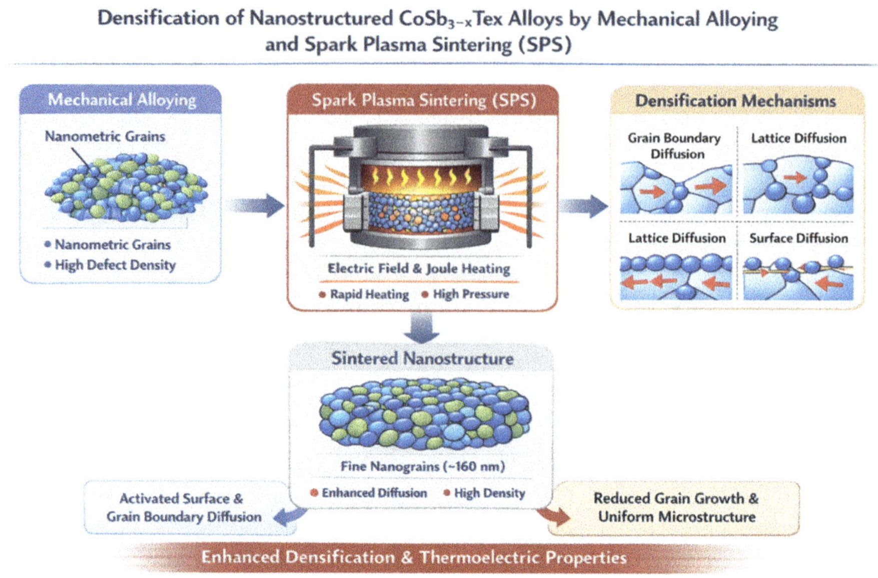

Fig. 10.5 Schematic representation of diffusion-controlled densification in nanostructured $CoSb_{3-n}Te_n$ alloys produced by MA and SPS, highlighting grain boundary-dominated diffusion, electric field-assisted transport, and retention of a fine nanograined microstructure

In classical sintering theory, densification occurs through the reduction of surface free energy, driven by atom diffusion along several possible paths, including lattice diffusion, grain boundary diffusion, and surface diffusion. The relative dominance of these mechanisms depends on the sintering temperature, particle size, and defect density. The group of Restivo [18] in 2010 demonstrated that MA can intentionally design powder microstructures that enhance surface and grain boundary diffusion, facilitating densification at reduced temperatures. They introduced the concept of sintering by activated surface (SAS), where the MA process produces lamellar, pod-like morphologies with metallic layers surrounding ceramic particles, thus generating high energy interfaces that serve as diffusion paths once exposed during initial sintering. This configuration promotes diffusion-assisted neck formation between particles and enhances densification kinetics without extensive grain growth. Diffusion mechanisms in these systems follow the principles established in classical sintering theory, where the effective diffusion coefficient is significantly influenced by defect concentration and particle size. For mechanically alloyed CoSb3-xTex systems, the fine grains and high defect densities result in accelerated atomic mobility, particularly along grain boundaries and dislocations. The SPS step introduces a pulsed electric current, which further enhances diffusion through localized Joule heating and electromigration effects. The combined mechanisms resulted in rapid densification and the development of a uniform microstructure. The diffusion enhancement observed in the samples was consistent with Fickian transport, in which the diffusion flux increases with decreasing particle size and with increasing temperature and concentration gradient. Low temperature activation of densification indicated the predominance of grain boundary diffusion during the initial stage, with a transition toward lattice diffusion as densification progressed beyond the intermediate regime [17].

The sintering kinetics observed by the group of Restivo [18] report that activation energies for diffusion decrease significantly when metallic additives such as copper are present, which lower the sintering temperature by facilitating diffusion through the formation of transient liquid phases. The MA process, followed by SAS, resulted in densification behavior consistent with an activated diffusion process controlled by surface reactions. The experimental sintering kinetics, described by an Arrhenius-type relation, showed a strong dependence of densification rate on both the metallic additive and the oxygen partial pressure. The associated apparent activation energies were consistent with diffusion-based models, in which shortened diffusion paths and increased defect densities reduce the energy barrier for mass transport during sintering.

In the thermoelectric CoSb3-xTex system reported by Liu et al. [33], diffusion governed both microstructural development and transport behavior. Te incorporation on Sb lattice sites was associated with systematic changes in lattice parameters with composition, consistent with substitutional solid solution formation. The solubility limit of Te in the CoSb3 structure, determined to be $x = 0.15$, arises from the balance between substitutional diffusion and the formation of secondary phases. The diffusion behavior of Te atoms influences the overall densification and defect distribution, which in turn affect the phonon scattering and thermal conductivity. As the Te content increases, lattice distortion enhances phonon scattering, reducing lattice thermal conductivity, while the refined microstructure produced by MA contributes to grain boundary scattering. These microstructural features were associated with diffusion-driven densification, with the resulting physical properties governed by the thermodynamic and kinetic constraints of the sintering process.

Classical sintering models formulated by Herring and Coble describe densification in terms of diffusion fluxes along defined transport paths. Activation energies extracted by Restivo et al. [18] were consistent with a grain boundary diffusion-controlled regime, particularly during the initial stage when interparticle necks form. Refractory metal additions were associated with reduced atomic mobility, resulting in diffusion barrier effects and corresponding changes in densification kinetics. Similarly, results reported by Liu et al. [17] demonstrate that grain boundary and lattice diffusion govern the microstructural evolution of CoSb3-xTex alloys, as indicated by the homogeneous distribution of nanometric grains after sintering.

10.3.2 Grain Growth Kinetics

The driving force for grain growth is fundamentally rooted in the reduction of the total grain boundary energy within a polycrystalline material. This energy reduction provides the thermodynamic impetus for grain boundary migration, leading to an increase in average grain size over time. The kinetics of this process is influenced by various factors, including temperature, initial microstructure, and the presence of second-phase particles that can exert pinning forces on migrating boundaries. The interaction between the driving force for grain growth and the resisting pinning force is central to understanding microstructural evolution during thermal processing. Li et al. [34] utilized a two-dimensional cellular automata model to systematically investigate the pinning effects of second-phase particles during grain coarsening, specifically examining complex scenarios involving particle precipitation, coarsening, growth, and dissolution. Simulations demonstrated that the average pinning force per particle is strongly dependent on particle size, a finding that contrasts with some classical two-dimensional theories which suggest pinning force is size-independent. An enhanced pinning force with larger particles was attributed to an increased number fraction of particles located at grain boundary junctions, such as triple and quadruple points, where a single particle can pin multiple boundaries simultaneously (Fig. 10.6).

Kinetic stabilization of nanocrystalline grains against coarsening at high homologous temperatures was achieved through Zener pinning by nanoscale dispersoids. Multicomponent nanocrystalline nickel alloys, including Ni-11W-3Ta-2Y and Ni-3Ta-2Y, were produced by high energy cryogenic mechanical alloying, with the Ni-11W-3Ta-2Y composition retaining a nanoscale grain size after annealing at 900 °C for 100 h [12]. This stability was primarily ascribed to the kinetic pinning of grain boundaries by stable Y_2O_3 nanoparticles and argon bubbles introduced during processing. The efficacy of these pinning agents was quantitatively evaluated using a grain boundary pinning model, which related the limiting mean grain radius to the average particle radius, the volume fraction of particles, and the fraction of particles situated at grain boundaries. Superior long-term stability was associated with Y_2O_3 dispersoids relative to YN in the reference alloy, as Y_2O_3 particles exhibited markedly reduced coarsening during extended annealing, resulting in a higher and more sustained pinning pressure. The stability of the microstructure directly translated to stability in mechanical properties, with the hardness of the Ni-11W-3Ta-2Y alloy remaining significantly above values predicted by the Hall Petch relationship alone, indicating substantial contributions from secondary phase strengthening.

Rare earth additions were used to enhance thermal stability through Zener pinning in high entropy alloys. Samarium additions to nanocrystalline CoCrFeNi were examined by Kotan et al. [35], with the as-milled grain size of approximately 16 nm in the base alloy increasing to about 1.35 μm after 1 h at 1100 °C, indicating limited intrinsic resistance to grain coarsening (Kotan et al. 2025, p. 1). In contrast, the Sm-modified alloy, $(CoCrFeNi)_{96}Sm_4$, maintained an average grain size of 110 nm under the same annealing conditions. This resistance to grain growth was attributed to the formation of additional Sm-based oxide phases, specifically Sm_2O_3 and $SmCrO_3$, which increased the total volume fraction of second phase particles in the microstructure. These finely dispersed oxides acted as effective pinning points for grain boundaries, with the authors explicitly linking the stability to the Zener pinning mechanism, whereby the capillary force of a boundary attempting to break away from an obstacle is balanced by the pinning force exerted by the particle. The

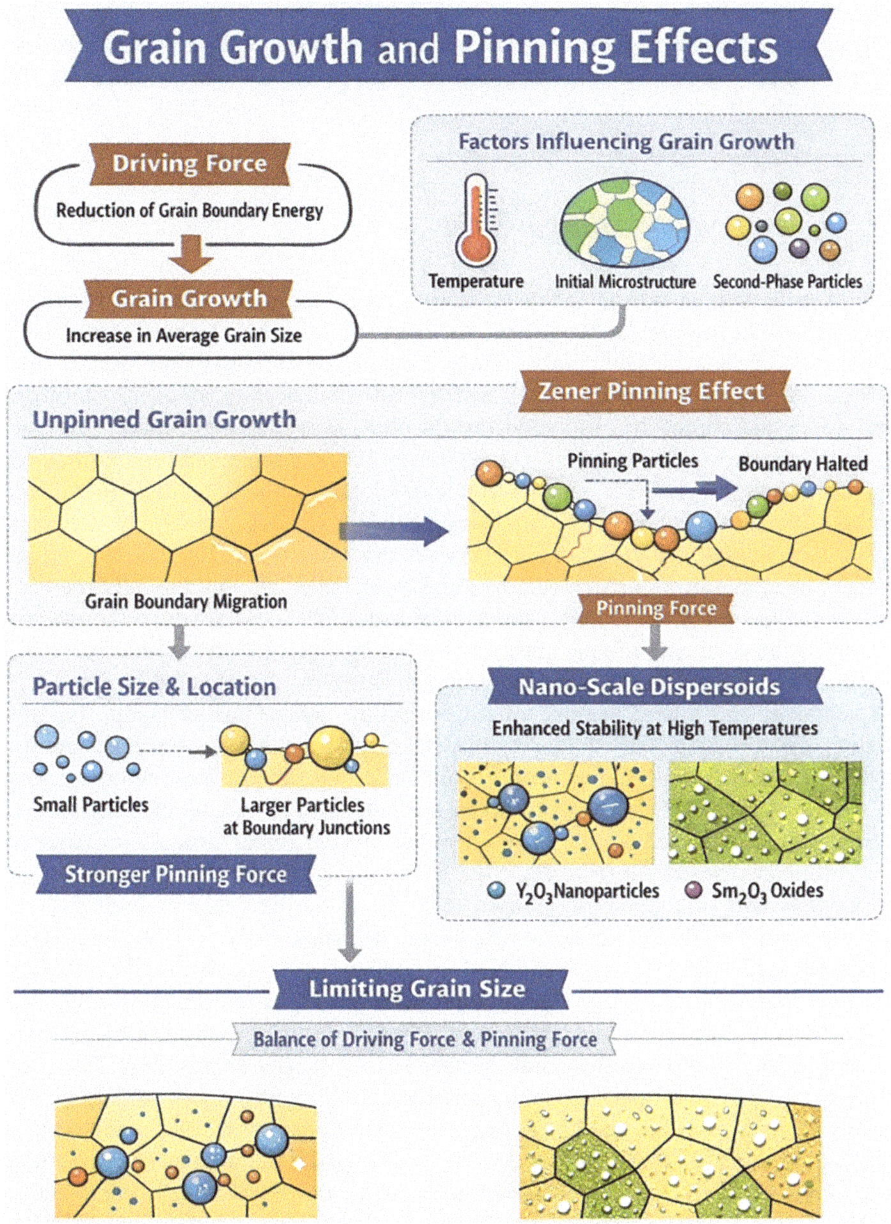

Fig. 10.6 Schematic 2D illustration of grain growth driven by grain boundary energy reduction and its inhibition by Zener pinning from second phase particles, highlighting the roles of particle size, location, and dispersion in controlling the limiting grain size at high temperatures

microstructural stability conferred by the Sm addition had a direct and substantial impact on mechanical properties, as the alloy retained 84% of its as-milled hardness after the high temperature anneal, compared to a severe drop in hardness for the unreinforced alloy.

10.3.3 Computational Approaches

The investigation of powder behavior during processes such as densification and grain growth in complex alloy systems has been significantly advanced through the application of sophisticated computational methods, including kinetic Monte Carlo (kMC), Molecular Dynamics (MD), and hybrid simulation techniques. Atomistic and mesoscale approaches provided insight into phase evolution, mixing mechanisms, and microstructural stability under nonequilibrium processing conditions. An extended Bellon-Averback kinetic Monte Carlo model was applied to ternary NiCoCr systems with negative mixing enthalpies, in which thermally activated vacancy diffusion and deformation-induced slip-plane shearing were treated as competing atomic scale processes governing mechanical alloying [36]. By tracking the evolution of Warren Cowley short-range order (SRO) parameters, the authors identified three distinct mixing mechanisms, chemically dominated, mechanically dominated, and mechanochemical mixing, dictated by the molar content of the element with the lowest average mixing enthalpy, which was chromium in this system. The kMC simulations successfully linked the competition between chemical and mechanical driving forces to the resultant steady-state phases, demonstrating that a high configurational entropy from mechanical mixing can stabilize single-phase random solid solutions, while chemical driving forces promote ordered or phase-separated structures (Fig. 10.7).

Monte Carlo simulations were used to examine the influence of hydrogen on microstructural evolution during severe plastic deformation of a TiVZrNbHf HEA combined with copper, with the model including metal and hydrogen diffusion together with shear steps representing high pressure torsion deformation [37]. The simulations revealed that the significantly different hydrogen affinities of the constituent phases created a substantial thermodynamic barrier against MA. The coupling of metal and hydrogen dissolution rendered the free energy of mixing highly positive for the hydride-containing composite, thereby inhibiting the formation of a single-phase solid solution and preserving a two-phase nanocomposite structure. The Monte Carlo results were consistent over a wide range of diffusion coefficients, with hydrogen reducing atomic scale mixing and thereby promoting microstructural stabilization, in agreement with complementary experimental observations.

Further demonstrating the utility of hybrid computational methods, Dias et al. [38] investigated the phase stability of a Cu Fe Ni Mo W multicomponent alloy using both standard Molecular Dynamics and a hybrid Molecular Dynamics/Monte Carlo (MD/MC) approach. A comparative analysis revealed that pure MD simulations tended to produce a single-phase solid solution, whereas the hybrid MD/MC

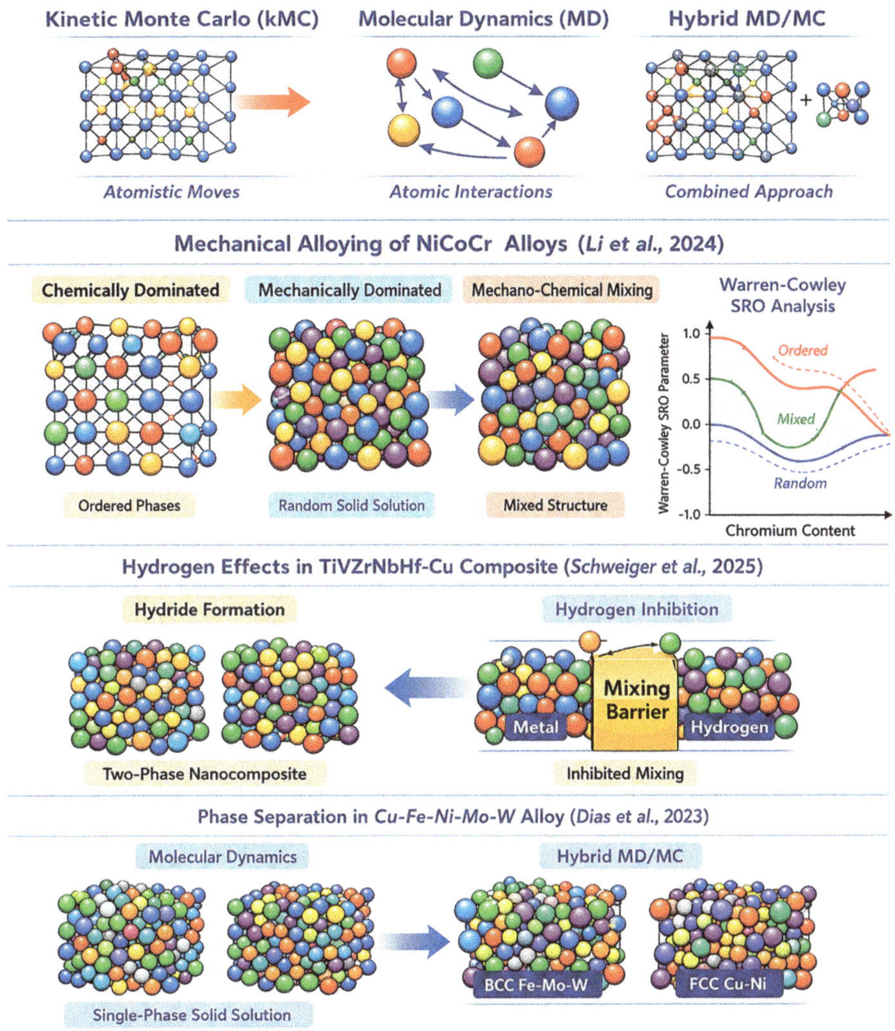

Fig. 10.7 Schematics illustrating atomistic and mesoscale simulation approaches (kMC, MD, and hybrid MD/MC) used to study mixing, phase stability, and microstructural evolution in complex multicomponent alloys under nonequilibrium processing conditions

method consistently predicted chemical segregation into a body-centered cubic (bcc) Fe-Mo-W phase and a face-centered cubic (fcc) Cu Ni phase. This phase separation was energetically favored at lower temperatures, as indicated by the calculated free energy difference between the single-phase and two-phase structures. The hybrid simulations, which allowed for atomic swaps beyond the physical limits of diffusion in MD, yielded microstructural predictions, specifically a two-phase mixture of bcc and fcc structures that aligned more closely with experimental results from arc melting and mechanical milling than those from MD simulations alone.

10.3.4 Multiscale Modeling

Multiscale modeling, as illustrated in Fig. 10.8, links atomistic simulations with continuum scale material response, enabling macroscopic behavior to be derived from atomic level interactions. This methodology is applicable to systems such as HEAs, oxide dispersion strengthened steels, and metal matrix composites, in which properties arise from coupled phenomena operating across multiple length and time scales. In a Cu40Zn24Ni24Ag8Hg4 HEA with antibacterial functionality, density functional theory combined with CALPHAD calculations was used to evaluate phase stability and to guide alloy design (Sharma et al. 2024). The density functional theory (DFT) calculations confirmed the greater stability of an FCC structure over a BCC structure, which was subsequently validated experimentally through XRD after MA. This integration of computational prediction with experimental verification exemplifies how atomistic and thermodynamic simulations can reliably inform processing parameters and final microstructure, ensuring the formation of a single-phase FCC structure essential for the alloy antibacterial performance [27].

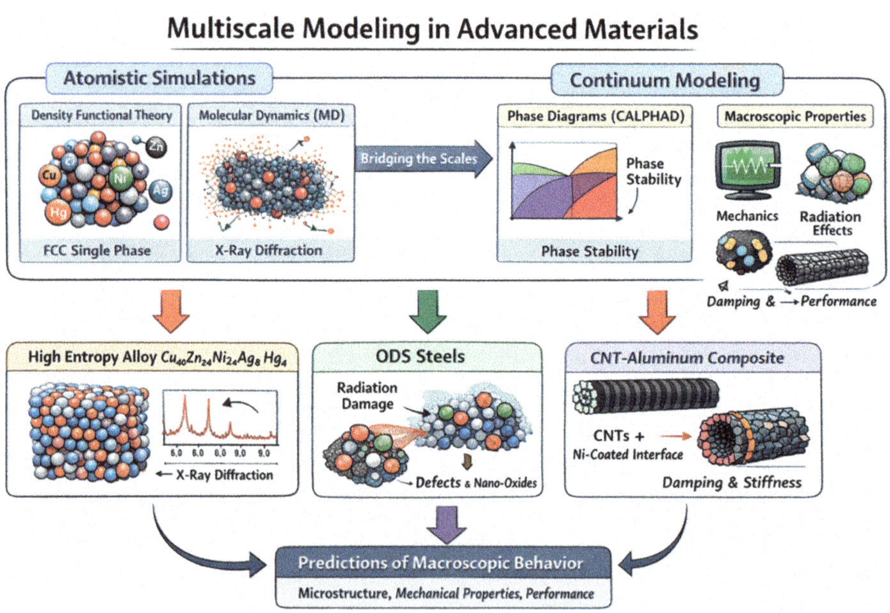

Fig. 10.8 Schematic illustration of multiscale modeling linking atomistic simulations (DFT and MD) to continuum models for predicting macroscopic properties in advanced materials, including HEAs, Oxide Dispersion Strengthened (ODS) steels, and CNT-reinforced aluminum composites

In the context of nuclear materials, multiscale modeling is indispensable for predicting radiation damage tolerance, a phenomenon inherently spanning from atomic displacement cascades to macroscopic swelling and embrittlement. Azeem et al. [39] reviewed the application of integrated computational and experimental methods to study radiation effects in ODS steels, which are candidate materials for next-generation nuclear reactors. Atomic scale simulations, particularly molecular dynamics (MD), are used to model primary radiation damage events, such as the formation of defects near oxide particle interfaces. Simulations reveal that nano dispersoids like Y_2O_3 act as efficient sinks for radiation-induced defects, thereby enhancing the material radiation resistance. The information garnered from MD simulations, such as the number and type of surviving defects, provides critical parameters for higher scale models, including rate theory or phase field models, which can then predict long-term microstructural evolution and mechanical property changes under irradiation.

Similarly, for carbon nanotube (CNT)-reinforced aluminum matrix composites, multiscale modeling connects interfacial atomic interactions to the macroscopic damping and stiffness behavior. Wang et al. [40] employed MD simulations to investigate the role of a nickel-coated interface between CNTs and the aluminum matrix. Simulations demonstrated that the interface exhibits a gradient in elastic modulus and energy dissipation capacity, a finding that could not be captured by traditional homogenized models like the rule of mixtures. Atomic scale information was used to formulate a modified rule of mixtures incorporating an effective interfacial modulus together with a gradient damping term to represent energy dissipation at heterogeneous interfaces (Wang et al. 2023). The MD results showed that introducing nickel atoms defeated the typical conflict between stiffness and damping, leading to a significant increase in the loss modulus, a key figure of merit for structural damping materials.

10.3.5 Predictive Design

The development of advanced alloys, such as Oxide Precipitation Hardened (OPH) steels and HEAs, presents significant challenges due to the complex, nonlinear relationships between their processing parameters, chemical compositions, and resulting mechanical properties. The increasing complexity of materials systems has led to the use of data-driven design methods combining large experimental datasets with ML models to estimate mechanical response without reliance on extensive physical testing (Fig. 10.9). Hybrid ML techniques were applied to predict ultimate tensile strength and elongation in OPH steels [26]. The methodology integrated experimental data from tensile tests on various OPH alloy variants, using chemical composition (Al, Mo, Fe, Cr, Ta, Y, O), MA conditions, and heat treatment parameters as inputs for the models. Three hybrid ML models were implemented, including a feedforward artificial neural network optimized by particle swarm optimization and two adaptive neuro fuzzy inference systems trained using

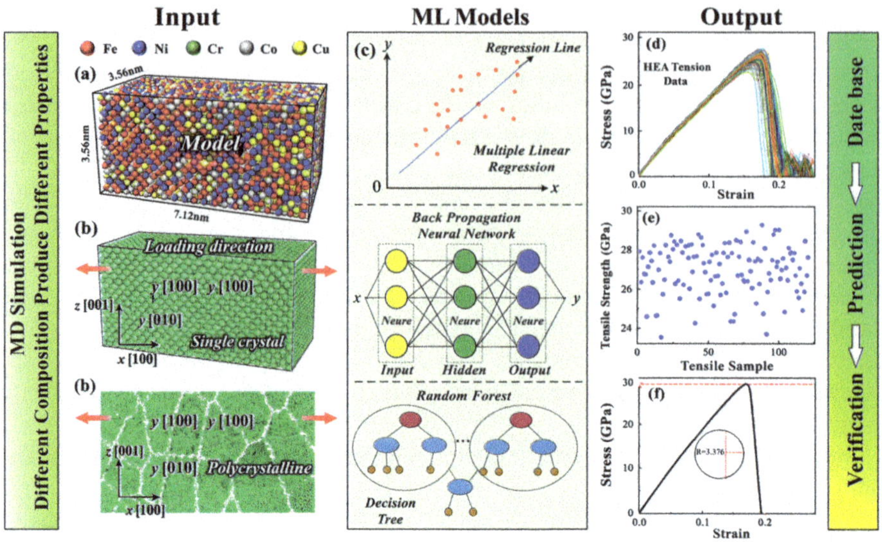

Fig. 10.9 (**a**) Tensile model diagram of FeNiCrCoCu HEA, (**b**) single crystal model diagram of HEA, (**c**) three ML models: multiple linear regression (MLR), back propagation neural network (BPNN) and random forest (RF), (**d**) stress-strain curves for 120 tensile models, (**e**) Tensile strength distribution of 120 tensile models, (**f**) stress-strain curves of HEAs simulated by MD to verify optimal mechanical properties [42]

fuzzy C means and subtractive clustering. These models reproduced the mechanical response of OPH steels with an accuracy of approximately 95%, with the ANFIS model based on subtractive clustering providing the highest fidelity in predicting ultimate tensile strength and elongation from the experimental dataset.

Khalaj et al. [41] further demonstrated the efficacy of ML for property prediction in OPH steels. They constructed and compared three models, Artificial Neural Network (ANN), ANFIS, and Support Vector Machine Regression (SVMR), using a dataset derived from experimental hardness tests (HV10). The inputs included compositional elements and processing conditions such as milling time, rolling temperature, and heat treatment parameters. It was reported that the ANFIS model, particularly when optimized with subtractive clustering, achieved the highest prediction accuracy with a coefficient of determination (R^2) of 0.99 and the lowest mean absolute error on the test set, approximately 32 HV. Sensitivity analysis indicated that heat treatment temperature exerted the strongest influence on hardness in OPH alloys, followed by heat treatment duration, rolling temperature, and milling time.

Expanding beyond ferrous alloys, the integration of ML with computational mechanics has also been successfully applied to multi-principal element systems. Yu et al. [43] combined ML with MD simulations to predict the mechanical properties and microstructural evolution of FeNiCrCoCu HEAs. They utilized three ML algorithms, Multiple Linear Regression (MLR), Back Propagation Neural Network (BPNN), and Random Forest (RF), to model the relationship between the content of principal elements and the tensile strength of the HEAs. Results indicated a strong linear correlation, leading them to select the MLR model, which achieved a prediction confidence of 84.9%. The MLR model predicted that an $Fe_{33}Ni_{32}Cr_{11}Co_{11}Cu_{13}$ HEA would possess the optimal tensile strength of 28.25 GPa. This prediction was subsequently verified using MD simulations, which yielded a tensile strength of 28.38 GPa, resulting in an error of less than 0.5% between the ML prediction and the simulated result. Coupling data-driven ML with physics-based molecular dynamics enabled rapid screening of alloy compositions with targeted properties, reducing reliance on purely experimental trial-and-error procedures.

10.4 Novel Materials Research: High Entropy Alloys, MAX Phases, and Amorphous Metals

10.4.1 HEAs: Design Principles, Stability, and Mechanical Performance

The design of HEAs represents a paradigm shift from traditional alloying strategies, leveraging multiple principal elements in near-equiatomic or non-equiatomic ratios to form concentrated solid solutions. A foundational aspect of HEA design involves the calculation of key thermodynamic parameters to predict phase stability. These parameters include the mixing entropy (ΔS_mix), mixing enthalpy (ΔH_mix), atomic size difference (δ), the Ω parameter, and the valence electron concentration (VEC). For instance, in the study of Ni-doped $Co_4Fe_2Ni_n(Al_{1.5}Mn_{1.5})$ alloys, these parameters were calculated to understand the phase formation during MA (Bilgin et al., 2025). The calculated Ω parameter was greater than one for all compositions, indicating solid solution stability, while VEC values between 7.44 and 7.70 suggested a potential coexistence of BCC and FCC phases, which was confirmed experimentally via XRD. Contrary to theoretical predictions based on equilibrium thermodynamics, the mechanically alloyed powders exhibited a multiphase structure consisting of BCC, FCC, and Laves phases (Fig. 10.10). This deviation was attributed to the nonequilibrium conditions inherent to the MA process, including severe plastic deformation, rapid cooling, and the introduction of significant lattice strain [25].

The stability of phases in HEAs is highly sensitive to the processing route, as demonstrated across multiple studies. MA, a common powder processing technique, often results in nonequilibrium phases that may not be predicted by CALPHAD software. In the $Fe_{40}Mn_{20}Cr_{20}Ti_{10}Al_{10}$ system, MA produced a dual-phase structure of BCC and a complex χ phase [44]. The 40-h milled powder was thermally stable up to 400 °C, beyond which an exothermic event observed via Differential Scanning Calorimetry (DSC) at approximately 500 °C correlated with the formation of an FCC phase, as verified by ex situ XRD of heat-treated samples. This phase evolution at elevated temperatures highlights the metastable nature of the mechanically alloyed

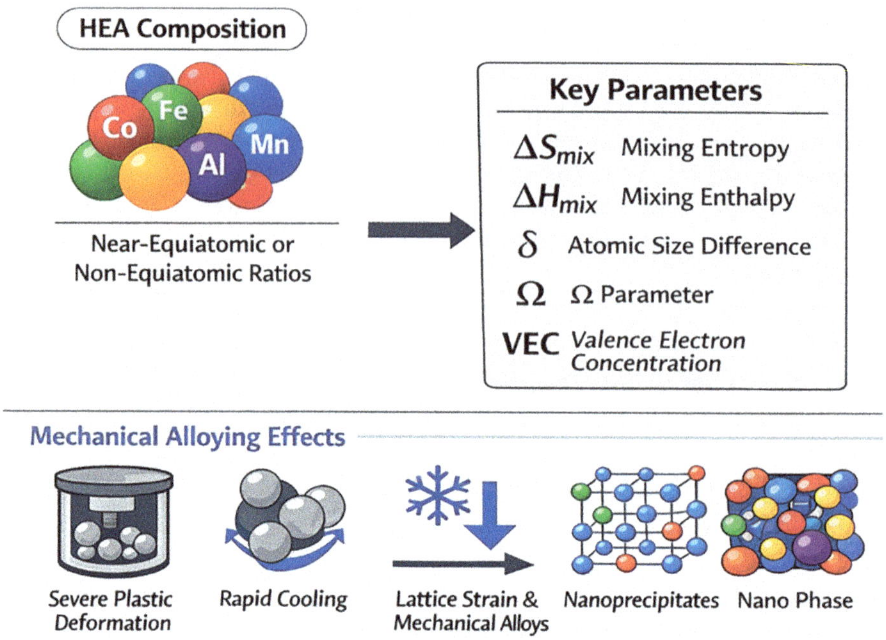

Fig. 10.10 Schematic illustrating the design, processing, phase formation, and structure–property relationships in mechanically alloyed HEAs, highlighting the role of thermodynamic parameters, nonequilibrium phases, and resulting mechanical and magnetic properties

state. Furthermore, subsequent consolidation via SPS at 900 °C led to a microstructure comprising a major BCC phase, a secondary FCC phase, and retained χ-phase nanoprecipitates, showcasing the complex interplay between thermodynamic predictions and kinetic constraints imposed by processing.

The optimization of MA parameters is critical for controlling phase formation, elemental homogeneity, and final powder characteristics. Liu et al. [33] systematically investigated the effects of milling speed and ball to powder weight ratio on the synthesis of $CoCrFeNiAl_{0.9}Nb_{0.1}$ HEA powder. They determined that a milling speed of 400 rpm and a ball to powder ratio of 17:1 for 25 h were optimal parameters. A fully alloyed powder was obtained with a nanoscale grain size of 10.8 nm, a lattice strain of 0.82%, a powder yield of 85%, and a homogeneous elemental distribution verified by energy-dispersive X-ray spectroscopy mapping. Increasing milling energy accelerated alloying, whereas excessively high rotation speeds or ball to powder ratios promoted cold welding, energy dissipation through ball-to-ball collisions, and localized temperature rises, resulting in grain coarsening.

The mechanical performance of HEAs is directly governed by their resultant microstructures, which are a consequence of both alloy design and processing. The sintered $Fe_{40}Mn_{20}Cr_{20}Ti_{10}Al_{10}$ alloy exhibited exceptional mechanical properties, including a high compressive yield stress of 1900 ± 200 MPa, a hardness of 6.4 ± 0.2 GPa, and a respectable failure strain of approximately 6.7%. Quantitative analysis attributed this high strength to a combination of strengthening mechanisms. Grain boundary strengthening, calculated using the Hall–Petch relationship, was a major contributor due to the fine microstructure. Significant contributions also came from dislocation strengthening, resulting from the high dislocation density introduced during MA, and precipitation strengthening from the stable χ phase nanoprecipitates. Solid solution strengthening provided a further, though smaller, contribution to the overall strength [44]. In contrast, the magnetic properties of the mechanically alloyed $Co_4Fe_2Ni_n(Al_{1.5}Mn_{1.5})$ alloys were less influenced by Ni content. Saturation magnetization (Ms) values remained around 111–113 emu/g, and intrinsic coercivity (Hci) was consistently between 169 and 171 Oe across different Ni concentrations. First Order Reversal Curve (FORC) analysis revealed that Ni addition narrowed the vertical diffusion in the FORC diagrams, suggesting a partial facilitation of domain wall movement. However, structural irregularities and the presence of nonferromagnetic phases, such as the Laves phase, counteracted any significant reduction in coercivity, underscoring the complex relationship between chemical composition, phase structure, and functional properties in HEAs [25].

10.4.2 MAX Phases

MAX phases, represented by the general formula Mn + 1AXn, constitute a unique family of nanolaminate materials that integrate the advantageous properties of both metals and ceramics. Their layered hexagonal crystal structure, characterized by strong M-X bonds of metallic covalent nature and weaker M-A bonds of metallic character, provides a distinct combination of stiffness, thermal and electrical conductivity, oxidation resistance, and machinability. These materials exhibit extraordinary thermal stability, resistance to damage, and conductivity, which distinguish them from traditional ceramics. The structural anisotropy of these compounds contributes to their exceptional mechanical performance under extreme conditions. For instance, the presence of alternating metal carbide or nitride layers and A group element layers allows easy deformation through dislocation glide along basal planes, while maintaining high stiffness and strength perpendicular to those planes [9, 45].

The layered configuration directly influences the functional behavior of MAX phases, with compositions such as Ti_2AlC, Ti_3SiC_2, Cr_2AlC, and V2AlC serving as the most studied examples. Historically, the discovery of these materials traces back to the 1960s, when Nowotny and co-workers identified ternary carbides with layered crystal structures, later formalized under the Mn + 1AXn notation. Interest in MAX phases increased after the synthesis of Ti_3SiC_2 in 1996, which exhibited a combination of metallic machinability and ceramic resistance to oxidation and thermal degradation [9]. These phases crystallize predominantly in the P63/mmc space group and are classified into 211 (M2AX), 312 (M3AX2), and 413 (M4AX3) structural types (Fig. 10.11). Variations such as intergrown and i-MAX structures have expanded the compositional diversity and tunability of these materials, enabling

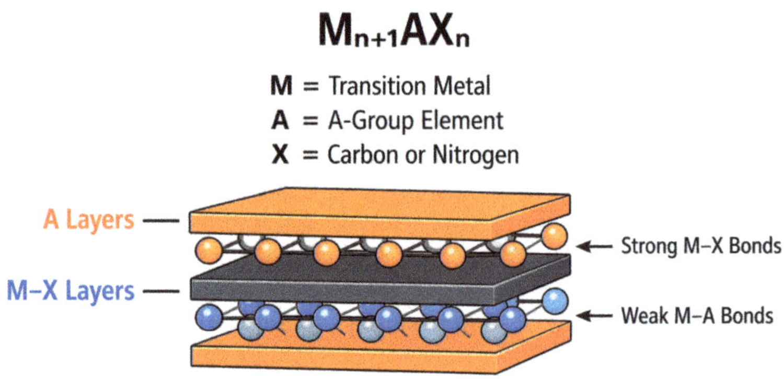

Fig. 10.11 Schematic representation of MAX phases ($M_{n+1}AX_n$), highlighting their layered hexagonal structure with strong M–X bonds and weaker M–A bonds, which underpins their unique combination of metallic and ceramic properties

manipulation of their mechanical, electronic, and thermal properties for specialized applications [9, 45].

Experimentally, the synthesis of MAX phases requires precise control of composition and thermal conditions to achieve phase purity. For example, in the mechanochemical synthesis of Ti3AlC2, Shahin et al. [46] demonstrated that increasing the aluminum content beyond stoichiometry enhances the phase purity due to compensation for Al loss during high temperature annealing. Their experiments involved ball milling Ti-Al-C mixtures with different Al contents followed by annealing at 900 °C and 1200 °C. XRD revealed that extended milling times promoted the formation of Ti3AlC2 and TiC through a mechanically induced self-propagating reaction, while excess Al facilitated the completion of the Ti3AlC2 phase with up to 88% purity after annealing at 1200 °C. Annealing promoted atomic diffusion and phase homogenization, while mechanical activation increased reactivity through defect introduction and particle size reduction. The combined treatments enabled the formation of dense, highly crystalline MAX phases under moderate conditions relative to conventional hot pressing or self-propagating high temperature synthesis.

Şenyurt et al. [15] examined the synthesis of Cr2AlC via mechanical activation followed by annealing, providing significant insight into the reaction mechanisms governing Cr-Al-C systems. A two-stage process, involving high energy ball milling (HEBM) and subsequent thermal treatment, demonstrated that MAX phase formation initiates near 700 °C and reaches completion at 1100 °C with a phase fraction exceeding 99% under optimal conditions. Detailed analysis using DSC, XRD, and TEM revealed that the formation proceeds through intermediate carbide phases before the establishment of the layered Cr2AlC structure. Furthermore, the addition of stearic acid as a process control agent prevented powder agglomeration and facilitated single-phase formation at higher annealing temperatures (1300 °C). The resultant Cr2AlC exhibited high density, uniform morphology, and the absence of residual carbide impurities, emphasizing the significance of milling duration and thermal treatment optimization in achieving phase purity.

10.4.3 Amorphous Metals

Amorphous metals, also referred to as metallic glasses, constitute a distinctive class of materials characterized by a lack of long-range atomic order, which imparts to them exceptional combinations of strength, corrosion resistance, and magnetic softness. Their synthesis relies on suppressing crystallization through rapid solidification or solid-state techniques such as MA, SPS, vacuum hot pressing (VHP), or cold spray deposition. Among these, MA has proven particularly effective for producing amorphous alloys and nanocrystal-embedded composites because it enables atomic interdiffusion and structural disordering at room temperature. The ability of an alloy system to form an amorphous phase is determined by its glass-forming ability (GFA), a function of thermodynamic parameters such as mixing enthalpy and atomic size mismatch, as well as kinetic factors that hinder crystallization [10, 47].

Research carried out on Al-based amorphous alloys demonstrates that alloying elements with large negative mixing enthalpies, including Fe, Ni, Ti, and Y, significantly promote amorphization during the MA process. In Al-Fe-Ni systems, the negative enthalpies of mixing for Al-Fe, Al-Ni, and Fe-Ni pairs favor solid-state amorphization, facilitating the formation of a homogeneous amorphous matrix. The group of Binh [47] synthesized Al82Fe14Ni4 alloys using planetary ball milling under argon at rotation speeds of 250 and 350 rpm, finding that a fully amorphous structure was achieved after 60 h of milling at the higher speed due to more intense collisions and enhanced atomic diffusion. The disappearance of sharp diffraction peaks and the emergence of a broad halo in the XRD patterns confirmed amorphization, while differential scanning calorimetry (DSC) indicated that powders milled at higher speeds exhibited increased thermal stability. Subsequent SPS consolidation at 600 °C produced dense bulk samples with a compressive strength of 710 MPa, demonstrating that MA followed by SPS is an effective route to generate amorphous Al-based materials with high mechanical performance (Fig. 10.12).

Compositional tuning was used to increase glass forming ability and to regulate thermal response. Al82Fe14Ni2Y2 and Al82Fe14Ti2Y2 alloys produced by mechanical alloying were examined to assess the influence of Ni and Ti substitution on glass formation and magnetic behavior. Dissolution of Ni and Al into the Fe lattice promoted the formation of Fe(Ni,Al) solid solutions, resulting in a fully amorphous structure after 60 h of milling [10]. In contrast, Ti additions slowed the amorphization process, as Al82Fe14Ti2Y2 retained partial nanocrystallinity even after 100 h. The slower dissolution kinetics of Ti reduced the amorphous fraction but improved the alloy's resistance to crystallization. Magnetic characterization indicated that both coercivity and saturation magnetization decreased markedly with milling time, consistent with the transition from crystalline to amorphous structures.

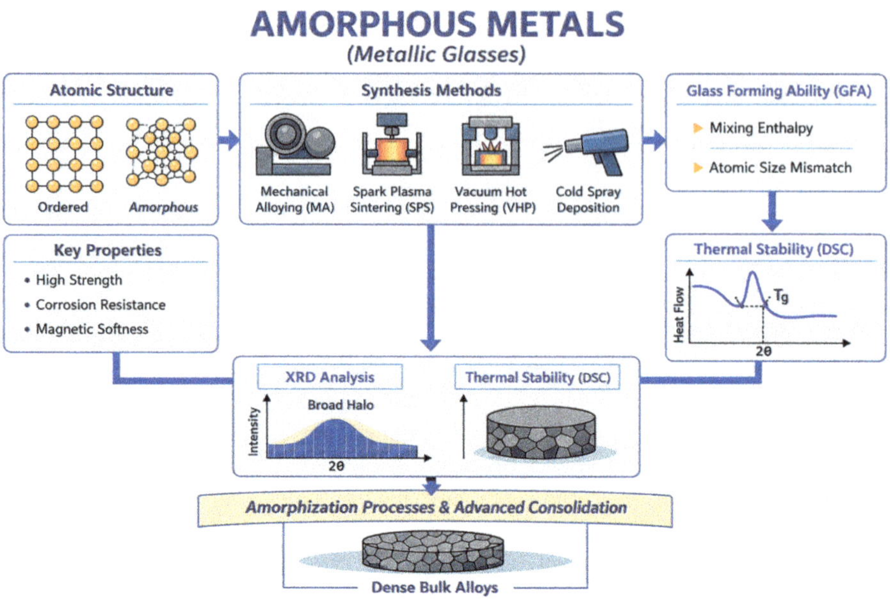

Fig. 10.12 Schematic representation of amorphous metals (metallic glasses), illustrating the absence of long-range atomic order, key synthesis routes based on MA and solid-state consolidation, glass-forming ability factors, and typical structural and thermal characterization methods leading to dense bulk amorphous materials

In addition to Al-based systems, the combination of MA and thermal consolidation has enabled the synthesis of bulk Cu-based amorphous composites embedded with nanocrystals. Lee et al. [48] developed (Cu60Zr30Ti10)91Ta9 powders by high energy ball milling for 5 h, producing Ta nanocrystal-embedded amorphous powders characterized by a supercooled liquid region of 43 K between the glass transition and crystallization temperatures. Subsequent vacuum hot pressing at 753 K for 30 min under a pressure of 1.2 GPa yielded dense nanocrystal-embedded bulk composites with relative densities up to 98.3%. The embedded Ta, Cu51Zr14, and Cu-Zr-Ti intermetallic nanocrystals were homogeneously dispersed within the amorphous matrix, enhancing both thermal stability and mechanical performance.

The integration of MA with advanced coating methods has further expanded the applications of amorphous metals. Aldhameer et al. [49] combined MA with cold spray deposition to fabricate Cu50(Ti50-xNix) ($x = 10$, 20, 30, and 40 at.%) metallic glass coatings on stainless steel substrates (SUS304). The metallic glass powders were produced through low energy ball milling for 100 h, followed by cold spray application, a solid-state coating process that preserves the amorphous nature of the feedstock. The Cu50Ti20Ni30 composition exhibited outstanding nanohardness, high wear resistance, and a low coefficient of friction ranging from 0.32 to 0.45. Moreover, antibacterial tests showed effective inhibition of *Escherichia coli* biofilm formation on the coated surfaces, attributed to the presence of nanocrystalline Cu-based phases within the amorphous matrix.

10.4.4 Hierarchical Composites

Hierarchical composites represent a strategic evolution in materials design, where the deliberate integration of multiple structural scales from the nanoscale to the macroscale generates emergent mechanical, thermal, and functional properties unattainable in conventional homogeneous materials. The concept relies on orchestrating reinforcing phases, interfaces, and matrix microstructures to promote synergistic strengthening, ductility retention, and thermal stability. Hierarchical architectures in metallic and ceramic matrix systems enable a balance between competing property pairs such as strength and ductility or electrical conductivity and thermal resistance, through multiscale constraint and coordinated deformation mechanisms (Fig. 10.13).

The hierarchical Cu-15W composite produced by MA and SPS exemplifies how bimodal particle distributions can stabilize microstructures under extreme conditions. This material exhibited an ultimate tensile strength of 709 MPa and exceptional thermal stability, maintaining nearly all its strength up to 700 °C due to the coexistence of nanoscale (4.6 nm) and submicron (27.5 nm) W particles within the Cu matrix. The smaller W nanoparticles enhanced dislocation storage, while the larger ones served as pinning sites for grain boundaries, collectively suppressing grain growth and dislocation motion. The coexistence of coherent and semicoherent interfaces across multiple particle size scales produced a hierarchical constraint network that preserved structural integrity and strength at elevated temperatures, with thermal and mechanical stability maintained concurrently [50].

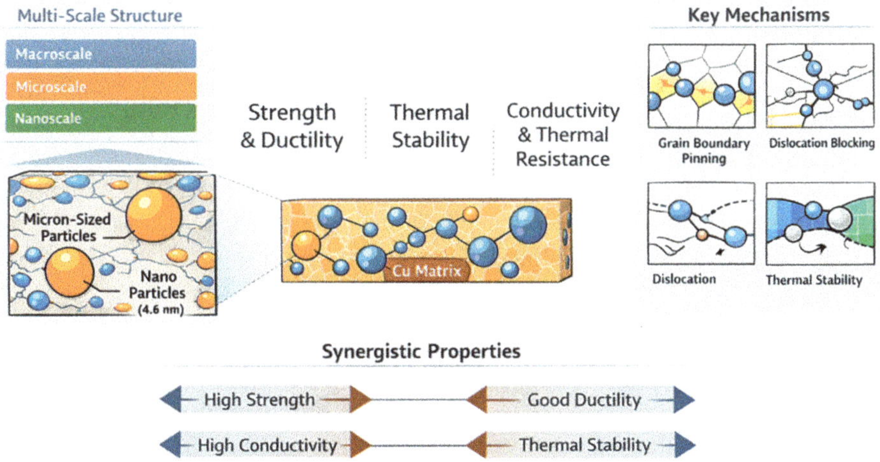

Fig. 10.13 Schematized representation of a hierarchical composite, illustrating the integration of nano-, micro-, and macroscale reinforcements within a metallic matrix. The multiscale architecture promotes synergistic mechanisms such as dislocation storage, grain boundary pinning, and interfacial constraint, enabling the simultaneous optimization of strength, ductility, thermal stability, and functional properties

An analogous hierarchical strategy was employed in the Mo-Cu composite system where Al doping induced a three-dimensional micro/nano heterostructure. Here, micron-sized Mo phases formed a percolated network, surrounded by submicron and nanoscale Mo-Cu domains, while Al-rich nanoparticles occupied grain boundaries and phase interfaces. The introduction of Al significantly reduced interfacial energy and promoted covalent bonding across phase boundaries, as confirmed by density functional theory calculations. These modifications resulted in a pronounced enhancement of mechanical and tribological properties: hardness increased by 24.4%, compressive strength more than doubled, and frictional wear decreased substantially. The hierarchical network thus acted not only as a structural reinforcement but also as a friction-mitigating architecture, revealing that chemical tuning of interfacial structures can complement hierarchical geometric design to yield superior multifunctional composites [50].

Hierarchical design principles extend beyond binary metallic systems into complex multicomponent alloys and composites. The integration of nano-TiC reinforcements into a CoCrFeMnNi HEA through laser powder bed fusion (L-PBF) enabled the formation of a hierarchical microstructure that reconciled strength and ductility. Incorporation of 1 to 2 wt% TiC nanoparticles refined the grain structure and increased dislocation density without inducing pronounced anisotropy. This hierarchical configuration produced a combination of yield strength up to 882 MPa and elongation up to 29% associated with the simultaneous contributions of grain boundary, Orowan, solid solution, and dislocation strengthening. The multiscale network of reinforcing particles and deformation-induced substructures promoted strain hardening and delayed strain localization, with additive manufacturing enabling hierarchical control of phase distribution across length scales [51].

Hierarchical reinforcement of copper matrices by the combined introduction of TiB_2 microclusters and nanoscale $B_{24}Cu$ precipitates produced composites that exceeded conventional balance between strength and electrical conductivity. The material achieved a tensile strength of 926 MPa and conductivity of nearly 80% IACS by combining in situ synthesized microscale TiB_2 networks with uniformly distributed, thermally stable nanoprecipitates. The hierarchical arrangement formed continuous low resistance pathways and promoted efficient load transfer, resulting in enhanced work hardening capacity and balanced performance [8]. In this regard, bioinspired design extended hierarchical composite concepts to ceramic–metal systems, with an Al_2O_3/Al-laminated composite produced by flake powder metallurgy replicating the brick-and-mortar microarchitecture of nacre. The aligned flake morphology created alternating stiff and compliant layers that enhanced strain hardening and delayed necking. The hierarchical interfaces between Al_2O_3 flakes and ultrafine-grained Al-promoted dislocation accumulation and stress redistribution, improving strength by 22% with only minor ductility loss. The multiscale arrangement provided intrinsic strengthening and extrinsic toughening, with structural hierarchy enabling concurrent increases in stiffness and fracture resistance [3]. In addition, titanium matrix composites containing hierarchical heterostructures applied these principles to lightweight materials for aerospace applications. Controlled spatial distributions of reinforcements generated heterogeneous strain fields that produced heterodeformation-induced strengthening and enhanced strain hardening. Layered, bimodal, and networked architectures combined hard and soft domains to balance plastic accommodation with stress redistribution, with strain gradients and geometrically necessary dislocation accumulation mitigating the conventional strength ductility trade-off and enabling high specific strength together with improved toughness under multiaxial loading [19].

Computational methods were developed to analyze and predict the mechanical response of hierarchical composites. An FE2-nested homogenization scheme coupled macroscopic and microscopic deformation by embedding representative volume elements within finite element calculations, thereby resolving the multilevel periodicity characteristic of laser powder bed-fused hierarchical structures. Experimental comparisons indicated that this formulation reproduced elastic behavior and macro-to-micro interactions, enabling numerical design of hierarchical architectures with prescribed mechanical response [52].

10.5 Conclusions

The chapter establishes a coherent scientific foundation in which powder behavior, microstructural evolution, and macroscopic properties are quantitatively connected through advanced characterization and modeling. X-ray diffraction, combined with Scherrer and Williamson-Hall analysis and neural network-assisted interpretation, provides a rigorous basis for tracking phase constitution, crystallite size, and lattice strain in mechanically alloyed and sintered systems. Scanning electron microscopy and automated image analysis extend this capability to powder morphology, melt pool geometry, and defect topology, enabling objective assessment of powder suitability and process stability. In this regard, in situ and operando diffraction, imaging, and multimodal sensing reveal transient phase transformations, melt pool dynamics, and residual stress development during additive manufacturing and electrochemical cycling, transforming powder processing from a static to a time-resolved discipline. Classical diffusion and sintering theories, augmented by models of activated surfaces and electric field-assisted transport, explain densification in nanostructured powders, while grain growth and Zener pinning models rationalize thermal stability in oxide- and rare-earth-stabilized alloys. Moreover, atomistic, Monte Carlo, and hybrid MD/MC simulations associate chemical mixing, hydrogen interactions, and phase separation with nonequilibrium processing routes, with multiscale schemes propagating these effects to continuum scale behavior. Machine learning-based property prediction translates experimental and simulated data into quantitative design rules for complex powder-derived materials.

References

1. Nagarjuna C et al (2023) Application of artificial neural network to predict the crystallite size and lattice strain of CoCrFeMnNi high entropy alloy prepared by powder metallurgy. Met Mater Int 29(7):1968–1975
2. Shen M et al (2024) Architectural and component design of CNTs/Al hierarchical composite for enhanced mechanical/thermal properties. J Mater Res Technol 30:120–133
3. Zhang Z et al (2021) Bioinspired hierarchical Al2O3/Al laminated composite fabricated by flake powder metallurgy. Compos A Appl Sci Manuf 140:106187
4. Brillon A et al (2022) Characterization of Al/B4C composite materials fabricated by powder metallurgy process technique for nuclear applications. J Nucl Mater 565:153724
5. Basem A et al (2024) Characterization of FeCoNiCr high-entropy alloys manufactured by powder metallurgy technique. J Mater Res Technol 30:88–100
6. Lim A et al (2025) In-situ/in-operando approach for topology optimization of structural components by means of thermal imaging, thermal analysis and CT scan. J Japan Soc Powder Powder Metall 72(Supplement):S1667–S1673
7. Singh K et al (2024) Investigating the potential of powder metallurgy for fabricating graphene nanoplatelets reinforced copper nanocomposites. J Mater Res Technol 31:1258–1269
8. Shi H et al (2025) A novel hierarchical structure of in-situ copper matrix composites reinforced with micro-clusters of TiB2 particles and nano-precipitates of B24Cu particles. Compos Part B 297:112311
9. Alam MS et al (2024) Advances of MAX phases: synthesis, characterizations and challenges. Eng Rep 6(8):e12911
10. Binh DN, Oanh NTH, Viet NH (2022) The effect of Ni and Ti additions on the glass forming ability and magnetic properties of Al-Fe-Y alloy prepared by mechanical alloying. J Non Cryst Solids 583:121478
11. Biyik S (2024) Effect of Y2O3 addition and milling time on the synthesis of nanocrystalline ag–ZnO composite powder via mechanical alloying. J Radiat Res Appl Sci 17(2):100867
12. Burton M-TS et al (2024) Enhanced microstructural stability and hardness of multi-component nanocrystalline nickel alloys processed via mechanical alloying. Metall Mater Trans A 55(5): 1338–1350
13. Jang K et al (2024) Impact of hierarchical dopant-induced microstructure on thermoelectric properties of p-type Si-Ge alloys revealed by comprehensive multi-scale characterization. Adv Funct Mater 34(40):2403785
14. Ssendagire K, Phiri I, Ryou S-Y (2025) Operando characterization of dendrite-free silver nanoparticle/lithium metal powder composite anodes. ACS Appl Mater Interfaces 17(20): 29757–29769
15. Şenyurt B, Ağaoğulları D, Akçamlı N (2025) Solid-state synthesis of Cr2AlC MAX phase from mechanically activated Cr/Al/C powder blends. J Alloys Compd 1026:180449
16. Bakas G et al (2022) A tool for rapid analysis using image processing and artificial intelligence: automated interoperable characterization data of metal powder for additive manufacturing with SEM case. Metals 12:1816. https://doi.org/10.3390/met12111816
17. Liu W-S et al (2007) Enhanced thermoelectric properties in CoSb3-xTex alloys prepared by mechanical alloying and spark plasma sintering. J Appl Phys 102(10):103717
18. Restivo TAG, Mello-Castanho SRH (2010) Microstructure design by mechanical alloying. J Eur Ceram Soc 30(14):2991–2996
19. Li S-L, Liu H-Y, Li S-F (2025) Titanium matrix composites featuring hierarchical heterostructures. cMat 2(2):e70007
20. Gaudez S et al (2025) Evolution of texture and residual stresses in 2205 duplex stainless steel during laser powder bed fusion. Mater Des 251:113658
21. Singh KK, Wakai A, Moridi A (2024) Advancements in operando X-ray techniques for metal additive manufacturing. Commun Mater 5(1):258

22. Si C et al (2021) Influence of mechanical alloying on the particle size, microstructure and soft magnetic properties of coarse Fe-based amorphous powders prepared by gas atomization. J Non-Cryst Solids 559:120675
23. Gao C et al (2025) Mechanical alloying and laser powder bed fusion: a potential preparation route for homogenized equiatomic TiTaNbZr high-entropy alloys with excellent wear and corrosion resistances. J Alloys Compd 1036:182116
24. Harikrishna K et al (2025) A novel sintering approach for interface diffusion layer control in Al–$Al_4CrFeMnTi_{0.25}$ high-entropy alloy composites: microstructure, strength, and corrosion resistance. J Alloys Compd 1039:183254
25. Bİlgİn S et al (2025) The role of mechanical alloying on the phase structure and magnetic properties of Ni-doped CoFeAlMn high-entropy alloys. J Alloys Compd 1041:183597
26. Khalaj O et al (2021) Hybrid machine learning techniques and computational mechanics: estimating the dynamic behavior of oxide precipitation hardened steel. IEEE Access 9:1
27. Sharma P et al (2024) Multiscale modeling-driven synthesis of Cu40Zn24Ni24Ag8Hg4 high entropy alloy with antibacterial properties. Nano Struct Nano Objects 40:101391
28. Chai R et al (2021) Effect of scan speed on grain and microstructural morphology for laser additive manufacturing of 304 stainless steel. Rev Adv Mater Sci 60(1):744–760
29. Mussatto A et al (2021) Influences of powder morphology and spreading parameters on the powder bed topography uniformity in powder bed fusion metal additive manufacturing. Addit Manuf 38:101807
30. Chandra S et al (2024) Powder-size driven facile microstructure control in powder-fusion metal additive manufacturing processes. Nat Commun 15(1):3094
31. Soundarapandiyan G et al (2023) In situ monitoring the effects of Ti6Al4V powder oxidation during laser powder bed fusion additive manufacturing. Int J Mach Tools Manuf 190:104049
32. Pandiyan V et al (2022) Deep learning-based monitoring of laser powder bed fusion process on variable time-scales using heterogeneous sensing and operando X-ray radiography guidance. Addit Manuf 58:103007
33. Liu X et al (2025) Effects of milling process parameters on the mechanical alloying behavior of CoCrFeNiAl0.9Nb0.1 high-entropy alloy powder. Mater Today Commun 48:113458
34. Li Z et al (2023) Influences of second phase particle precipitation, coarsening, growth or dissolution on the pinning effects during grain coarsening processes. Metals 13:281. https://doi.org/10.3390/met13020281
35. Kotan H, Koç RC, Batıbay AB (2025) Remarkable thermal stability of nanocrystalline CoCrFeNi high entropy alloy achieved through the incorporation of rare-earth element samarium. Intermetallics 178:108608
36. Li L, Yao Y, Hassani M (2024) Mechano-chemical competition in driven complex concentrated alloys. Materialia 38:102300
37. Schweiger L et al (2025) Achieving complex nanostructures: the role of hydrogen in controlling mechanical alloying and microstructure evolution in the TiVZrNbHf-cu system. Adv Sci 12(33):e07168
38. Dias M et al (2023) Hybrid molecular dynamic Monte Carlo simulation and experimental production of a multi-component cu–Fe–Ni–Mo–W alloy. Intermetallics 161:107960
39. Azeem MM, Jamil I, Khan MB (2022) Multiscale modeling of radiation damage in oxide dispersed strengthened steel alloys: a perspective. Int J Eng Work 9(12):193–201
40. Wang F et al (2023) Damping of aluminum-matrix composite reinforced by carbon nanotube: multiscale modeling and characteristics. Sci China Technol Sci 66(4):1062–1074
41. Khalaj O et al (2021) Potential role of machine learning techniques for modeling the hardness of OPH steels. Mater Today Commun 26:101806
42. Yu J et al (2023) Combining machine learning and molecular dynamics to predict mechanical properties and microstructural evolution of FeNiCrCoCu high-entropy alloys. Nanomaterials 13(6):968

43. Yu J et al (2023) Combining machine learning and molecular dynamics to predict mechanical properties and microstructural evolution of FeNiCrCoCu high-entropy alloys. Nanomaterials 13:968. https://doi.org/10.3390/nano13060968
44. Jain H et al (2025) Structure and properties of Fe40Mn20Cr20Ti10Al10 high-entropy alloy fabricated by mechanical alloying and spark plasma sintering. J Mater Sci 60(3):1705–1725
45. Alam MS et al (2024) Advancements in MAX phase materials: structure, properties, and novel applications. RSC Adv 14(37):26995–27041
46. Shahin N, Heidarpour A, Kazemi S (2024) The purity evolution of Ti3AlC2 MAX phase synthesized by mechanochemical route: the effect of Al content and high-temperature annealing. Eng Rep 6(8):e12814
47. Binh DN, Oanh NT, Viet NH (2022) Al-Fe-Ni metallic glasses via mechanical alloying and its consolidation. Appl Sci 12:10561. https://doi.org/10.3390/app122010561
48. Lee P-Y et al (2025) Synthesis of nanocrystal-embedded bulk metallic glass composites by a combination of mechanical alloying and vacuum hot pressing. Materials 18:360. https://doi.org/10.3390/ma18020360
49. Aldhameer A et al (2022) Mechanical alloying integrated with cold spray coating for fabrication Cu50(Ti50−xNix), x; 10, 20, 30, and 40 at.% antibiofilm metallic glass coated/SUS304 sheets. Nanomaterials 12:1681. https://doi.org/10.3390/nano12101681
50. Song H et al (2025) Achieving ultrahigh strength and high thermal stability of Cu-15 W composite by hierarchical structure with bimodal particle size distribution. J Nucl Mater 615:155992
51. Torabi Parizi M et al (2022) Trimodal hierarchical structure in the carbonaceous hybrid (GNPs +CNTs) reinforced CoCrFeMnNi high entropy alloy to promote strength-ductility synergy. Mater Sci Eng A 850:143446
52. Gao T et al (2025) Unveiling the mechanics of micro-LPBF manufactured hierarchical composites: a novel FE2-nested homogenisation approach. Virtual Phys Prototyp 20(1):e2456693

GPSR Compliance
The European Union's (EU) General Product Safety Regulation (GPSR) is a set of rules that requires consumer products to be safe and our obligations to ensure this.

If you have any concerns about our products, you can contact us on

ProductSafety@springernature.com

In case Publisher is established outside the EU, the EU authorized representative is:

Springer Nature Customer Service Center GmbH
Europaplatz 3
69115 Heidelberg, Germany

www.ingramcontent.com/pod-product-compliance
Ingram Content Group UK Ltd.
Pitfield, Milton Keynes, MK11 3LW, UK
UKHW022004280926
476074UK00003B/13
* 9 7 8 3 0 3 2 2 7 1 2 9 7 *